Textbook on Semiconductors

NJATC

Textbook on Semiconductors

NJATC

THOMSON
DELMAR LEARNING

Australia Canada Mexico Singapore Spain United Kingdom United States

Textbook on Semiconductors
NJATC

Vice President, Technology and Trades SBU:
Alar Elken

Executive Director, Professional Business Unit:
Gregory L. Clayton

Product Development Manager:
Patrick Kane

Development Editor:
Angie Davis

Channel Manager:
Beth A. Lutz

Marketing Specialist:
Brian McGrath

Production Director:
Mary Ellen Black

Production Manager:
Larry Main

Senior Project Editor:
Christopher Chien

Art/Design Coordinator:
Francis Hogan

Printed in the United States of America
2 3 4 5 XX 06 05 04

For more information contact
Delmar Learning
Executive Woods
5 Maxwell Drive, PO Box 8007,
Clifton Park, NY 12065-8007
Or find us on the World Wide Web at
www.delmarlearning.com

Library of Congress Cataloging-in-Publication Data:

NJATC.
Textbook on Semiconductors / NJATC.
p. cm.
ISBN 1-4018-5688-8
1. Semiconductors—Analysis.
2. Semiconductors—Design and construction. 3. Electric circuits—Analysis. I. Title.
TK7871 .85 C33 2003
621.3815′2—dc22
2003021989

Contents

CHAPTER 3
Power Supplies 32

CHAPTER 4 Transistors 58

CHAPTER 5 JFETs, MOSFETs, UJTs 78

CHAPTER 14
Differential and Operational Amplifiers 224

CHAPTER 15
Oscillators 256

CHAPTER 17 Integrated Circuits 296

CHAPTER 18 Microprocessors and Systems Components 310

Preface

The National Joint Apprenticeship and Training Committee (NJATC) is the training arm of the International Brotherhood of Electrical Workers and National Electrical Contractors Association. Established in 1941, the NJATC has developed uniform standards that are used nationwide to train thousands of qualified men and women for demanding and rewarding careers in the electrical and telecommunications industries. To enhance the effectiveness of this mission, the NJATC has partnered with Thomson Delmar Learning to deliver the very finest in training materials for the electrical profession.

Textbook on Semiconductors is designed to provide the necessary background required to understand the concepts and theory associated with semiconductor theory that has been developed using all of the principles learned throughout the years from the 18th Century to the present. In 1745 Cuneus and Muschenbrock from Leyden, Netherlands developed a storage method for electrical charges by placing in a jar several metal foils with insulation between their layers and attaching a wire to each of the end foils—the first electrical capacitor. Between 1746 and 1752, Ben Franklin demonstrated static electricity (recall his kite flying hobby) and theorized that electrical charges were either positive or negative. In 1785, Charles Augustin Coulomb experimented with the magnitude of opposing charges and gave us the unit of charge known today as the "coulomb". Not long thereafter (1800), Count Alessandro Volta used bimetallic arcs dipped in brine to show a source of electricty—known today as the battery—and, as you can guess, the unit of potential called the "volt". History also records the relationship of electricity and magnetism to a Danish gentleman, Hans Christian Oersted, around 1820. Between 1822 and 1827, Andre Marie Ampere developed the algebraic formulas for the relationships between electricity and magnetism and contributed the term "ampere" or "amps". Georg Simon Ohm (1826) discovered that conductors of different materials worked differently in electrical circuits and gave us the unit of resistance we know today as the "ohm". This brings us to the topic of the text—semiconductors.

By definition, a semiconductor is a material that has a resistance somewhere between an insulator and a conductor. At first glance, this concept seems insignificant, but the advantages of controlling the amount of resistance in a circuit will become clear once you study this text. The basic operation of a transistor allows a varying low current to proportionally control a large current flow. During the early years of radio development, the only electronic components available were the vacuum tube diode and triode. But in 1948 three inventors from Bell Labs (William Bradford Shockley, John Bardeen, and Walter Houser Brattain) together invented the solid-state (semiconductor) transistor. Shortly thereafter, the semiconductor diode was derived.

Today, semiconductor electronics is the backbone in nearly every electronic device we use. No longer are radios, telephones, and televisions based on vacuum tube technology. Solid-state chips containing millions of transistors, capacitors, and resistors can easily fit on the tip of your finger. In the electrical construction and maintenance environment a large portion of today's controls and instrumentation are designed with semiconductor components. Understanding the basics of semiconductors is essential in order for proper installation and maintaining of electrical systems. This text explains the basic concepts of the most commonly used semiconductor devices found today. It is not intended to be an engineering manual but to provide common troubleshooting concepts and basic awareness when working with semiconductors. The mixture of theory supported by graphical representations, examples and mathematical problems will give the electrical apprentice and journeyman the knowledge to aid in the installation and maintenance of semiconductor based systems.

The eighteen chapters in this text are essential building blocks by which the student can progress to the next step. You will find the text focuses on the electronics of the construction and maintenance industry. Therefore, the electrical student will not be studying unrelated information such as biomedical electronics, or VCR repair, but instead the circuits and semiconductor components used in their industry. Often the electrical industry asks "Why do we need to learn semiconductor electronics?" By simply looking around every construction jobsite, it is easy to see that many of the electrical systems are no longer controlled by relays and coils but instead by semiconductor electronics. Personal computers and solid-state controllers continue to creep into every electrical application; and with this comes the requirement for basic knowledge of how to install and troubleshoot the systems.

The first three chapters of this book introduce the semiconductor device and the basic PN junction upon which 90% of all modern electronics are based. The diode is discussed in detail and basic power supply circuits and concepts are covered in detail.

Chapters 4 and 5 introduce the bipolar junction transistor, the junction field-effect transistor and the unijunction transistor. All are important components, with the bipolar junction transistor being the basis for the overwhelming majority of integrated circuitry and other such modern electronic devices. The operation of all these devices and their basic applications are covered in these chapters.

Chapters 6, 7, and 8 present the theory, design, application and analysis of the electronic amplifier. The amplifier is found in radios, televisions, control circuits, and variable frequency drives. In fact, as a circuit the amplifier is the most common of them all. The text describes the principles and explains how to analyze amplifiers in their various forms.

Chapter 9 visits the fundamentals and operation of three of the most common digital (on or off) devices in use today—the silicon controlled rectifier (SCR), the triac, and the diac. The SCR and the triac are particularly common in lighting control circuits, variable frequency drive circuits, DC power supplies, inverters, and uninterruptible power supplies. The electrician who specializes in power electronics will find this chapter to be especially useful in his/her career.

Chapters 10 and 11 introduce the circuits and equipment used in one of the fastest growing technologies of the 21st century—fiber optics and optical data transfer. Because of its high data rates and its immunity to noise, optical data transmission is fast becoming the most widely used method for the broad-band needs of modern data communication systems such as the Internet.

Chapters 12 and 13 introduce the number systems and calculations that underlie modern computers. In chapter 12 you will learn the basis of modern computers by studying number systems such as the decimal, binary, octal, and hexadecimal. Chapter 13 teaches you how a computer is used to add, subtract, multiply, and divide in the various number systems.

Chapter 14 introduces the first of the various types of integrated circuits that are used in modern electronics—the operational amplifier (Op-amp). The Op-amp is an analog device that is used in many modern electronic circuits such as controls and radio transmitters and receivers.

Chapters 15 and 16 provide introductions into fundamentals that will be of interest to those who decide to turn their career into communications systems such as radio and television. Various types of oscillators are introduced and the way they operate is discussed. Then, in Chapter 16, the principles of modulation and information transfer are explained.

Chapter 17 discusses another type of integrated circuit—the NE555 timer. This timer is one of the mainstays in electronic timing circuits and has held that position for many, many years. Understanding its operation will be critical to you should your career lead you into industrial controls.

Chapter 18 introduces you to the heart of the modern digital computer—the microprocessor. As time goes by, modern power systems and power electronics are using the microprocessor to control virtually all of the various functions that have been manual or discretely controlled in the past.

Taken together, this text will serve you well as an introduction to the wide world of electronics for the electrician.

The NJATC can provide a complete line of electrical and telecommunication training materials, including CBT programs and courses. Visit the NJATC online at www.njatc.org to review the finest electrical training curriculum the industry has to offer. The subject of semiconductor theory is both interesting and essential for the electrical and electronics student. Take the time to progress through the semiconductor theory material, perform the calculations, and review the chapter objectives before moving forward to the next section. Your understanding of semiconductor theory will provide all of the essentials to move to the next level of expertise in the electrical and electronics fields. Should you decide on a career in the electrical industry, the International Brotherhood of Electrical Workers and the National Electrical Contractors Association (IBEW-NECA) training programs provide the finest electrical apprenticeship programs the industry has to offer. If you are accepted into one of their local apprenticeship programs you'll be trained for one of four career specialties, journeyman lineman, residential wireman, journeyman wireman or telecommunications (Voice/Data/Video) installer/technician. Most importantly, you'll be paid while you learn. To learn more, visit www.njatc.org.

NJATC ACKNOWLEDGEMENTS

NJATC Technical Editor

William R. Ball, NJATC Staff

Contributing Writers

Ed Swearingen, Instructor, Alaska JATC
Chris MacCreery, Training Director, Battle Creek Elect. JATC
Paul LeVasseur, Training Director, Bay City JEATC
Gary Strouz, Training Director, Houston Electrical JATC

ADDITIONAL ACKNOWLEDGEMENTS

This material is continually reviewed and evaluated by Training Directors who are also members of the NJATC Inside Education Committee. The invaluable input provided by these individuals allows for the development of instructional material that is of the absolute highest quality. At the time of this printing the Education Committee was comprised of the following members.

Inside Education Committee

Dennis Anthony—Phoenix, AZ; John Biondi—Vineland, NJ; Dan Campbell—Tangent, OR; Peter Dulcich—Syracuse, NY; John Gray—San Antonio, TX; Gary Hunziker—Sacramento, CA; Dave Kingery—Salt Lake City, UT; Bill Leigers—Richmond, VA; James Lord Jr.—Atlanta, GA; Bud McDannel—West Frankfort, IL; Bill McGinnis—Wichita, KS; Jerry Melson—Bakersfield, CA; Tom Minder—Fairbanks, AK; Bill Newlin—Dayton, OH; Jim Paladino—Omaha, NE; Dan Sellers—Collegeville, PA and Jim Sullivan—Winter Park, FL.

PUBLISHER ACKNOWLEDGEMENTS

John Cadick, P.E., Contributor

A registered professional engineer, John Cadick has specialized for almost four decades in electrical engineering, training, and management. In 1986 he founded Cadick Professional Services (forerunner to the present-day Cadick Corporation), a consulting firm in Garland, Texas. His firm specializes in electrical engineering and training, working extensively in the areas of power system design and engineering studies, condition based maintenance programs, and electrical safety. Prior to the creation of Cadick Corporation, John held a number of technical and managerial positions with electric utilities, electrical testing firms, and consulting firms. Mr. Cadick is a widely published author of numerous articles and technical papers. He is the author of the *Electrical Safety Handbook* as well as *Cables and Wiring.* His expertise in electrical engineering as well as electrical maintenance and testing coupled with his extensive experience in the electrical power industry makes Mr. Cadick a highly respected and sought-after consultant in the industry.

Monica Ohlinger

The publisher would like to thank Monica Ohlinger of Ohlinger Publishing Services for her diligent work in development on the text.

Semiconductor Principles and Introduction to Diodes

OUTLINE

OVERVIEW

Virtually all electric and electronic circuits share the use of the so-called **passive circuit elements**. Prior to the 1960s, active devices such as transistors and vacuum tubes were used only in electronic circuitry. Since that time, improved semiconductors have found multiple applications in high-powered electric power circuits such as variable-frequency drives, UPS systems, power conditioners, high-voltage DC transmission systems, and many others. In this chapter, you will learn the fundamental operating principles of semiconductors and you will be introduced to the most basic of semiconductors—the diode.

OBJECTIVES

After completing this chapter, the student should be able to:

1. Explain the fundamental concepts of semiconductor materials.
2. Describe the basic operating principles of the PN junction.
3. Describe electron flow through a diode.
4. Identify diode symbols to indicate forward and reverse bias.
5. Interpret characteristic curves for semiconductor diodes.
6. Determine if diodes are operating properly.
7. Predict the output waveforms for diode circuits.
8. Calculate component and circuit values for current and voltage in circuits that have diodes.

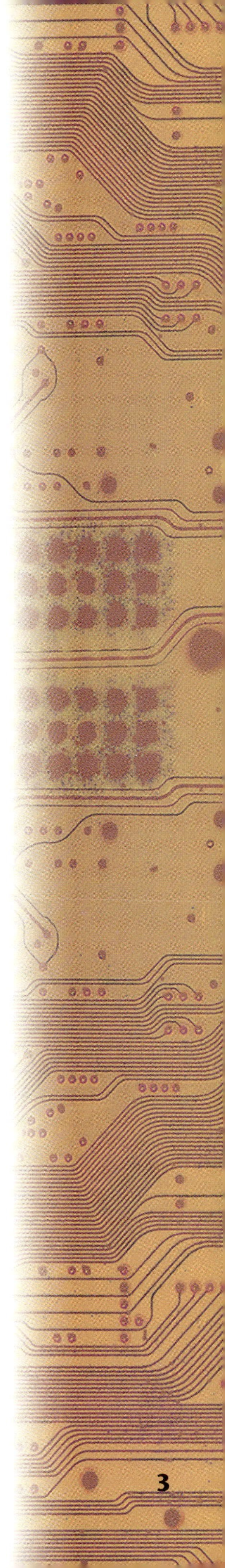

GLOSSARY

Amplitude modulation A system of attaching information to a single-frequency carrier wave. The information is included by varying the amplitude (magnitude) of the carrier wave.

Avalanche voltage See Reverse breakdown voltage.

Bias A current or voltage applied to a semiconductor device to obtain a specific result such as conduction.

Depletion layer The layer that forms between the P and N material in a PN junction. Also called the depletion region.

Diode A two-terminal semiconductor device that passes current of one polarity and blocks current of the opposite polarity.

Doping material A material added to a semiconductor to cause either a P-type material (electron deficiency) or an N-type material (electron excess).

Frequency modulation A system of attaching information to a single-frequency carrier wave. The information is included by varying the frequency of the carrier wave.

Passive circuit element Those electric devices that do not add energy to an electric circuit. Examples include resistors, inductors, and capacitors.

Peak inverse voltage See Reverse breakdown voltage.

PN junction The interface formed when a P-type material is conjoined with an N-type material.

Rectify To change alternating current (AC) to direct current (DC).

Reverse breakdown voltage The reverse bias voltage required to cause a PN junction to fail.

Semiconductor Any of various solid crystalline substances, such as germanium or silicon, having electric conductivity greater than insulators but less than good conductors.[1]

[1]Excerpted from *American Heritage Talking Dictionary.*

INTRODUCTION

1.1 Historical Background

The history of semiconductors is based on the work of Michael Faraday (1831) and James Clerk Maxwell (1864). These men, among others, showed that electric current could be produced by magnets in motion and that this electric current had electromagnetic wave properties.

In 1883, Thomas Edison discovered that electron flow can be generated between a hot filament and a positively charged plate when they are isolated in a vacuum chamber. In 1896, Guglielmo Marconi transmitted and received radio signals over a distance of two miles. These two events provided the foundation for modern-day tubes and later semiconductors.

1.2 Vacuum Tubes

Tubes have been used in radio and television for many years. **Amplitude modulation** (AM) radio was being used in the early 1900s, and **frequency modulation** (FM) was introduced in the 1930s. Electronic black-and-white television systems were developed by the early 1940s, and color TV was commercially available by the mid-1950s. High-power applications for tubes continued to dominate the communications field until full development of the transistor and integrated circuits.

1.3 Transistors

The invention of the transistor in 1948 permitted machinery and devices requiring electronic components to become smaller and more portable. The rows and rows of vacuum tubes and mechanical relays that once made computers room-sized installations have now been replaced by transistors and integrated circuits, which have enabled some computers to be reduced in size to less than 3 × 5 inches. In addition, the use of semiconductors has dramatically reduced the power requirements and heat dissipation of electronic equipment. What used to require 240 V and 15 A now takes less than 5 V and 250 mA.

KEY REVIEW ELEMENTS

1.4 Current Flow

The electrons in the outer ring of the atom are called valence electrons. The valence ring can contain a maximum of eight electrons. These electrons are more or less unstable and can be moved from atom to atom. The number of valence electrons determines how well the material conducts electric current.

There are two different theories for describing electric current flow: hole flow and electron flow. Both are used and both can be proven mathematically. The electron is the particle that moves; that is, it has motion. Holes simply give the appearance of movement (motion) due to the movement of electrons.

1.5 Conductors, Semiconductors, and Insulators

Electrical Conductivity Principles

As you know from earlier lessons in DC and AC electricity, all materials can be classified as conductors, insulators, or semiconductors. Material classification depends on the ability to conduct an electric current. The ability to conduct electricity depends on the number of free or relatively loose electrons in the material. All **semiconductor** devices such as diodes, transistors, and integrated circuits (ICs) are made of semiconductor materials.

Conductors

Conductors are of materials whose atoms have only one or two electrons in the outer valence ring. These atoms (copper, silver, gold, etc.) are good conductors because the outer electrons can be "pulled away" very easily. Electrical conductivity of conductors tends to go down as temperature goes down.

Insulators

Insulators are materials with very high resistance to current (electron flow). This means that their electrons strongly resist being "pulled away" from the outer valence ring of the atom. Insulators have almost full valence rings with seven or eight electrons. Wood, paper, and glass are good examples of insulators. Electrical conductivity of insulators tends to go up as temperature goes up.

Semiconductors

Semiconductors have four electrons in the outer valence ring. They are better conductors than insulators, but they do not conduct as well as conductors. Electrical conductivity of semiconductors reacts to temperature in the same way that it does in insulators; that is, conductivity goes up as temperature goes up.

SEMICONDUCTOR CHARACTERISTICS

Semiconductor materials have characteristics of both insulators and conductors. Two of the most common semiconductors are silicon and germanium. These materials form crystals. In their pure (intrinsic) form they do not have enough free electrons to be useful, so they are modified or doped by adding an impurity to help the conducting process.

After the intrinsic crystal is doped, it is called an extrinsic semiconductor. **Doping material** comes in two types. The first has an excess of free electrons and, when added to a pure semiconductor crystal, the resulting crystal is called an N-type semiconductor material. The second has a deficiency of free electrons (an excess of free holes) and, when added to a pure semiconductor crystal, the resulting crystal is

FIGURE 1–1 N-type and P-type materials

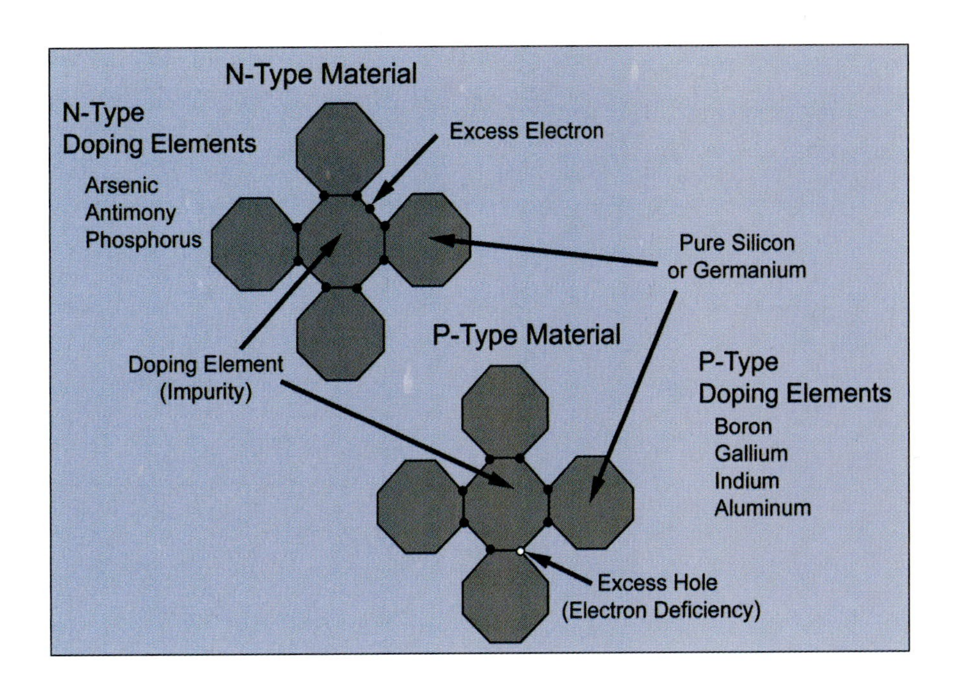

called a P-type semiconductor. Figure 1–1 shows the resulting crystal effect with the different N- and P-type doping materials.

THE PN JUNCTION

1.6 Forming the PN Junction

As you can see from Figure 1–1, current will flow in either P-type (hole flow) or N-type (electron flow) material. When these two types of materials are joined together, they form the enormously useful PN junction. The **PN junction** is vital to diode and semiconductor operation. During the manufacturing process of the PN junction, an interaction takes place between the two types of materials. Some of the excess electrons move into the P material and combine with the excess holes, and some of the excess holes move into the N material and combine with the excess electrons. This combination creates negative and positive ions—negative ions in the P material and positive ions in the N material. Figure 1–2 shows this PN junction effect. The area affected by the combining holes and electrons is called the depletion region or **depletion layer**.

FIGURE 1–2 The PN junctions

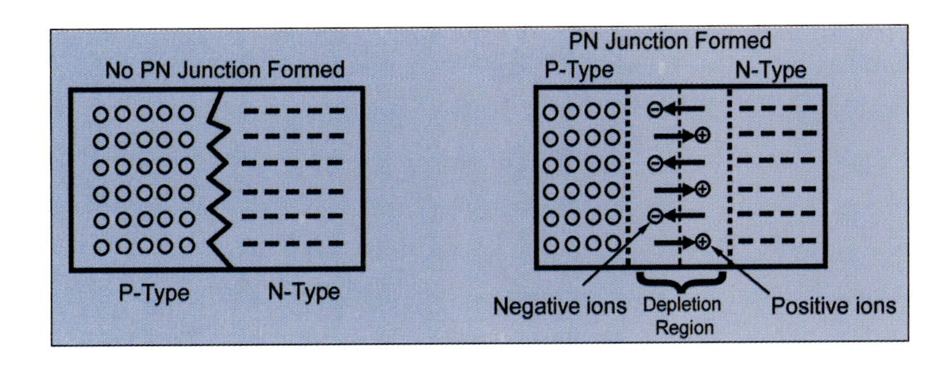

The negative and positive ions in the depletion region create an internal barrier voltage of .5 V to .7 V (silicon) and .25 V to .3 V (germanium) that prevents further current flow from the N- or P-type material. This internal voltage barrier is a constant that comes from the semiconductor material itself. This means that it is present in all PN junctions of diodes and transistors. As with any voltage, this barrier voltage has a polarity. The negative ions (in the P material) are negatively charged and thus repel any further movement from the excess electrons in the N material. The positive ions (in the N material) are positively charged and thus repel any further movement from the excess holes in the P material. Figure 1–3 shows how this barrier voltage works.

1.7 Biasing the PN Junction

The behavior of the PN junction can be changed by applying a bias voltage. The **bias** voltage changes the width of the PN junction's depletion region and, therefore, changes its resistance. Figure 1–4 shows the result of forward or reverse bias on the PN junction. The PN junction conducts when it is forward biased and does not conduct when it is reverse biased.

Forward Bias

A PN junction is forward biased by making the P material positive with respect to the N material. Depending on whether the PN junction is made of silicon (Si) or germanium (Ge), the minimum amount of external voltage required to forward bias the junction is .5 V to .7 V (Si) or .25 V to .3 V (Ge). Power semiconductors may have greater forward voltage drops, depending on current magnitude. When the junction is forward biased, the depletion layer decreases and the PN junction conducts.

Reverse Bias

Reverse biasing occurs when the N material is made positive with respect to the P material. In this condition, the depletion layer widens and essentially no current flows.

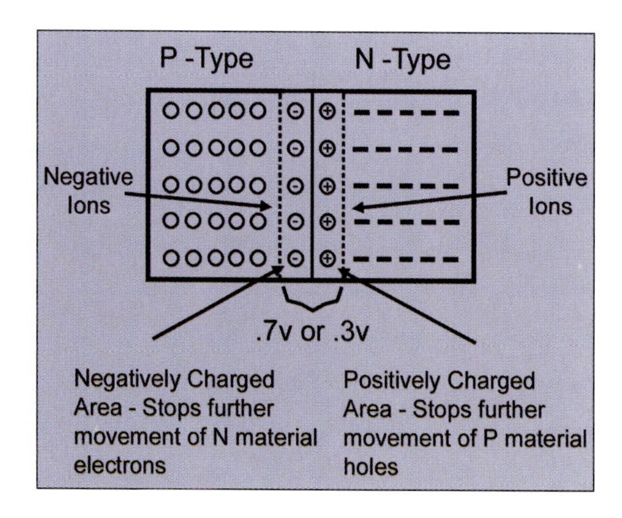

FIGURE 1–3 The ion barrier voltage

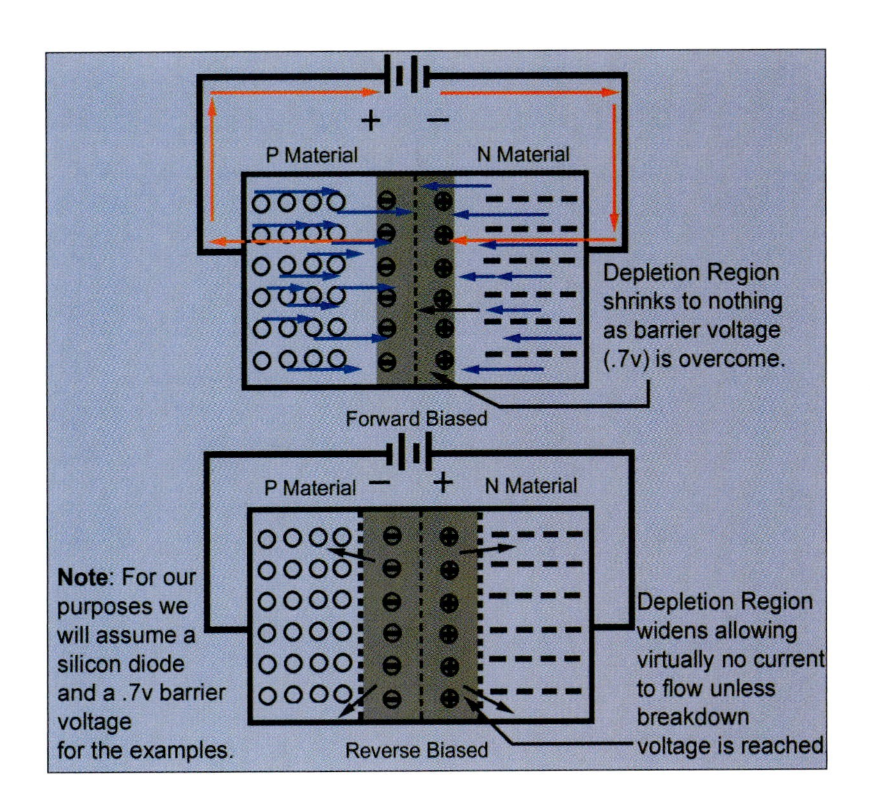

FIGURE 1–4 Biasing the PN junction

1.8 Reverse Breakdown Voltage

Excessive reverse bias will break down and possibly destroy the PN junction. When the voltage is large enough, the PN junction's resistance drops rapidly, a very high current develops, and the PN junction is destroyed. The voltage at which this happens is called **reverse breakdown voltage, peak inverse voltage**, or **avalanche voltage**.

There are certain types of semiconductor devices that are designed to operate in the reverse breakdown area. The Zener diode is the most common of these devices. The Zener diode is used to regulate voltage in a circuit, even when there is a widely varying current. You will learn more about these devices later on in your study. Normally, however, all semiconductor devices that have PN junctions operate at less than peak inverse or avalanche voltages. Figure 1–5 shows normal conduction and reverse breakdown (avalanche) voltage and current relationships during forward and reverse bias conditions.

THE DIODE

1.9 Basic Principles

A **diode** is a two-terminal or two-lead device that has a PN junction and acts as a one-way conductor. When forward biased, the diode conducts. When reverse biased, the diode conduction is so small that it is usually considered as zero. There are other types of diodes (Zener, light emitting, etc.); for now, however, when we use the term we are referring to a simple PN junction two-lead device.

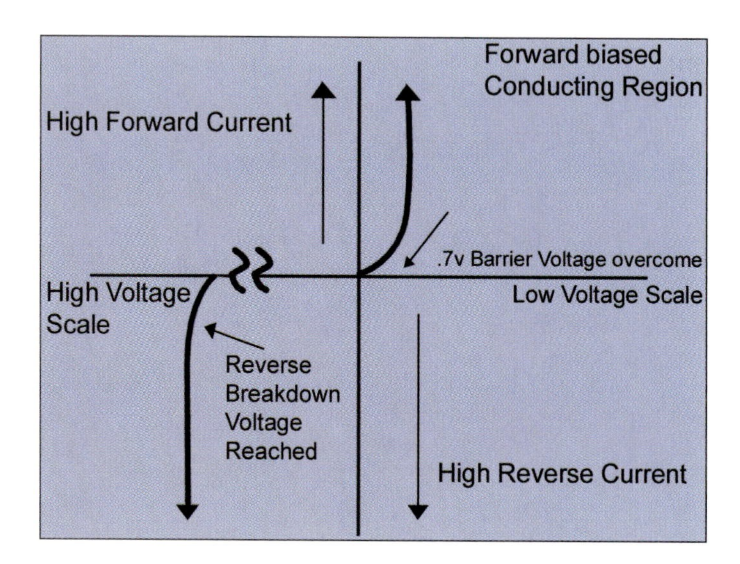

FIGURE 1–5 PN junction voltage–current characteristic curve

1.10 Diode Construction and Characteristics

The Anode and the Cathode

Two examples of diodes and the schematic symbol are shown in Figure 1–6. The N material is called the cathode and the P material is called the anode. The diode will conduct when it is forward biased. Remember, electrons flow against the arrow (from the negative to the positive potential).

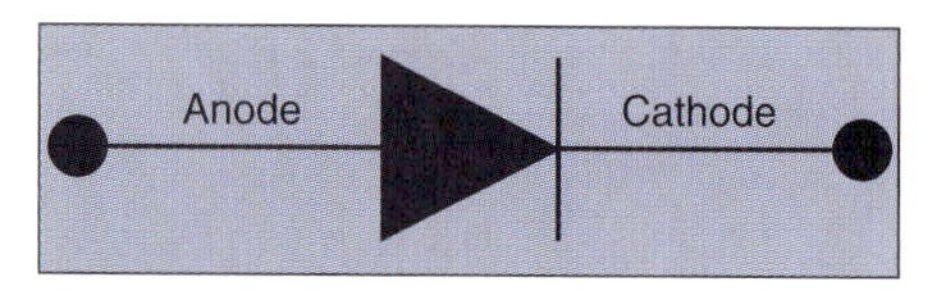

(a)

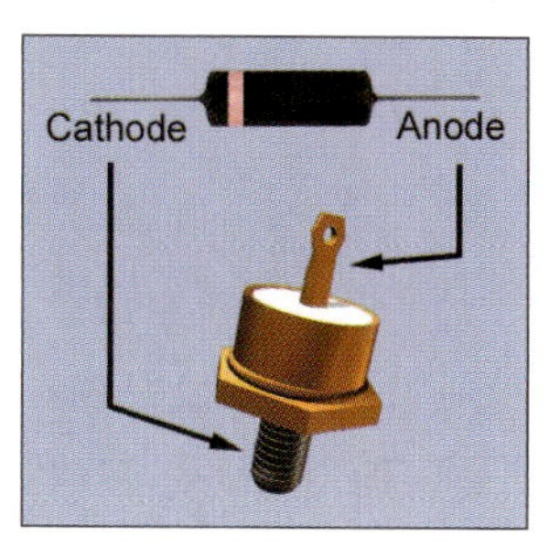

(b)

FIGURE 1–6 The diode; a) schematic, b) typical examples

Forward and Reverse Biasing

Figure 1–7 shows a diode with forward bias connections in a simple series circuit. Note that the circuit is drawn in three different ways. The first way shows the diode connection with respect to the negative and positive terminals of the voltage supply. The second and third ways show the diode connection with respect to a voltage potential and ground. Note that in each case the anode is positive with respect to the cathode. As your career progresses, you will probably see all of these types of circuit connections.

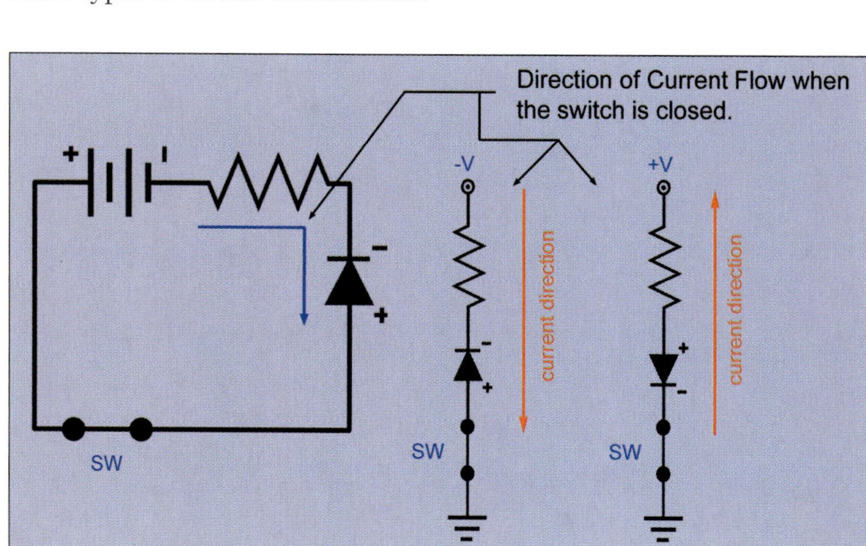

FIGURE 1–7 A forward biased diode allows current flow

FIGURE 1–8 A reverse biased diode blocks current flow

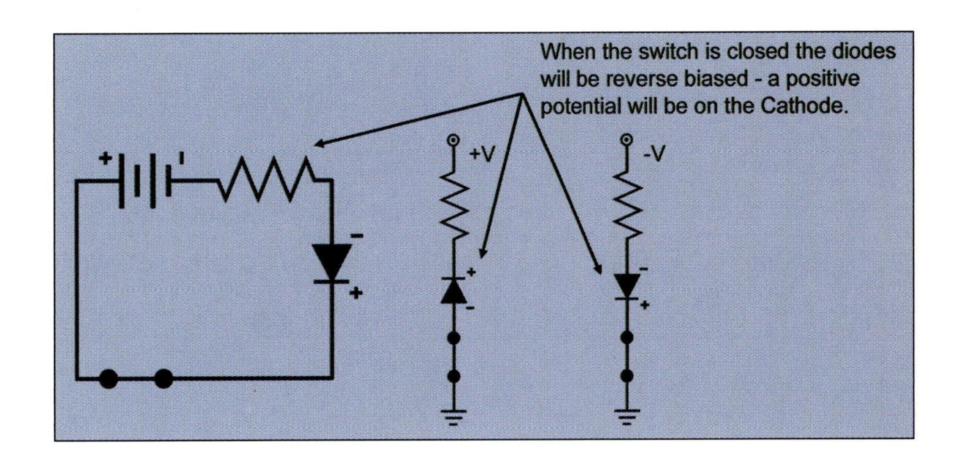

Figure 1–8 shows a diode with reverse bias connections in a simple series circuit. Again, note that the circuit is drawn in three different ways. The first way shows the diode connection with respect to the negative and positive terminals of the voltage supply. The second and third ways show the diode connection with respect to a voltage potential and ground. In all three diagrams, the anode is negative with respect to the cathode and, therefore, the diode will not conduct.

Rating Characteristics

Diodes have several characteristics that must be considered when using them in circuits. The most common are shown in Table 1–1.

Diodes have other characteristics as well; however, these three are among the most important.

Diode Characteristic Curves

Diodes also have characteristic curves. These are plotted on charts to show how the diode will react at different temperatures and with

Table 1–1 Diode Characteristics (Ratings)

Characteristic (Rating)	Description
Peak Reverse Voltage (V_{RRM})	The maximum reverse voltage allowed for the diode. Exceeding this voltage will cause the diode to be destroyed. Sometimes called the peak inverse voltage or the avalanche voltage.
Average Forward Current (I_0)	Maximum allowed DC forward current. Exceeding this current will cause the diode to be destroyed.
Forward Power Dissipation ($P_{D(max)}$)	Maximum power that can be dissipated by the diode when it is forward biased. Exceeding this power limit will cause the diode to be destroyed.

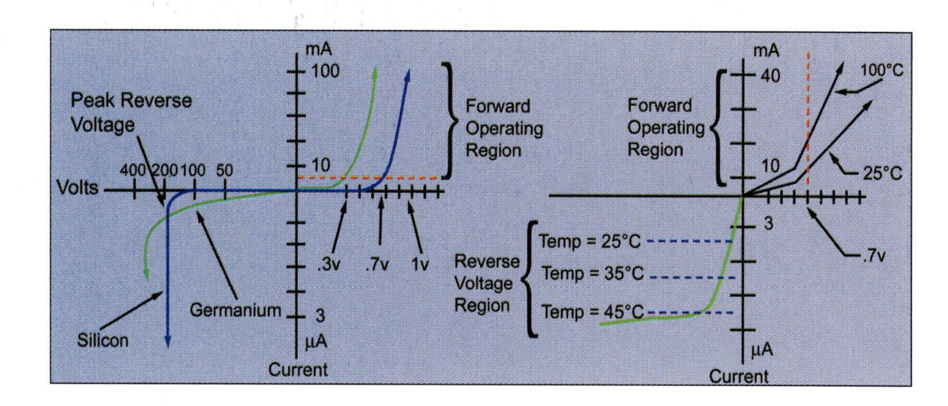

FIGURE 1–9 Diode characteristic curves

different circuit voltages and currents applied. Figure 1–9 shows examples of typical diode characteristics.

1.11 Diode Testing

Diodes can be tested by using an ohmmeter. This type of test can determine if the diode is operating properly and which lead is the cathode (negative) and which is the anode (positive). Refer to Figure 1–10.

The following steps outline those in Figure 1–10.

1. Start by placing one lead of the ohmmeter on each lead of the diode and note the resistance (high or low).
2. Reverse the ohmmeter leads and again note the resistance (low or high).

If the ohmmeter reads high for one connection and low for the other, the diode is probably good. Furthermore, when the diode reads low, the positive lead is connected to the anode.

If the meter reads high (or low) in both directions, the diode is bad. Diodes usually fail short. That means that the PN junction fuses and the diode becomes shorted.

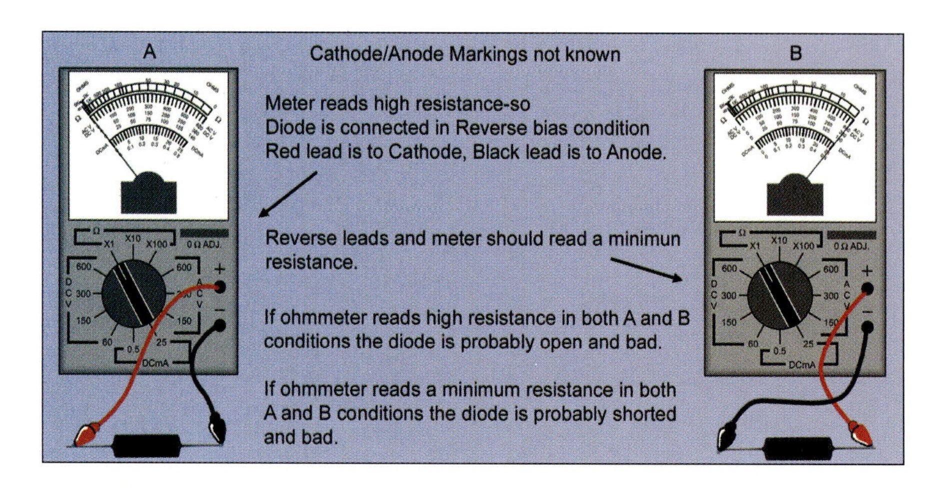

FIGURE 1–10 Testing diode

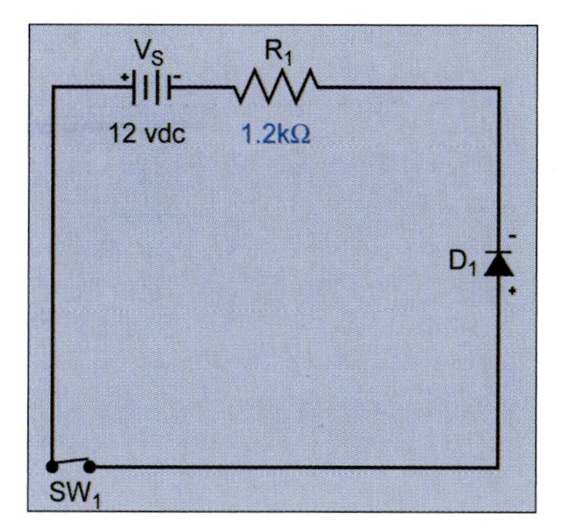

FIGURE 1–11 Simple DC diode circuit

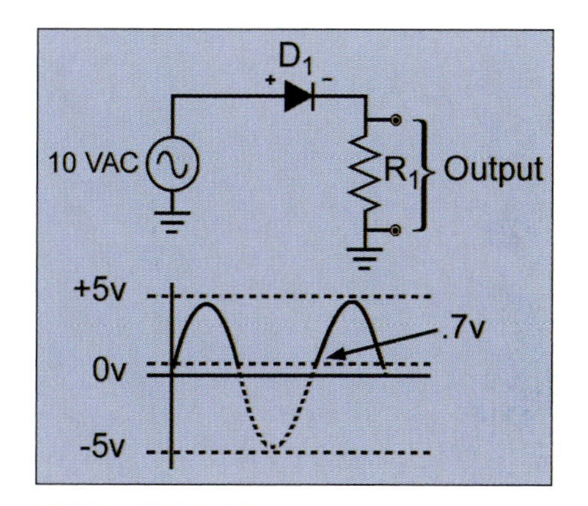

FIGURE 1–12 Diode used to rectify AC

DIODE CIRCUIT ANALYSIS

Diode circuit analysis is more difficult than it might appear at first. Remember that many of the circuit analysis tools that you have used previously depend on the circuit being bilateral—all of the elements must work the same both ways. Clearly, diodes do not behave in this way. Therefore, tools such as superposition and Kirchhoff's laws must be used very carefully.

1.12 Calculating DC Voltage and Current

The following illustrates the effect of the .7-V barrier voltage in a simple series circuit.

In Figure 1–11, calculate the voltage drops for R_1 and D_1 (silicon) and the total circuit current.

Using Ohm's law:

$$V_T = V_{D1} + V_{R1}$$

Where

$$V_{D1} = 0.7 \text{ volts (silicon)}$$
$$V_{R1} = V_T - V_{D1} = 12 - 0.7 = 11.3 \text{ V}$$

Ohm's law:

$$I_T = \frac{V_{R1}}{R_1} = \frac{11.3}{1{,}200} = 9.42 \text{ mA}$$

Note: This same example can be calculated using a germanium diode with a barrier voltage of .3 V.

1.13 Diodes and AC Circuit Analysis

As you have seen, the diode has polarity, the cathode is negative, and the anode is positive. This polarity can be used in an AC circuit to **rectify** the output signal of the circuit. This happens because a diode placed in a circuit with an AC power supply will conduct (be forward biased) only half of the time. Figure 1–12 shows a diode connected to an AC power source in a series circuit and the resulting output waveform.

Practical Applications

A practical application of this 'rectifying' effect can be found in meters. In some meters, diodes are placed in series to allow current in only one direction. (See Figure 1–13.) In other meters, diodes are

FIGURE 1–13 Diode and meter movement in series

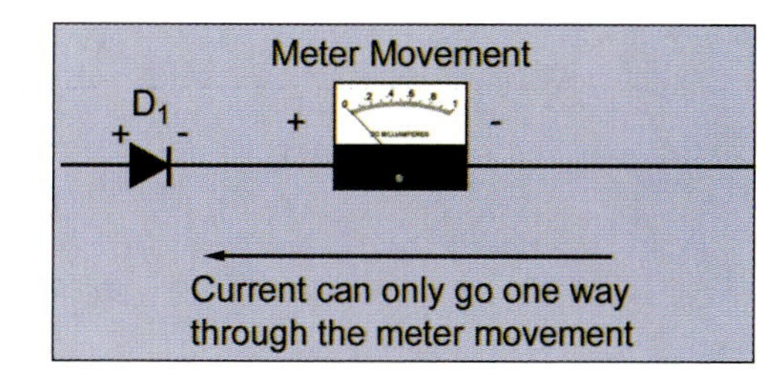

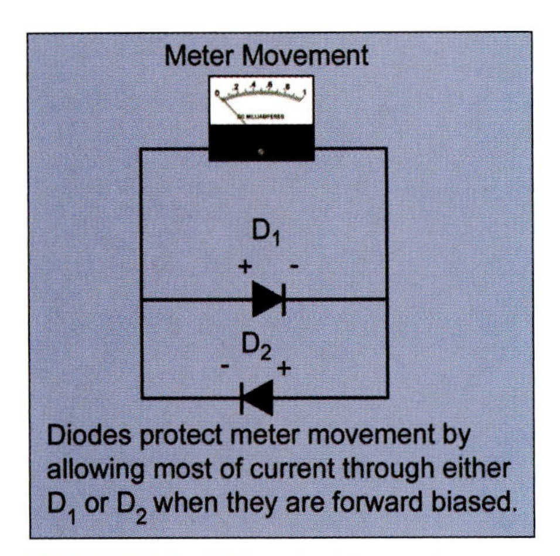

FIGURE 1–14 Diodes and meter movement in parallel

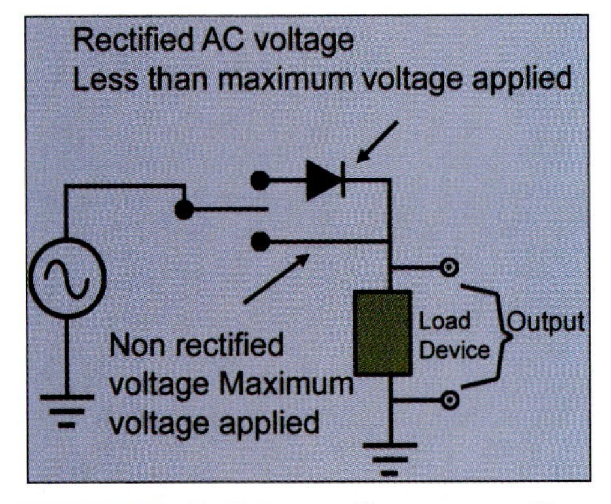

FIGURE 1–15 Half-wave rectifier application

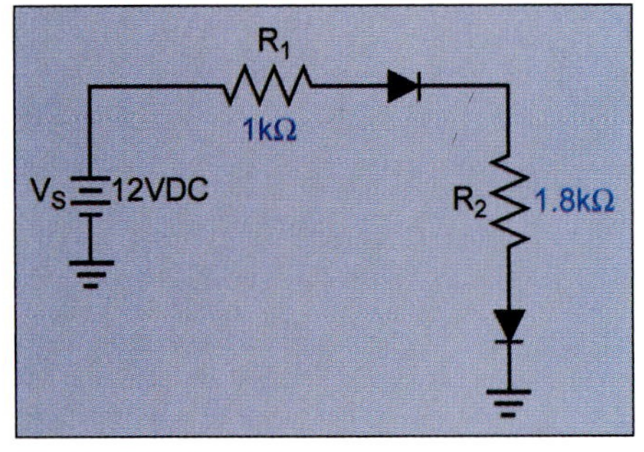

FIGURE 1–16 Multiple diode circuit

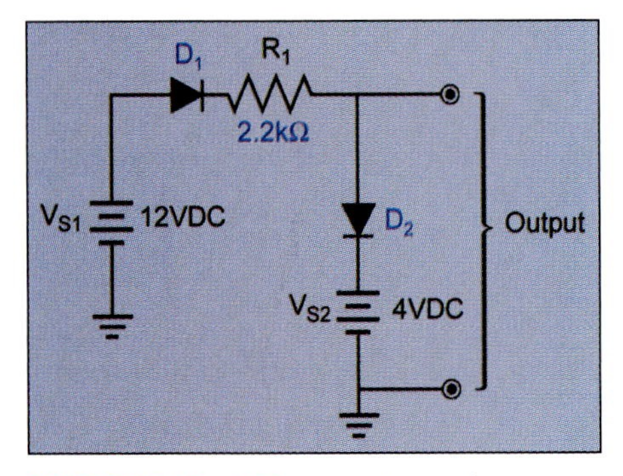

FIGURE 1–17 Diode and multiple power sources

placed in parallel (shunted) to the meter movement. This limits the current through the movement. (See Figure 1–14 for an example of this application.)

Figure 1–15 shows how a half-wave rectifier being used as a voltage reduction circuit. With the diode in the circuit (switch up), only half of the voltage waveform is let through. Although the current has been rectified, it still has some reduced magnitude. (See Figure 1–12.) When the switch is thrown down, the load receives the entire waveform and, thus, the full system voltage.

You can combine diodes to vary the circuit's output. Refer to Figure 1–16 for the following problem.

EXAMPLE 1

Calculate the voltage drops across R_1 and R_2. Calculate I_T for the circuit.

Using Ohm's law:

$$I_T = \frac{[V_S - (0.7 + 0.7)]}{R_T} = \frac{(12 - 1.4)}{2{,}800} = 3.79 \text{ mA}$$

and

$$V_{R1} = I_T \times R_1 = 3.79 \text{ mA} \times 1 \text{ k}\Omega = 3.79 \text{ V}$$

$$V_{R2} = I_T \times R_2 = 3.79 \text{ mA} \times 1.8 \text{ k}\Omega = 6.81 \text{ V}$$

When checking to see if your calculations are correct, add all the voltage drops. Their sum should equal the voltage applied. $V_T = 6.81 + 3.79 + .7 + .7 = 12$ V

The next example uses multiple power sources in a circuit with diodes. Refer to Figure 1–17 when calculating the following problem.

EXAMPLE 2

Calculate the total circuit current and voltage drop across R_1.

$$I_T = \frac{V_T}{R_1} = \frac{12 - 0.7 - 4 - 0.7}{2{,}200} = 3 \text{ mA}$$

$$V_{R1} = I_T \times R_1 = 2.2 \text{ k}\Omega \times 3 \text{ mA} = 6.6 \text{ V}$$

SUMMARY

Diodes are made from a P-type and N-type material. This material is joined to form a PN junction. This junction has polarity and can be forward or reverse biased. There is a depletion region where this PN junction forms. This depletion region forms a barrier that requires a voltage to cause current to flow through the junction. This barrier voltage is .5–.7 in silicon diodes and .25–.3 in germanium diodes.

One of the main applications for diodes is in rectifiers. A rectifier circuit takes an AC signal and converts it into pulsating DC. Half-wave, full-wave, and bridge are three of the various types of rectifier circuits. These circuits are discussed in chapter 3. Another major application for diodes is in motor control and voltage control circuits.

Diodes operate much like a switch. When they are reverse biased, they essentially are OFF. That is, they conduct only very small amounts of current. When they are forward biased, they turn ON and conduct current to their rated value.

Diodes have critical operational characteristics. These characteristics are shown in Table 1–1. In practice, most diode maximum reverse voltage ratings are in multiples of 50 (i.e., 50, 100, 150, etc).

REVIEW QUESTIONS

1. Discuss the depletion layer in a PN junction.
   a. What is its origin?
   b. How does it affect the operation of the junction?
2. What are three critical characteristics of diodes that must be considered to avoid destroying the semiconductor?
3. Describe the characteristics of a P-type material and an N-type material.
4. Draw a diode circuit with a resistor, diode, and battery.
   a. Forward biased
   b. Reverse biased
5. What are the forward voltage drops for silicon and germanium PN junctions?
6. Look at Figure 1–17. What is the output voltage?
7. If you wish to analyze the circuit of Figure 1–17 using the Superposition Theorem, you must assume a small current flow through the diodes when they are reverse biased. Why?
8. The load device in Figure 1–15 sees less voltage when the switch is up than when it is down. Why?
9. Look at Figure 1–7. How much current will flow if the battery voltage is 0.5 V and the diode is a silicon diode?
10. Based on your answer to review question 9, what kind of material would be used for diodes in small signal applications—germanium or silicon?

chapter 2

Zener and Other Diodes

OUTLINE

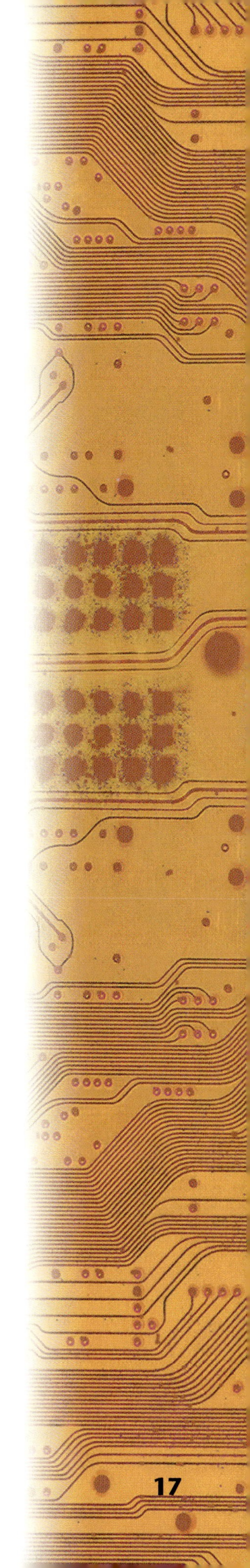

OVERVIEW

There are many variations of the simple diode. In this chapter, you will learn about four types. Some of these types are used primarily in electronic circuitry, while others, such as the Zener diode, are used in both electronic and power applications. The information in this chapter will build on your previous training and, in turn, will serve as a background for future chapters.

OBJECTIVES

After completing this chapter, the student should be able to:

1. Describe the operation of Zener diodes.
2. Determine current flow in Zener diode circuits.
3. Interpret characteristic curves for Zener diodes.
4. Determine if Zener diodes are operating properly.
5. Calculate component and circuit values for current, voltage, and power in circuits that have Zener diodes.
6. Identify the schematic symbols for other types of diodes.
7. Describe how other types of diodes are used.

GLOSSARY

Antiparallel Two circuit elements connected in parallel with opposite polarities.

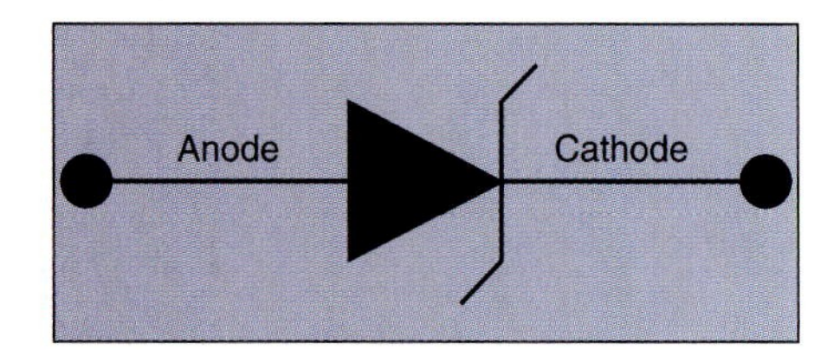

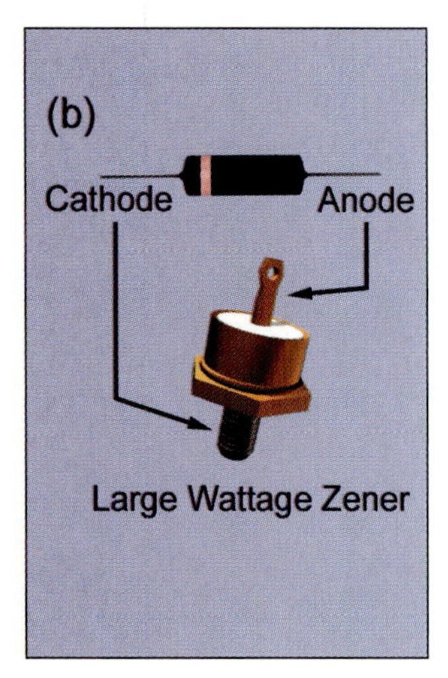

FIGURE 2–1 Zener diode; a) schematic symbol, b) typical examples

INTRODUCTION

The Zener diode is a special and very useful type of diode and is, perhaps, the most prevalent and useful form of diode made today. Zener diodes are unique in that they operate as a Zener in the reverse bias mode, and when forward biased, they operate as a normal semiconductor (rectifier) diode. Figure 2–1a is the schematic diagram of the Zener diode. Figure 2–1b shows a typical circuit and drawings of a low- and high-powered Zener.

2.1 Zener Diode Characteristics

Figure 2–2 is a typical characteristic for a 15-volt Zener diode. As you can see, the Zener diode voltage–current (VI) curve is somewhat different than the diodes that you studied previously. When the reverse bias voltage reaches a level called the "Zener voltage," current flows in the diode and the voltage drop remains essentially constant. This characteristic makes Zener diodes particularly useful in circuits such as voltage stabilizers or regulators.

The area to the left of the breakdown or Zener voltage is called the Zener region. Depending on the specific diode, Zener voltages range from 1.8 volts at a ¼ watt to 200 volts at 50 watts. The wattage ratings refer to the maximum power dissipation allowed for the Zener before it will be damaged. Zener diodes rated at over 1 watt usually come in stud-mount packages as shown in Figure 2–1b. Heat sinks can be used to dissipate heat and increase the power capability of the diode. The most common Zener values are 3.3 volts to 75 volts.

Notice that as the reverse voltage is increased, the (reverse) leakage current remains almost constant until the breakdown voltage is reached, then the current increases dramatically. (Recall that this is called the avalanche point.) This breakdown voltage is the Zener voltage for Zener diodes. In practical terms, before the Zener (breakdown)

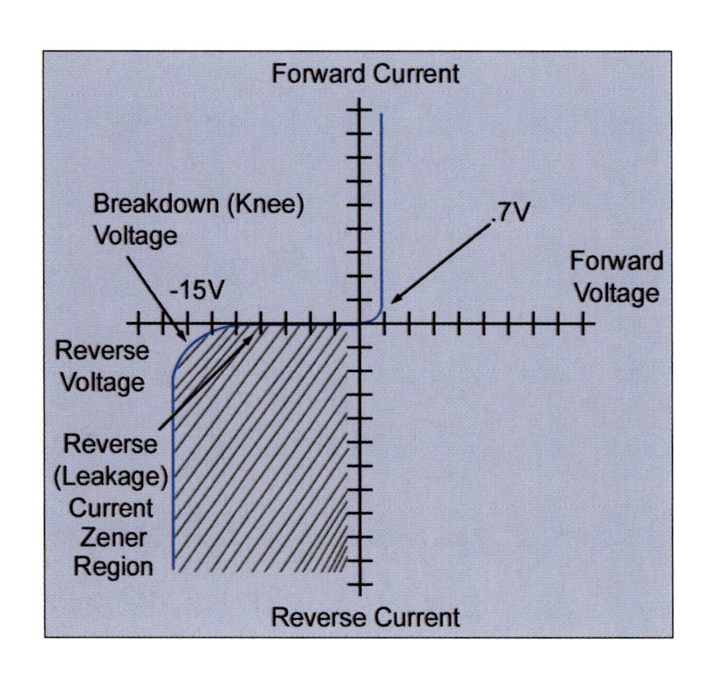

FIGURE 2–2 Zener diode characteristic curve

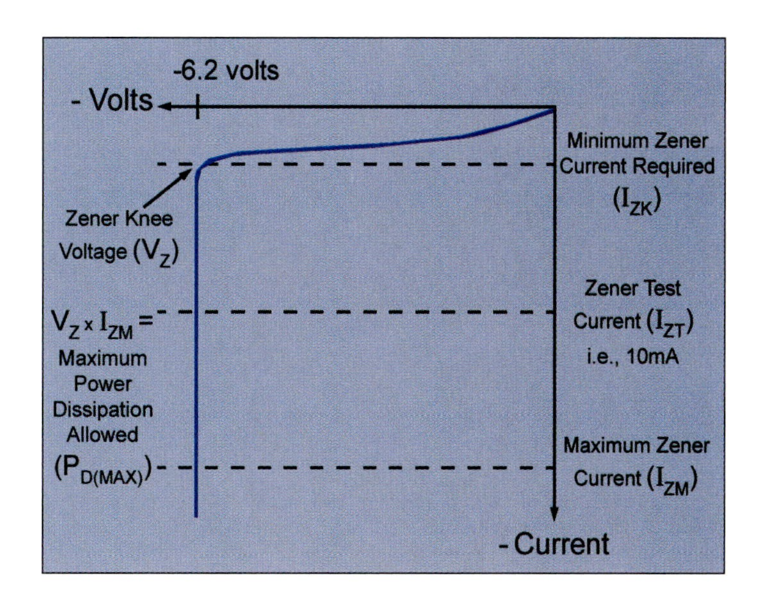

FIGURE 2–3 Zener diode current region

Table 2–1 Zener Diode Characteristics

Rating	Description
I_{ZM}	The maximum Zener current allowed before damage will occur
I_{ZT}	The Zener current at which the diode was tested to determine its rating. For example, a 6.2 V Zener has its voltage (6.2 V) tested at the factory at a value of 10 mA.
$V_{D(MAX)}$	The maximum power dissipation allowed before damage will occur

voltage is reached, the Zener diode performs like an open circuit. Although it is imperative for the normal PN junction diode or rectifier to operate below this voltage to prevent diode damage, the Zener diode is intended to operate at this voltage.

The V_{BR} (breakdown voltage) is a manufactured characteristic of the Zener. It is also called the knee voltage. Zener voltage is usually shown as V_Z. Other critical characteristics are shown in Table 2–1. (See Figure 2–3.)

CIRCUIT ANALYSIS

The Zener diode and, in fact, all diodes are not bilateral devices. This means that you must be very careful in trying to analyze such circuits using standard methods. Methods such as superposition, Kirchhoff's voltage law, and others will work reliably only if the system is evaluated for current flow in the forward direction or for voltages across the diode in excess of V_Z.

2.2 Zener Diode Circuit Operation

Figure 2–4a is a simple Zener diode circuit with a variable DC power supply. Note that the Zener is in parallel with the load and reverse biased.

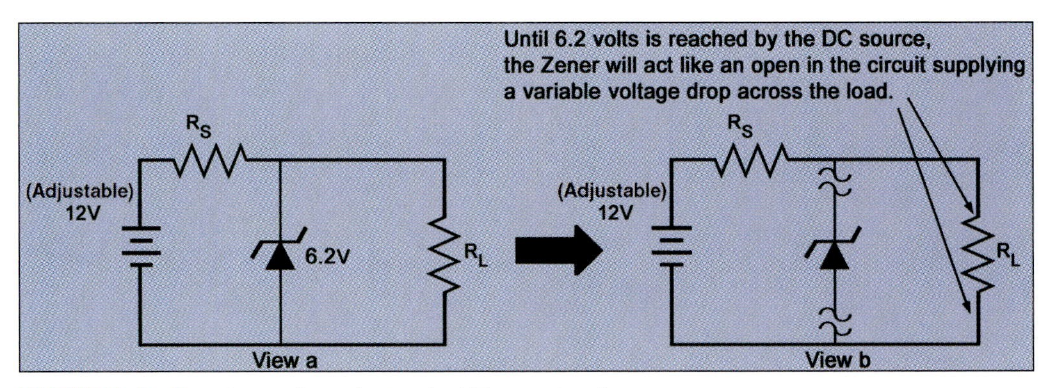

FIGURE 2–4 Operation of a typical Zener circuit

This is the normal configuration for the Zener diode. Voltage regulation is performed by the Zener in the following way:

1. As the DC voltage is increased from 0 V, the Zener will act like an open circuit.
2. When the Zener voltage reaches 6.2 V or higher, the Zener avalanches and draws current. It also maintains a constant 6.2 V across the load, as shown in Figure 2–4b.
3. As you will see, the operation of this circuit is very dependent on the values of R_S and R_L.

Refer to Figure 2.5. Start by calculating the current flow through the Zener diode with switch S_{W2} open. This is a simple application of Kirchhoff's voltage law (KVL).

$$V_{Source} + V_{RS} + V_Z = 0$$

$$-12 + V_{RS} + 6.2 = 0$$

$$V_{RS} = 5.8 \text{ V}$$

Using Ohm's law, the current through R_S can be calculated as:

$$I_{RS} = \frac{V_{RS}}{R_S} = \frac{5.8}{400} = 14.5 \text{ mA}$$

Notice that these values will stay the same for each of the following three examples as long as enough current is supplied to the Zener diode to allow it to stay in its Zener region. This can be checked later. First, consider the following three examples.

EXAMPLE 1

In Figure 2–5a, both switches are closed.

Because the Zener and the load resistance are in parallel, you know that $V_Z = V_{RL}$.

Therefore:

$$I_{RL} = \frac{V_Z}{R_L} = \frac{6.2 \text{ V}}{2.2 \text{ k}\Omega} = 2.82 \text{ mA}$$

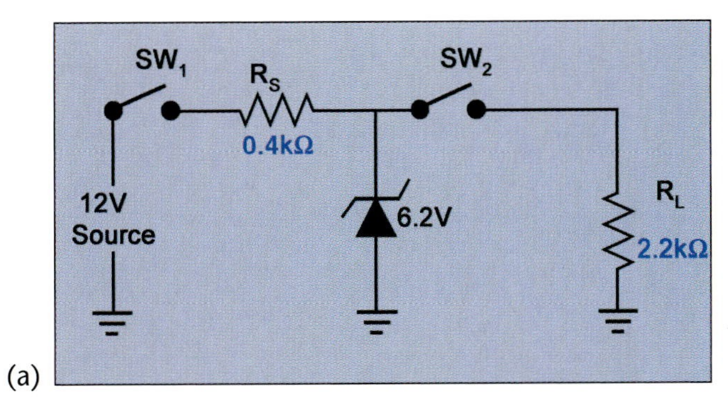

(a)

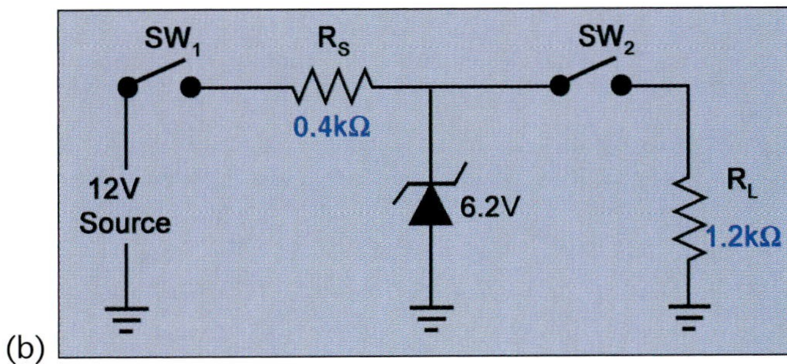

(b)

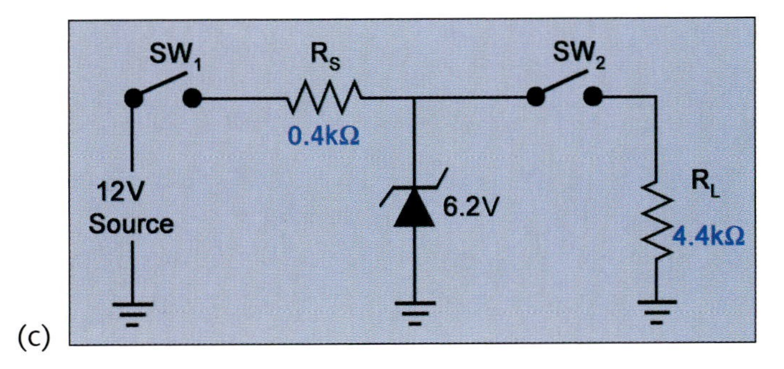

(c)

FIGURE 2–5 Zener circuit analysis; a) 2.2 kW load, b) 1.2 kW load, c) 4.4 kW load

And, by Kirchhoff's current law (KCL), the Zener current is calculated as:

$$I_Z = I_{RS} - I_{RL} = 14.5 - 2.82 = 11.7 \text{ mA}$$

EXAMPLE 2

In Figure 2–5b, both switches are closed.

Again, because the Zener and the load resistance are in parallel, you know that $V_Z = V_{RL}$.

Therefore:

$$I_{RL} = \frac{V_Z}{R_L} = \frac{6.2 \text{ V}}{1.2 \text{ k}\Omega} = 5.16 \text{ mA}$$

And, by Kirchhoff's current law (KCL), the Zener current is calculated as:

$$I_Z = I_{RS} - I_{RL} = 14.5 - 5.16 = 9.34 \text{ mA}$$

EXAMPLE 3

In Figure 2–5c, both switches are closed.

And as before, because the Zener and the load resistance are in parallel, you know that $V_Z = V_{RL}$.

Therefore:

$$I_{RL} = \frac{V_Z}{R_L} = \frac{6.2\text{ V}}{4.4\text{ k}\Omega} = 1.41\text{ mA}$$

And, by Kirchhoff's current law (KCL), the Zener current is calculated as:

$$I_Z = I_{RS} - I_{RL} = 14.5 - 1.41 = 13.1\text{ mA}$$

Note two very important points:

1. As the load resistance increases, the Zener current increases.
2. The maximum Zener current is 14.5 mA.

2.3 Checking an Installation

For any Zener circuit, the Zener current must stay between two limits:

1. The minimum Zener current must be sufficient to bias the Zener into its Zener region.
2. The maximum Zener current must be low enough so that the Zener power does not exceed its maximum.

Assume that the minimum Zener current for the diode used in the previous examples is 8 mA. This means that the minimum load that can keep the Zener operating properly is:

$$R_L = \frac{6.2}{14.5 - 8} = 954\ \Omega$$

Any value of load resistance less than 954 Ω will result in a Zener current that is less than the required value and the Zener will not regulate.

At the other end, you need to calculate the total power dissipation of the Zener at its highest current.

$$P_Z = V_Z \times I_Z = 6.2\text{ V} \times 14.5\text{ mA} = 89.9\text{ mW}$$

In this particular circuit, even a ¼-watt Zener (250 mW) will be sufficient.

2.4 Other Zener Circuit Considerations

Insufficient Reverse Bias

A Zener diode may be taken out of its Zener range by either one of two conditions:

1. Insufficient current, as discussed above
2. Insufficient supply voltage, as in Figure 2–6

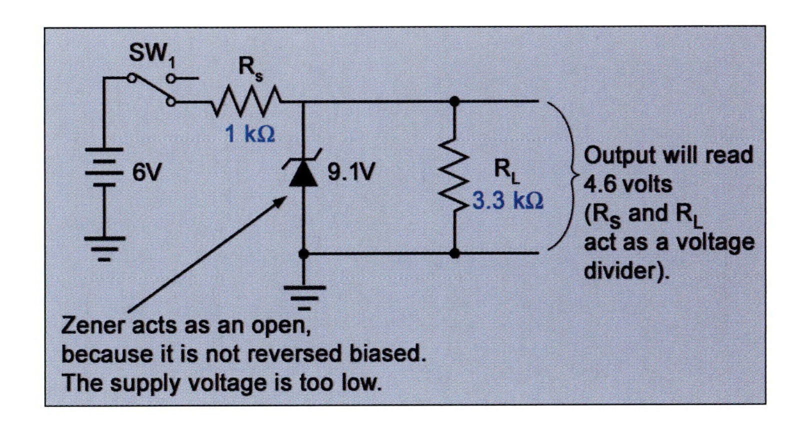

FIGURE 2–6 A Zener with reverse bias less than its Zener voltage

Actually, both of these conditions amount to the same thing. The circuit conditions cause the available voltage and current to be below the Zener level. In this condition, the Zener behaves like any reverse-biased diode. In Figure 2–6, the voltage of the source is not high enough to properly reverse bias the Zener. Therefore, the Zener behaves like any diode, and the voltage drop across the resistors is easily calculated using the voltage divider theorem.

In the earlier example (Figure 2–5), you saw that when the load resistance drops below 954 Ω, the Zener will drop out of its Zener range. This is because, at that load level, too much voltage is dropped across the R_S resistor. This makes the voltage available for the Zener too low to properly bias it.

A Forward-Biased Zener

A forward-biased Zener diode behaves exactly the same way that a normal diode behaves, as in Figure 2–7.

This circuit can be solved by a simple application of KVL:

$$-6 + iR_{1k\Omega} + V_{Z(\text{forward})} = 0$$

$$-6 + i \times 1000 + .7 = 0$$

$$i = \frac{5.3}{1{,}000} = 5.3 \text{ mA}$$

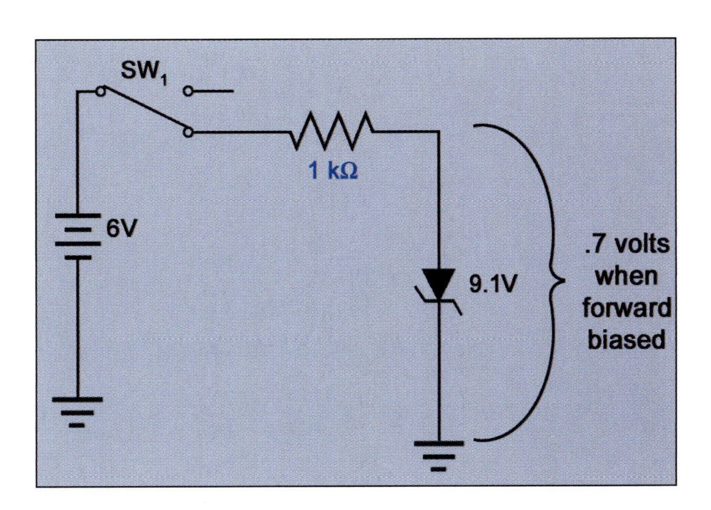

FIGURE 2–7 A forward-biased Zener diode

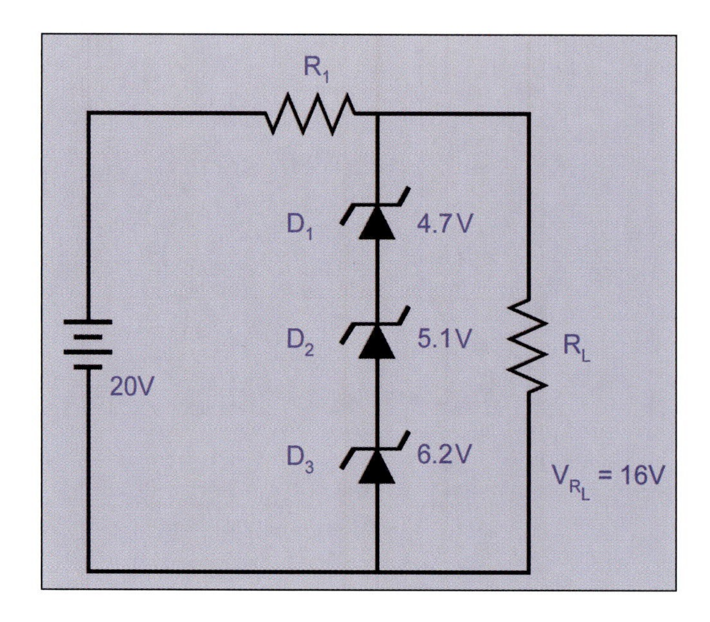

FIGURE 2–8 Reverse-biased Zener diodes in series

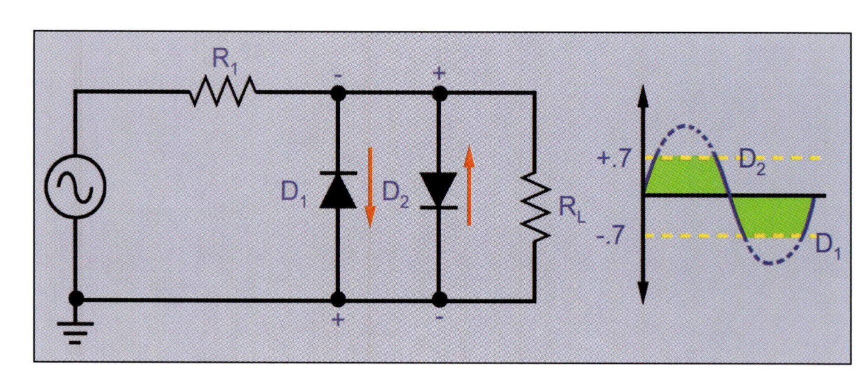

FIGURE 2–9 Regular diodes with an AC source

Reverse-Biased Zener Diodes in Series

Figure 2–8 shows an example of how Zener diodes connected in series provide an additive constant voltage drop across the circuit load. Refer to Figure 2–4. The Zener diode acts as a constant voltage source to the parallel load. The same principle applies in Figure 2–8, only the series Zener voltage drops are additive. All the voltage drops added together provide a constant voltage to the parallel load.

2.5 Diodes in an AC Circuit

Non-Zener

Figure 2–9 is a circuit and waveform for **antiparallel** diodes connected across the output of an AC supply.

Note that because of the internal .7-voltage drop of the diodes, both alternations of the AC cycle are "clipped" or blocked. The diode conducts only after its forward biasing voltage is reached (approximately .7 V). This means that until the forward bias is greater than .7 V, the diode acts as an open circuit. After .7 V is reached, the diode holds its voltage and, therefore, the voltage across R_L at .7 V. When D_1 is conducting, D_2 is reverse biased (open), and vice versa.

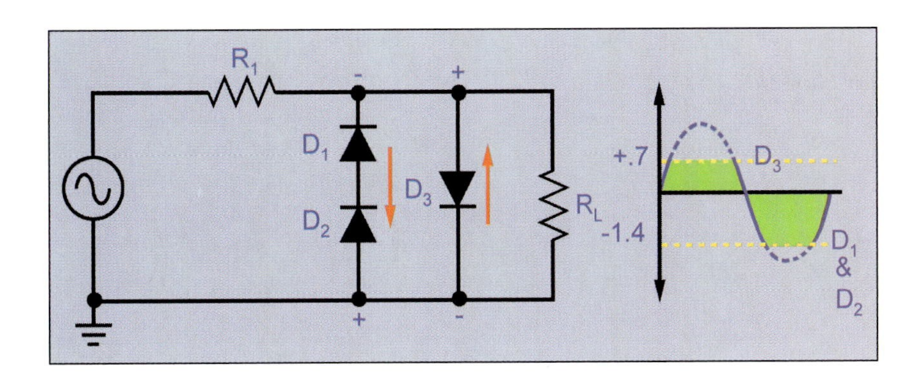

FIGURE 2–10 Clipping circuit with two diodes in one parallel leg

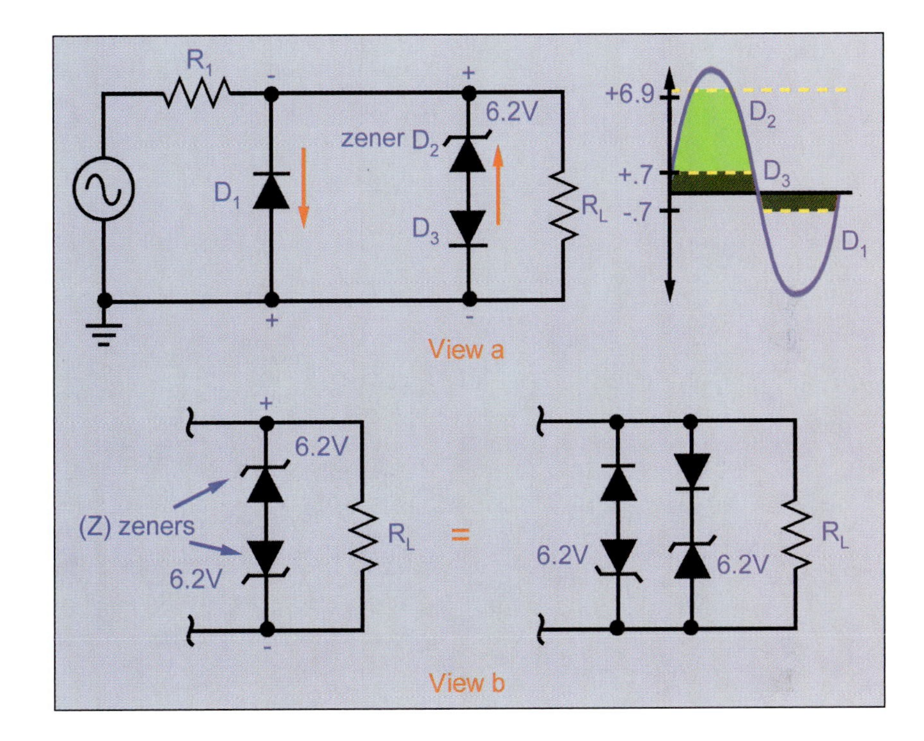

FIGURE 2–11 Clipping with a regular diode and a Zener diode

An example of multiple diodes in series and their clipping affect on the parallel load is shown in Figure 2–10. Note that the difference in this circuit from that of Figure 2–9 is that there are two diodes in series in the first parallel branch. These diodes' internal voltage drops are additive and provide a total drop of 1.4 V before they are forward biased and conduct. This is shown in the effect of the output signal voltage across the load.

The waveform shows that, in the positive direction, .7 V is developed across the load as before. In the negative direction, 1.4 V (or two diode voltage drops) is developed across the load.

Zener Diodes

Figure 2–11 shows an interesting variation on clipping. Note that in 2–11a, the Zener diode D_2 and regular diode D_3 are in series and are reversed to each other. In this connection, when the Zener is reverse biased, the diode will be forward biased, and vice versa.

On the negative half-cycle, the D_2D_3 combination will not conduct. All of the current will go through D_1 because it is forward biased, thus providing a .7-V output. During the positive half-cycle, the D_2D_3 combination will not conduct until the voltage has reached the Zener voltage (6.2 V) plus the diode forward voltage (.7 V). The resulting waveform is shown in Figure 2–11a.

Figure 2–11b shows two circuits that are functionally equivalent in regards to the output voltage to R_L when an AC voltage is applied. The back-to-back Zener diodes in the left circuit function the same in both the positive and negative cycles. This is because, in the positive or negative cycle, one Zener will be forward biased (.7 V) and the other Zener will be reverse biased (Zener voltage of 6.2 V). The final output will be +6.9 volts to −6.9 volts to R_L.

In the right circuit, the Zeners are in a series with the diodes. The diodes create a voltage drop of .7 V as the opposing Zener diode did in the left circuit. In the positive or negative cycle, the Zeners will operate at the Zener voltage of 6.2 V; therefore, the sum of the Zener and diode equal 6.9 V. Again, the final output will be +6.9 volts to −6.9 volts to R_L. This example assumes the Zeners and diodes have internal voltage drops of .7 volt.

OTHER TYPES OF DIODES

This next section shows how heavier doping of the PN junction with impurities can change the way certain diodes respond to a forward biasing current and voltage. This section will also discuss another special type diode—the PIN diode. This diode does not have a PN junction. Other special diodes such as photodiodes, light-emitting diodes, and laser diodes are not addressed in this lesson, but will be covered in a later semiconductor lesson.

2.6 Tunnel Diodes

Tunnel diodes are doped (impurities added to the P- and N-type materials) about 1,000 times more heavily than the normal PN junction diode. This doping causes the tunnel diode to be very useful in microwave and ultrahigh-frequency (UHF) devices. These high-frequency applications are possible because the tunnel diode has a very unique voltage and current relationship in the forward-biased condition—the relationship is inverse. That is, when the voltage and current reach peak value, any additional voltage will cause a decrease in forward current. This relationship is also called a "negative resistance."

Figure 2–12 shows the schematic symbol for a tunnel diode and a typical characteristic curve. Tunnel diodes are normally operated in this negative resistance region. They also can operate at higher temperatures than silicon or germanium diodes. Another use for tunnel diodes is in high-frequency oscillators.

2.7 Schottky or Shockley Diodes

These diodes are also known as hot-carrier or surface-barrier diodes, and they are used in switching circuits and microwaves. They are also

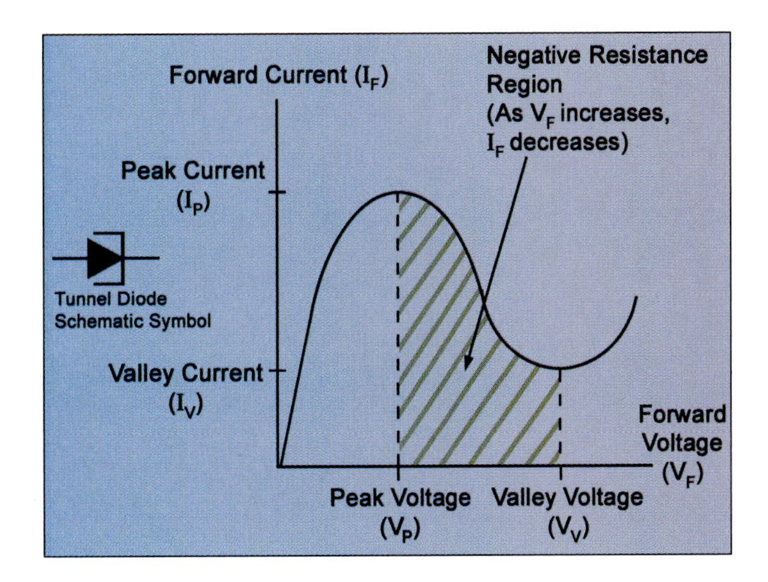

FIGURE 2–12 Tunnel diode characteristic curve

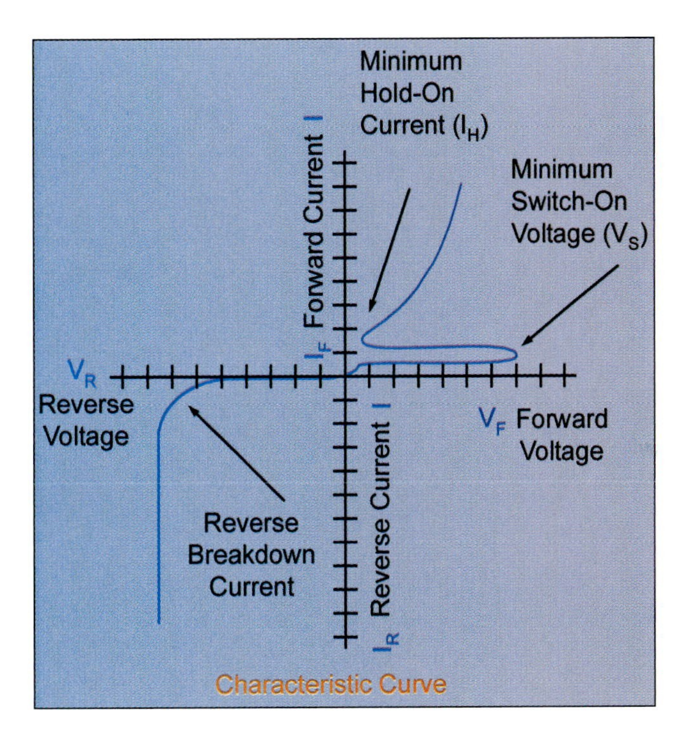

FIGURE 2–13 Characteristic curve of a Schottky diode

frequently used as a trigger device for SCRs (silicon-controlled rectifiers). When forward biased, they have a low breakover voltage of .3 volt. Their main advantage is that they reverse polarity (forward to reverse bias) almost instantaneously, allowing the diode to be used in very high-frequency ranges. This high-frequency response is made possible by one of this diode's unique characteristics—it uses metallic flakes (sort of a doping) on top of the silicon semiconductor-type material to produce its "junction." Figure 2–13 shows the characteristic curve for a Schottky diode.

The Schottky can also have two PN junctions, PNPN, but only two electrodes (anode and cathode). See Figure 2–14 for the schematic symbols and pictures of typical diode packaging. The usual forward-biased resistance for the Schottky is small, just a few ohms. This is considerably

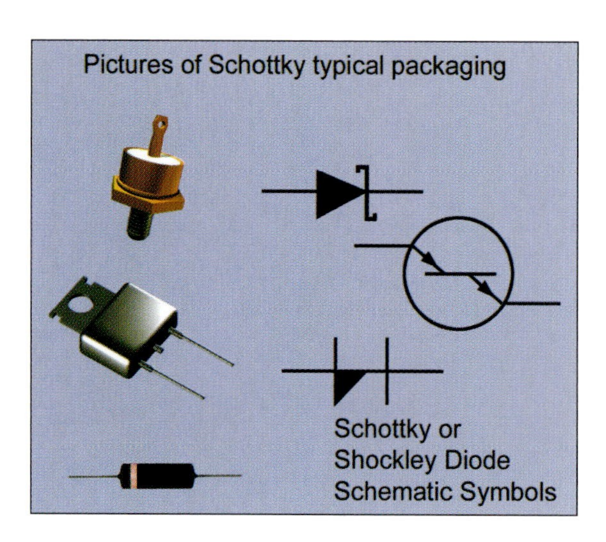

FIGURE 2–14 Schottky schematic symbols and typical examples

less than the forward-biased resistance of normal rectifier diodes. The Schottky is rated by its threshold (trigger) voltage. This is the voltage required to forward bias or turn on the diode. A Schottky can have a third electrode. If the third electrode, called a gate, is used to "fire" or turn on the diode, the Schottky is called a silicon-controlled rectifier (SCR). We will discuss SCRs in a later lesson.

2.8 PIN Diodes

The most common use of the PIN diode is in radio frequency (rf), microwave, and modulator circuits, and as switches. The PIN diode differs from the PN junction diode because it does not have a PN junction. Instead, it is made of three materials: P-type, N-type, and a pure (intrinsic) silicon slice. The intrinsic silicon slice is between the P- and N-type materials, so there is no PN junction. In fact, the I in PIN comes from the "intrinsic" material slice or layer between the P- and N-type materials. The intrinsic silicon has a very high resistance (recall that intrinsic silicon or germanium acts much like an insulator unless doped with impurities). The thicker the intrinsic material layer, the higher the resistance and the greater the breakdown voltage. See Figure 2–15 for an example of the construction of a PIN diode.

Another distinct characteristic of the PIN diode is that it does not have an actual knee voltage. Unlike the PN junction diode, there is no abrupt "turn-on" point, but rather a gradual increase in forward bias voltage. This causes a corresponding increase in forward bias current. See Figure 2–16 for a typical PIN diode characteristic curve.

The PIN family of diodes has two basic uses. One type is used in high-frequency (can be over 300 MHz) circuits as signal carriers and switches. The second PIN-type diode is used in a range of high-power diodes. This is possible because of the intrinsic layer that supports a high breakdown voltage needed for high-power applications. LEDs and laser diodes will be discussed later in this book.

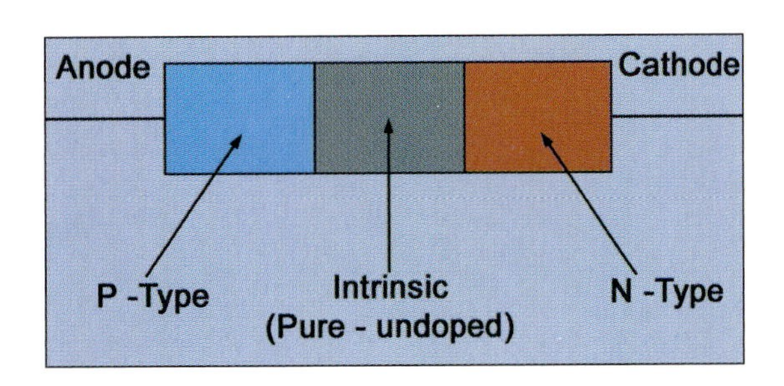

FIGURE 2–15 Construction of a PIN diode

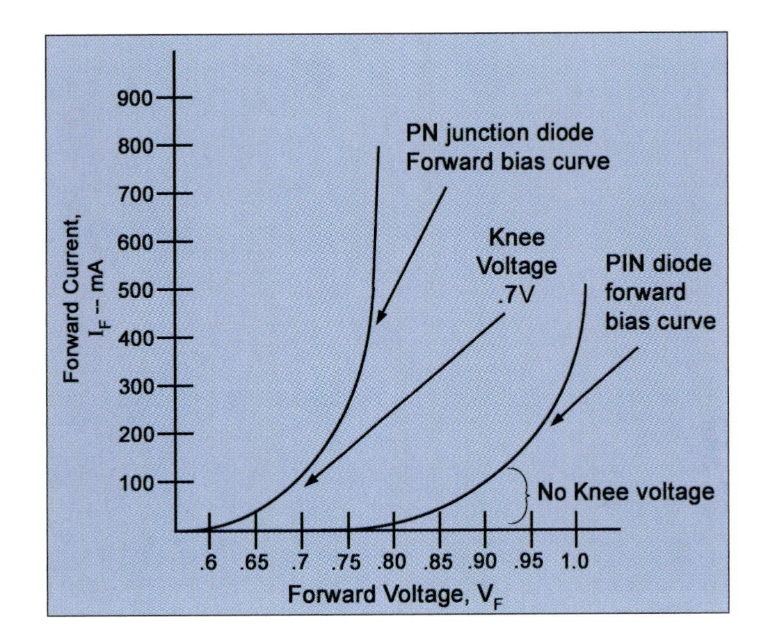

FIGURE 2–16 PIN diode characteristic curves

SUMMARY

Zener Diodes

The Zener is one of the special types of PN junction diodes: it is designed to operate in the reverse-biased condition. When reverse biased, the Zener is normally used as a voltage regulator. Zeners are designed to maintain a specified voltage drop over a wide range of currents. Zeners come in sizes (specified voltages) ranging from 1.8 volts at ¼ watt to 200 volts at 50 watts. Multiple Zeners of various breakdown voltages can be placed in series to increase the designed regulated voltage (voltage drop). In the forward-biased condition, the Zener acts much like a regular PN junction diode and has a forward voltage drop of approximately .7 volt. Critical Zener characteristics and ratings are given in Table 2–1.

Other Diode Types

Three other common non-PN junction diodes are the tunnel, Schottky, and PIN. All of these can be used with high-frequency and microwave applications.

The tunnel diode has a unique characteristic called "negative resistance." This means that voltage increases beyond peak forward voltage cause the forward current to decrease. Negative resistance allows the tunnel diode to be used as an oscillation device in high-frequency applications.

The Schottky diode is also known as the Shockley or hot-carrier diode. This family of diodes is used as a switching (on/off) device in high-frequency applications, especially microwaves, and they are frequently used as trigger devices for SCRs (silicon-controlled rectifiers). Some Schottky diodes use a metal doping on top of the silicon semiconductor material to produce its "junction." A Schottky diode that has a third electrode or gate is called a silicon-controlled rectifier (SCR).

The PIN diode is used extensively in modulator and radio frequency (rf) circuits. The PIN diode does not have a PN junction, but uses a third intrinsic or pure silicon layer between the P-type and N-type materials. The intrinsic silicon slice is where the I comes from in PIN. This slice also acts as an insulator and sets up a high-resistance barrier that provides for greater breakdown voltage ratings. The thicker the intrinsic layer, the higher the breakdown voltage. PIN diodes do not have specific 'knee' voltages but rather increase forward conduction more gradually than PN junction diodes.

REVIEW QUESTIONS

1. Describe the most common uses for the types of diodes discussed in this chapter.
2. Look at Figure 2–6. If the battery is increased to 12 volts, what value of load resistors will allow the Zener to operate in its Zener range? Assume that the Zener requires a minimum of 5 mA to operate correctly.
3. In Figure 2–9, what precautions would have to be observed to keep the two diodes from burning out due to forward current flow?
4. List at least four practical applications of a Zener diode regulator.
5. How is a PIN diode made? Of what value is the intrinsic layer?
6. The tunnel diode is normally biased to operate in its negative resistance region. How does this differ from regular diodes or Zener diodes?
7. You have a need for a regulated 11.3-V DC voltage. You have four Zener diodes with ratings of 11.2, 6.2, 5.1, and 4.7 volts. What is the best combination for your regulator?
8. In Figure 2–1b, explain what the differences might be in the application of the two types of Zener diodes that are shown.
9. Draw a circuit that will limit (clip) and AC voltage to a maximum of ±4.7 V.
10. Look at Figure 2–4. Assume the Zener $P_{D(MAX)}$ is 2 watts. What is the smallest size R_S that could be used in this circuit?

chapter 3

Power Supplies

OUTLINE

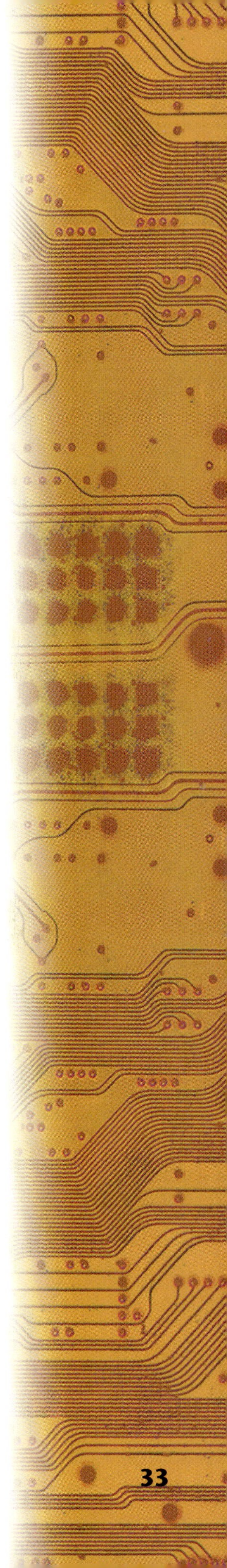

OVERVIEW

All physical processes require energy. The rate at which this energy is expended is called power. In our society, most electric power is distributed using alternating current (AC). Many electric and electronic circuits require a direct current (DC) form of power to operate properly.

In this chapter, you will learn the fundamentals of turning AC into DC. You learned something about this earlier when you saw that a diode will pass current only when it is forward biased. Thus, when an AC voltage is applied to a diode, a pulsating DC is produced. From this simple concept, by using ever more complex circuitry, AC power can be rectified into a DC that is virtually pure—almost as pure as the output from a battery.

OBJECTIVES

After completing this chapter, the student should be able to:

1. Draw schematic symbols for half-wave, full-wave, and bridge rectifiers.
2. Explain the operation of capacitors as filters.
3. Explain the operation of chokes as filters.
4. Describe the operation of voltage regulators and dividers.
5. Predict the output waveforms for different rectifier circuits.
6. Discuss the operation of voltage doublers.

GLOSSARY

Choke A coil or inductor used as a filter, often in an AC-to-DC power supply.

Effective value The magnitude of a DC waveform that generates as much power as a measured AC waveform.

Peak value The magnitude of a waveform as measured from the zero value to the peak value.

Peak-to-peak The peak-to-peak magnitude of an AC signal.

Power supply An electric circuit that is used to convert electric voltages and currents from one form to another form suitable for a particular application. Example: change the AC mains voltage into a DC voltage suitable for powering an electronic computer.

RMS value The same as effective value. RMS stands for root-mean-square.

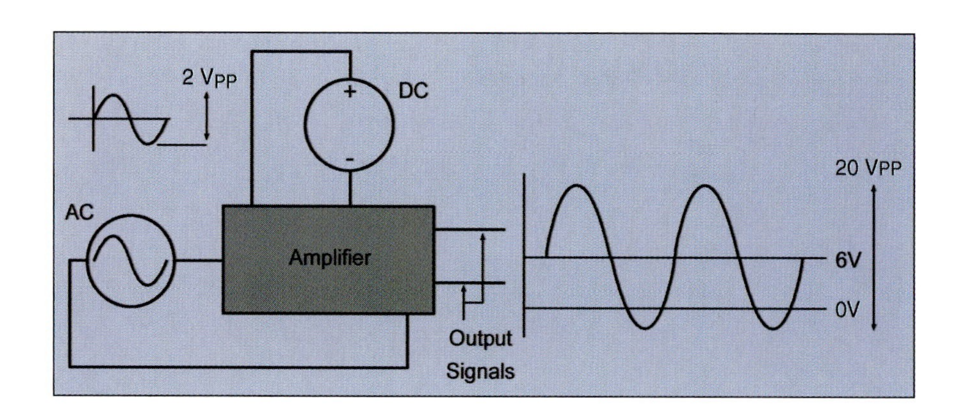

FIGURE 3–1 AC signal processing with DC control voltage

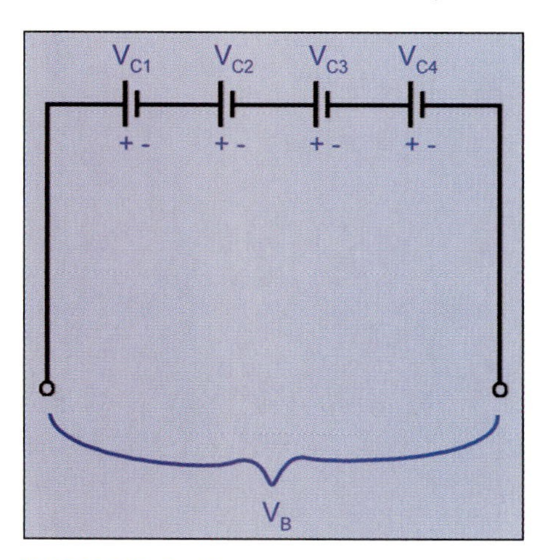

FIGURE 3–2 Battery power supply with batteries in series

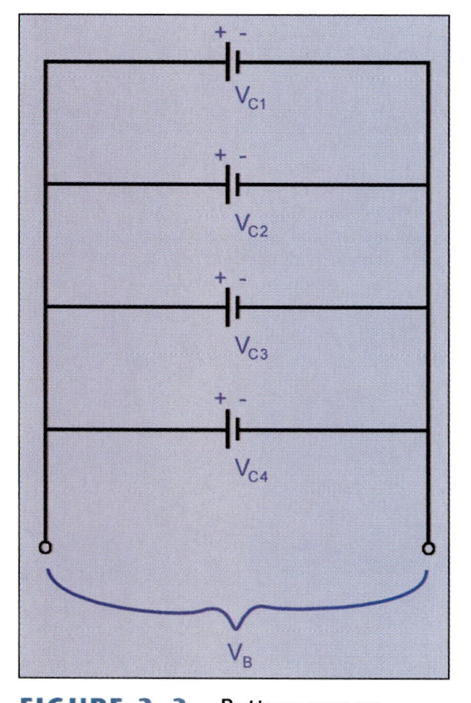

FIGURE 3–3 Battery power supply with batteries in parallel

INTRODUCTION

Earlier you learned that semiconductor devices can be forward biased by a very small voltage, typically .3 or .7 volt. Other semiconductor devices are 'turned on' with voltages ranging from 1.8 to 50 volts. These low DC voltages are used to control the semiconductor circuits in the form of forward or reverse biases on the semiconductor devices. Other applications, such as oscillators, receivers, modulators, and transmitters, use AC voltages that 'ride on' the DC control voltage to produce the 'signals' or waveforms that are processed by the various semiconductor devices.

Consider Figure 3–1 for example. In this circuit, a 2 V_{PP} signal is fed into a signal amplifier. The semiconductor circuits, possibly transistors, in the amplifier are biased with a DC supply voltage. The bias allows the transistors to control their output voltages by exactly reproducing the shape of the input magnified or amplified by a factor of 10. Notice that, as shown, the output also has a 6 VDC component.

POWER SUPPLY SYSTEM CHARACTERISTICS

3.1 Series Voltage Sources

Different **power supplies** have many different output voltages and currents, but most of them use the same basic operating principles. Figure 3–2 shows a simple battery power supply. This configuration uses batteries in series, with each cell (V_{C1} through V_{C4}) supplying 25 percent of the total supply voltage.

3.2 Parallel Voltage Sources

Figure 3–3 shows a power supply with the batteries connected in parallel. Recall that in this type of circuit, the current is additive instead of the voltage. Each cell (V_{C1} through V_{C4}) supplies 25 percent of the total source current.

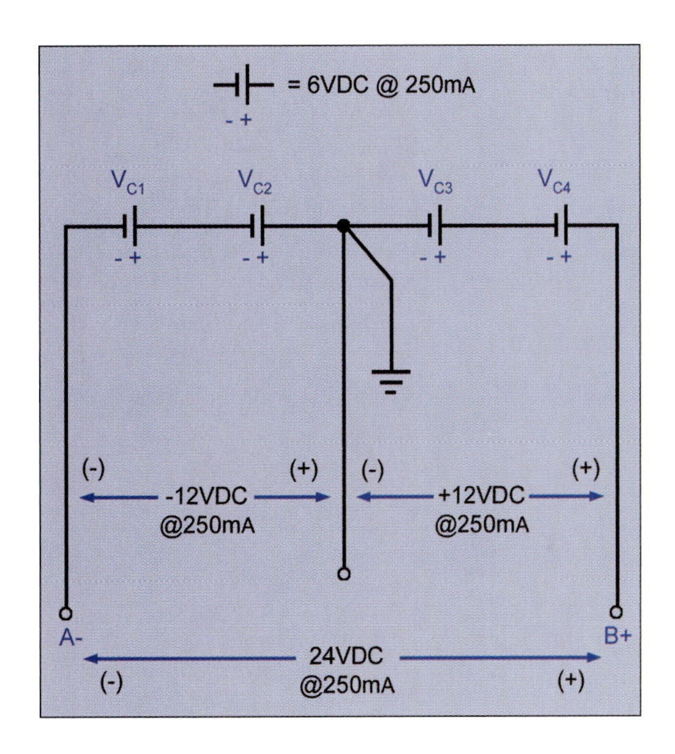

FIGURE 3–4 Dual-voltage power supply (also called a bipolar power supply)

3.3 Dual-Voltage Power Supply

By placing a reference ground between cell V_{C2} and V_{C3} in Figure 3–4, point A becomes negative to the ground reference and point B is positive to the same reference point. This allows the supply source to provide two polarities to a circuit. This is necessary with some semiconductor circuits to properly bias transistors and diodes. A dual-voltage power supply is often called a bipolar supply.

3.4 The Power Supply Block Diagram

The power supplies shown in this chapter use 120 VAC, 60 Hz input (common household electric power in North America) and produce a range of DC voltage forms. The most common power supplies have three major operational sections: the transformer, the rectifier, and the filter. In other lessons on DC and AC theory, you learned how transformers and filters operate. In the last two lessons, you analyzed the rectifier capabilities of diodes. In this lesson, all three operations are put together to produce a power supply.

This lesson will build on the five steps involved in a complete power supply system. These steps are:

- Transform the AC voltage.
- Rectify the AC to pulsating DC.
- Filter the pulsating DC output.
- Regulate the output.
- Divide the output.

Figure 3–5 shows a simple block diagram of the first four steps listed above, and the following paragraphs describe each process in more detail.

Transform the AC Voltage

This process is done by the transformer in Figure 3–5. The maximum DC output voltage will depend upon the amount of AC supplied to the rectifier and the filter. On the other hand, excessive pulsing DC voltage into the filter will cause the filter to run hot and probably result in excessive heat loss. By using a transformer, the voltage can be changed to an appropriate level to produce the desired DC output.

Rectify the AC to Pulsating DC

The box labeled "Full-Wave Rectifier" has this job. In fact, you recall that a rectifier is a circuit that changes AC into pulsating DC. Not all rectifiers are necessarily full wave; however, the full-wave rectifier is more efficient, and so most large power supplies use full-wave rectifiers.

Filter the Pulsating DC Output

Most electric and electronic circuits that use DC require a smooth DC similar to that produced by a battery. By methods described in detail later in this chapter, the filter changes the pulsating DC to a purer DC by removing at least part of the AC component.

Regulate the Output

As the load current changes, the output voltage of a power supply tends to change. This is caused by the increased internal voltage drop inside the power supply. A regulator, such as a Zener diode, will maintain the output voltage at some predetermined value so that the load is always supplied at the needed voltage level. Note from Figure 3–5 that the regulator will also help with the filtering process.

Divide the Output

Some circuits require multiple DC voltage levels and/or multiple DC voltage polarities. A voltage divider circuit can be used to provide different voltages from the same supply. Such a circuit may involve a simple resistance voltage divider, a capacitor-type voltage doubler, or even multiple Zener diodes from which regulated voltages are taken. The voltage divider may be located before or after the regulator in a power supply. No divider is shown in Figure 3–5.

FIGURE 3–5 Basic power supply block diagram

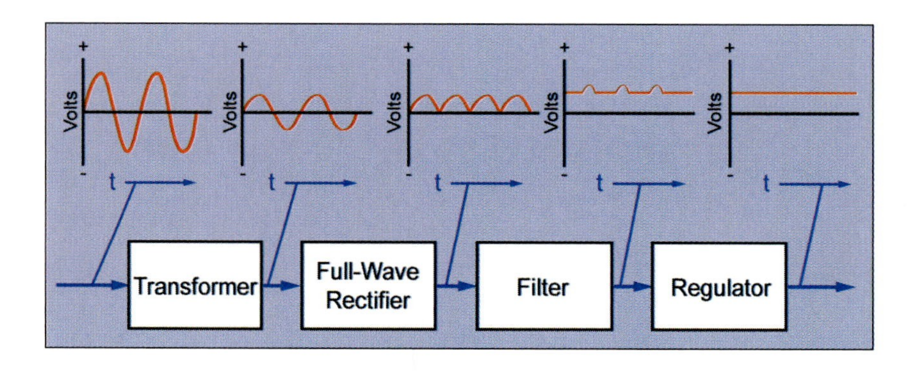

HALF-WAVE RECTIFIERS

3.5 Rectifying Characteristics

Recall from the past lessons that one of the main purposes of a diode is to rectify AC voltage. Also recall that there are different forms of measurements for these rectified voltages—peak-to-peak, peak, average, and rms. The **rms value** is also called the **effective value**. The rms rating is used to determine the effective output of an AC machine. We use this because the machine produces the peak only for an instant then returns through its cycle. When the rms values of voltage and current are used, the values identify the same amount of power as a like value of direct current.

An AC waveform is rectified by turning it into pulsating DC. Figure 3–6a shows an AC waveform before and after it is rectified by a PN junction diode. Also recall that for the AC signal to be processed, the diode must be forward biased. This forward bias voltage is typically .7 volt. Figure 3–6b shows a typical half-wave rectifier circuit that would produce the output waveform shown in Figure 3–6a. Note that in this case, the power is being supplied by a transformer. As stated earlier, the voltage may need to be increased or decreased. This is done by step-up and step-down transformers respectively. The two types are shown schematically in Figure 3–6c. The transformer in the schematic of Figure 3–6b is called a step-down transformer because the input is 120 VAC and the output is 15 VAC.

The voltage is reduced by the induction of the primary (input) coil into the secondary (output) coil of the transformer. Transformers may either step up (increase) or step down (decrease) the primary side voltage of the transformer to the secondary side output. This is done by increasing or decreasing the amount of wire coils (number of turns in the

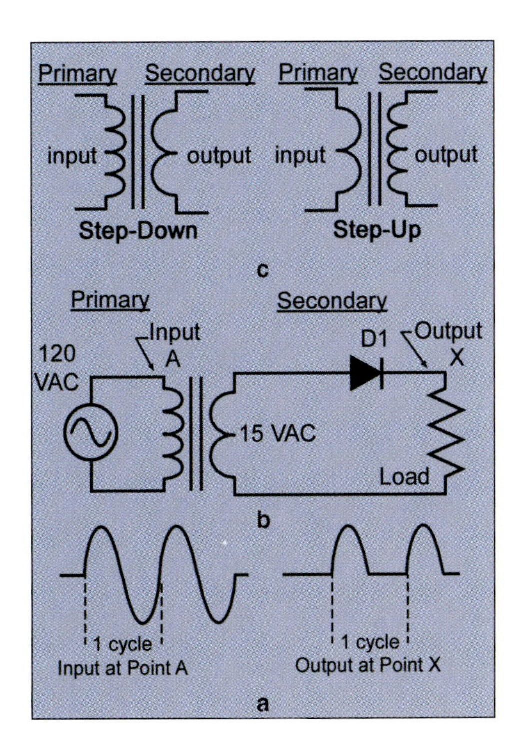

FIGURE 3–6 Elements of a half-wave rectifier; a) input and output waveforms, b) simple circuit, c) step-up and step-down transformers

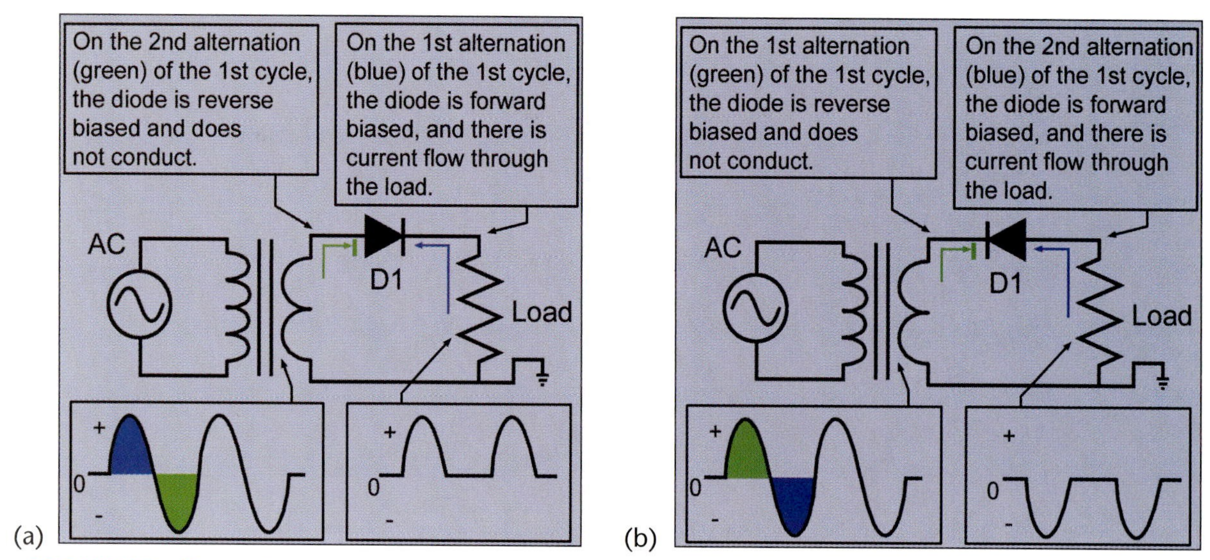

FIGURE 3–7 Half-wave rectifier; a) positive output, b) negative output

coil) of the transformer output (secondary) as compared to the transformer's input (primary) side windings. There is one class of transformers called isolation transformers that have the same number of turns on primary and secondary. The voltage does not change but the transformer does allow "isolation" from the primary to the secondary.

Figure 3–7a shows the polarity of the output voltage and path of current flow through the rectifier circuit. Note that by reversing the diode as shown in Figure 3–7b, the current reverses direction through the diode and through the load. This changes the voltage polarity across the load. In Figure 3–7a, only the positive peaks are passed to the load, and in Figure 3–7b, only the negative peaks are passed.

3.6 Circuit Analysis

Review of AC Measurements

The various values for AC voltage and current levels are almost always given in terms of the rms or root-mean-square value. This includes transformer output voltages. There are, in fact, several values for voltage and current that can be specified for an AC or a pulsating DC voltage or current. The values and their meanings are listed in Table 3–1. Their location and labels for an AC sine wave are shown in Figure 3–8.

Using the information, the values for the typical 120 V house supply are as follows:

$$V_{PP} = 2 \times 120 \times \sqrt{2} = 339.4 \text{ V}$$

$$V_P = \sqrt{2} \times 120 = 169.7 \text{ V}$$

$$V_{AVE} = 0 \text{ V}$$

$$V_{RMS} = 120 \text{ V}$$

Remember that voltages and currents in AC power systems are almost always measured and discussed based on their rms value.

Table 3–1 Key Values for Various Types of Waveforms

Symbol	Description	Value: Sine wave[1]	Value: Half-wave rectified[2]	Value: Full-wave rectified	Value: DC
V_{PP}	**Peak-to-peak** value. The distance from the highest peak to the lowest peak.	$2 \times V_P$	V_P	V_P	V_P
V_P	The distance from the zero line to the highest **peak value**, usually measured up.	$V_{RMS} \times \sqrt{2}$	$V_{RMS} \times 2$	$V_{RMS} \times \sqrt{2}$	V_P
V_{AVE}	The arithmetic average of all values. V_{AVE} is also the DC (not rms) value that would be measured by a non-rms, DC responding meter.	0	$\frac{V_P{}^2}{\pi}$	$\frac{2 \times V_P}{\pi}$	V_P
V_{RMS} V_{EFF}	The effective value or heating value of the waveform. A DC value that will produce the same amount of heat in a resistor.	$\frac{V_P}{\sqrt{2}}$	$\frac{V_P}{2}$	$\frac{V_P}{\sqrt{2}}$	V_P

[1]A sine wave has no average value because it is above the 0 line as much as it is below.
[2]$\pi = 3.14159\ldots$

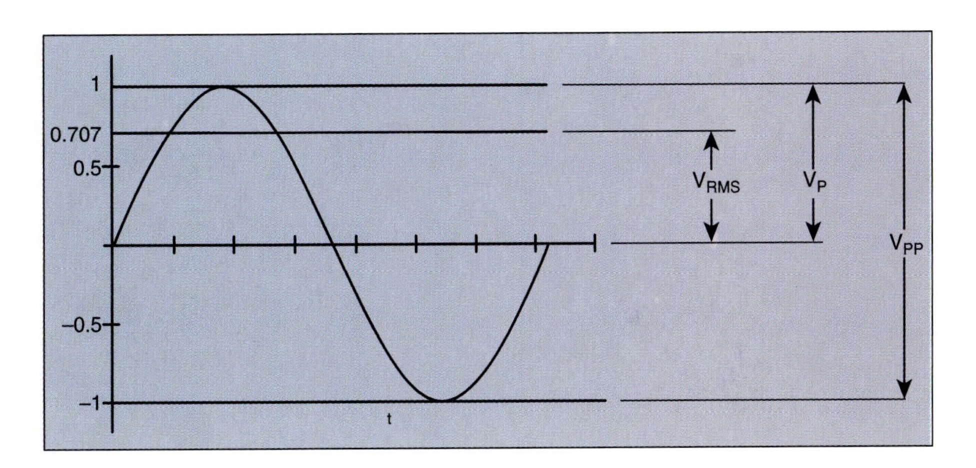

FIGURE 3–8 A sine wave and its various voltage values

EXAMPLE 1

Refer to Figure 3–9 and solve for the load V_{EFF}, V_{AVE}, and I_P and for the output waveform configuration.

The voltage out of the secondary side of the transformer is 24 VAC peak-to-peak.

From Table 3–1, the effective voltage of the transformer's secondary side is:

$$V_{EFF} = \frac{V_P}{\sqrt{2}} = \frac{V_{PP}/2}{\sqrt{2}} = \frac{12}{\sqrt{2}} = 8.49 \text{ V}$$

The half-wave output to the load has a peak voltage equal to the peak of the transformer output minus the forward drop across the diode:

$$V_{P\text{-Load}} = V_{P\text{-xfmr}} - V_D = 12 - .7 = 11.3 \text{ V}$$

FIGURE 3–9 Half-wave rectifier circuit analysis

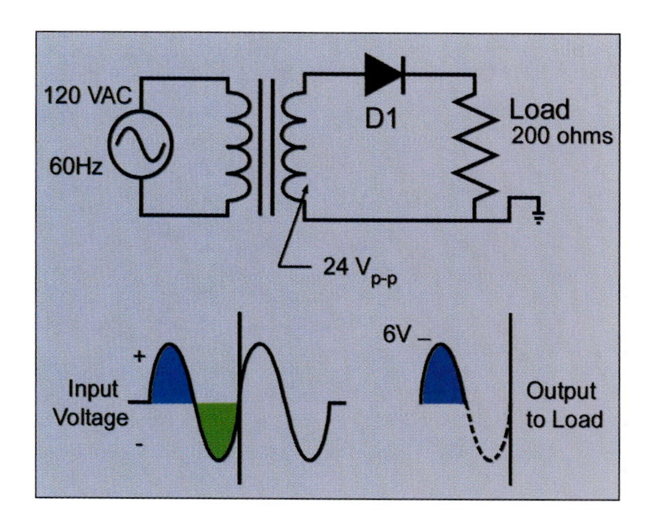

From Table 3–1, the average and rms voltage for the load are:

$$V_{AVE} = \frac{V_P}{\pi} = 3.60 \text{ V}$$

$$V_{RMS} = \frac{V_P}{2} = 5.65 \text{ V}$$

The peak current through the load can now be calculated as follows:

$$I_P = \frac{V_P}{R_L} = \frac{11.3}{200} = 56.5 \text{ mA}$$

From Table 3–1, the average and the rms value of the current are:

$$I_{AVE} = \frac{I_P}{\pi} = \frac{56.5 \text{ mA}}{\pi} = 17.98 \text{ mA}$$

$$I_{RMS} = \frac{I_P}{2} = \frac{56.5 \text{ mA}}{2} = 28.25 \text{ mA}$$

The voltage waveform is shown in Figure 3–10.

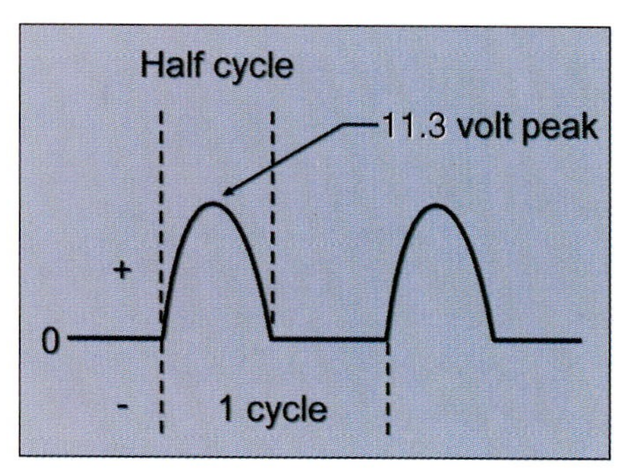

FIGURE 3–10 Half-wave load voltage waveform

FULL-WAVE RECTIFIERS

3.7 Rectifying Characteristics

A full-wave rectifier uses two diodes and a center tap on the secondary of the transformer to produce a positive output voltage on both positive and negative alternations of the AC input voltage cycle. This waveform is pulsating direct current (DC) with half the peak voltage of the secondary (less the diode forward voltage drop) because of the center tap. Figure 3–11 shows a simple full-wave rectifier and accompanying output waveform.

FIGURE 3–11 Simple full-wave rectifier circuit

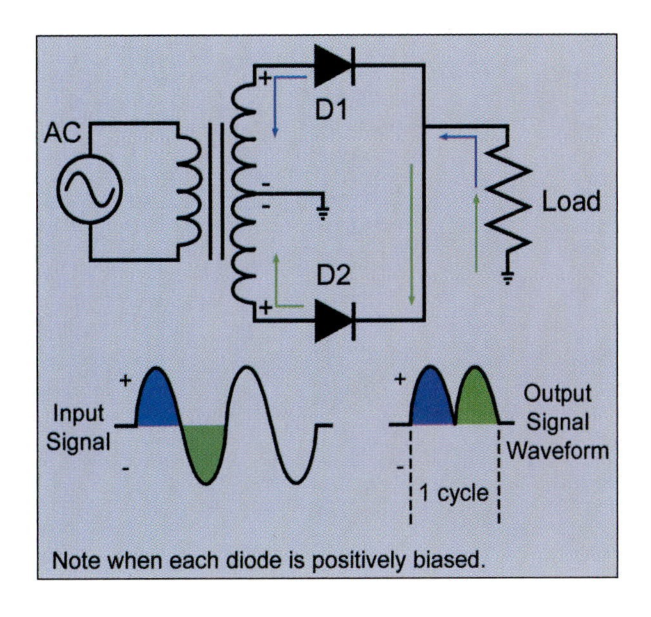

FIGURE 3–12 Solid-state rectifier case styles

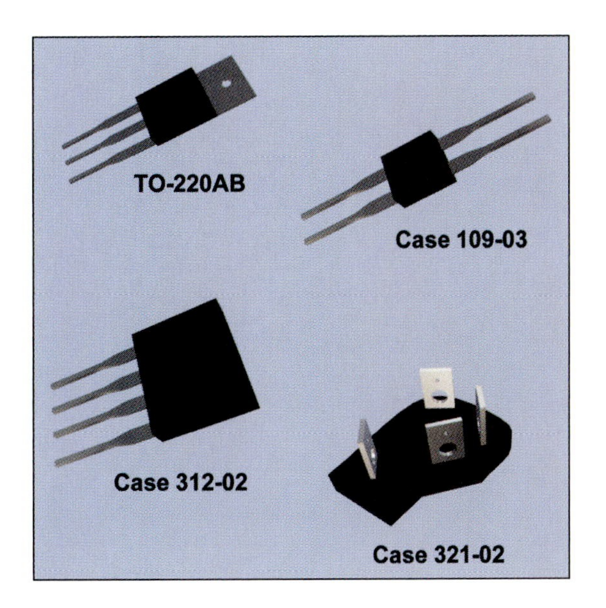

Note that by using a center tap referenced to ground, each half-cycle produces an equal but opposite voltage across the diodes. When D_1 is conducting, D_2 is reverse biased with respect to ground and the load. The opposite is true when D_2 conducts—D_1 is reverse biased with respect to ground and the load. With this arrangement, each diode conducts and allows a voltage drop across the load on alternating half-cycles. An output waveform is produced that has a positive peak for each half-cycle of the AC input voltage or current.

Figure 3–12 shows various rectifier case styles. Case TO-220AB is a half-wave 2-diode rectifier. The remaining cases are full-wave bridge 4-diode rectifiers. The full-wave bridge will be discussed later.

3.8 Circuit Analysis

EXAMPLE 1

Using Figure 3–13, solve for the load voltages: V_P, V_{EFF}, V_{AVE}, and I_P.

Solution:
The secondary of the transformer is p-p 24 VAC and it is center tapped.

Note that the peak voltage supplied to the load from the secondary is only half value because of the center-tap reference. The effective load voltage is:

$$V_{P\text{-Load}} = 6 - 0.7 = 5.3 \text{ V}$$

Recall that a full-wave rectifier has twice as many DC output pulses as the half-wave rectifier. From Table 3–1, the average and rms load voltages load are:

$$V_{AVE} = \frac{2 \times V_P}{\pi} = \frac{10.6}{\pi} = 3.37 \text{ V}$$

$$V_{RMS} = \frac{V_P}{\sqrt{2}} = \frac{5.3}{\sqrt{2}} = 3.75 \text{ V}$$

The peak load current through the load can now be calculated.

$$I_P = \frac{V_P}{R_L} = \frac{5.3}{200} = 26.5 \text{ mA}$$

Again from Table 3–1, the average and rms load currents are:

$$I_{AVE} = \frac{2 \times I_P}{\pi} = 16.87 \text{ mA}$$

$$I_{RMS} = \frac{I_P}{\sqrt{2}} = 39.95 \text{ mA}$$

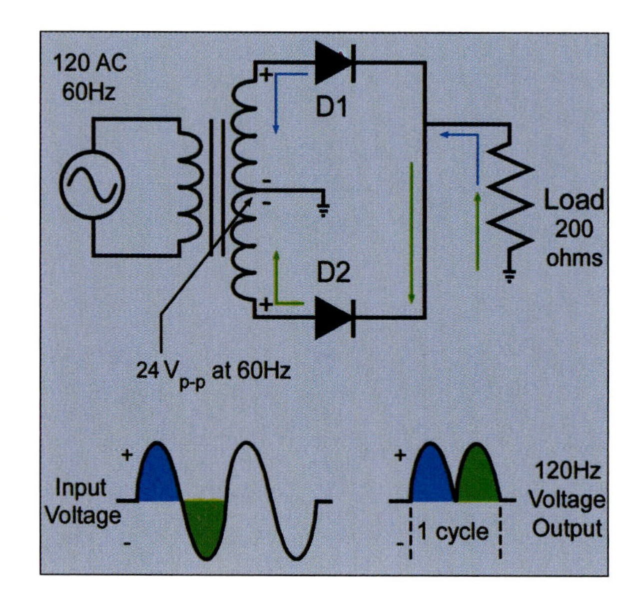

FIGURE 3–13 Full-wave rectifier circuit

Note that with half the voltage, the full-wave rectifier supplies almost the same average and rms current to the load; moreover, the full-wave pulsating DC is at twice the input or line frequency, has less ripple, and will allow better filtering. Also, the full-wave "center-tap transformer" rectifier peak voltage and current are 50 percent less than the half-wave rectifier. This allows both the voltage and current rating of the full-wave rectifier diodes to be 50 percent less in comparison to the half-wave rectifier diode.

Of course, this analysis does not mean that you can arbitrarily use half the voltage when designing a full-wave rectifier as compared to a half-wave. It does, however, show that the extra pulse provided for each cycle makes a full-wave rectifier a more efficient and generally better approach.

FULL-WAVE BRIDGE RECTIFIER

3.9 Rectifying Characteristics

One of the most useful and common rectifiers is the full-wave bridge rectifier (bridge rectifier). It does not use a center-tap transformer, but uses four diodes to rectify the incoming AC voltage or current. Figure 3–14 shows a simple bridge rectifier circuit. Notice that, like the full-wave rectifier just discussed, the pulsating DC output frequency is double that of the AC input or line frequency.

Another characteristic of the bridge rectifier is that the voltage drop across the diodes will be approximately double that of the center-tap, full-wave rectifier because there are two diodes in each conducting path.

3.10 Circuit Analysis

The current path in Figure 3–14 is shown in blue for the first half-cycle of the AC input voltage. The second half-cycle current path is shown in green. As the input AC builds to a positive (blue + symbol), the electrons flow from the negative terminal of the secondary through D_2 to the load. Electrons continue to flow through the load and on through D_3 back to the positive terminal. This completes the first half-cycle of the AC input voltage.

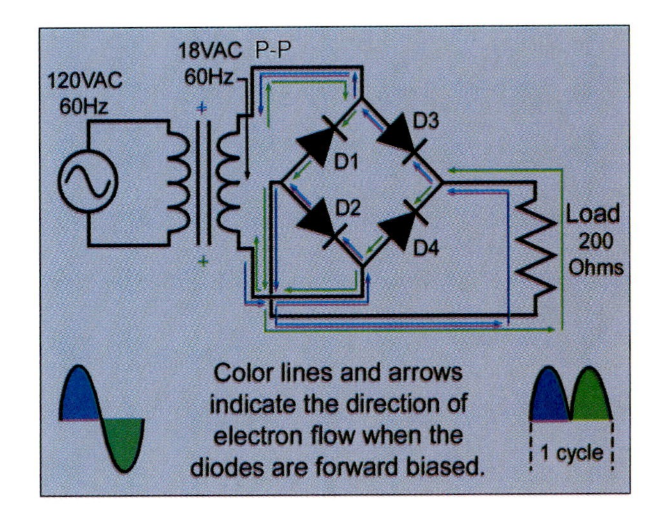

FIGURE 3–14 Bridge rectifier circuit

The second half-cycle builds a positive potential on the secondary side of the transformer (green + symbol). Electrons now reverse direction and flow from the negative secondary terminal through D_1 to the load. The electrons continue through the load and on through D_4 back to the positive terminal. This completes the second half-cycle of the AC input. Notice that the voltage or current always flows the same direction through the load, no matter which input half-cycle (negative or positive) is generating the signal.

EXAMPLE 1

Refer to Figure 3–14 and solve for the following load voltages: V_P, V_{EFF}, V_{AVE}, and I_P.

Solution:
The peak voltage across the load is:

$$V_{P-Load} = V_{P-xfmr} - (V_{D1} + V_{D4}) = \frac{V_{PP-xfmr}}{2} - (V_{D1} + V_{D4}) =$$

$$9 - 1.4 = 7.6 \text{ V}$$

From Table 3–1, the average and effective output voltages are:

$$V_{AVE} = \frac{2 \times V_P}{\pi} = \frac{2 \times 7.6}{\pi} = 4.84 \text{ V}$$

$$V_{RMS} = \frac{V_P}{\sqrt{2}} = \frac{7.6}{\sqrt{2}} = 5.37 \text{ V}$$

Using Ohm's law, the peak, average, and rms load currents are:

$$I_P = \frac{V_P}{R_L} = \frac{7.6}{200} = 38 \text{ mA}$$

$$I_P = \frac{V_{AVE}}{R_L} = \frac{4.84}{200} = 24.2 \text{ mA}$$

$$I_P = \frac{V_{RMS}}{R_L} = \frac{5.37}{200} = 26.85 \text{ mA}$$

Figure 3–15 summarizes some of what you have learned so far. There are three basic types of rectifiers: the half-wave, the full-wave center tap, and the full-wave bridge. Each schematic and output waveform is shown in the three views presented.

Regulated power supplies using transistors will be discussed in later chapters. Finally, there are other types of common power supplies that exist but will not be discussed in this text, such as 3-phase rectification and phase inverters.

FILTERS

A filter is a device that helps to change the pulsating DC into a smoother DC output. It is called a filter because it filters or eliminates much of the AC part of the waveform and leaves only DC. Virtually all

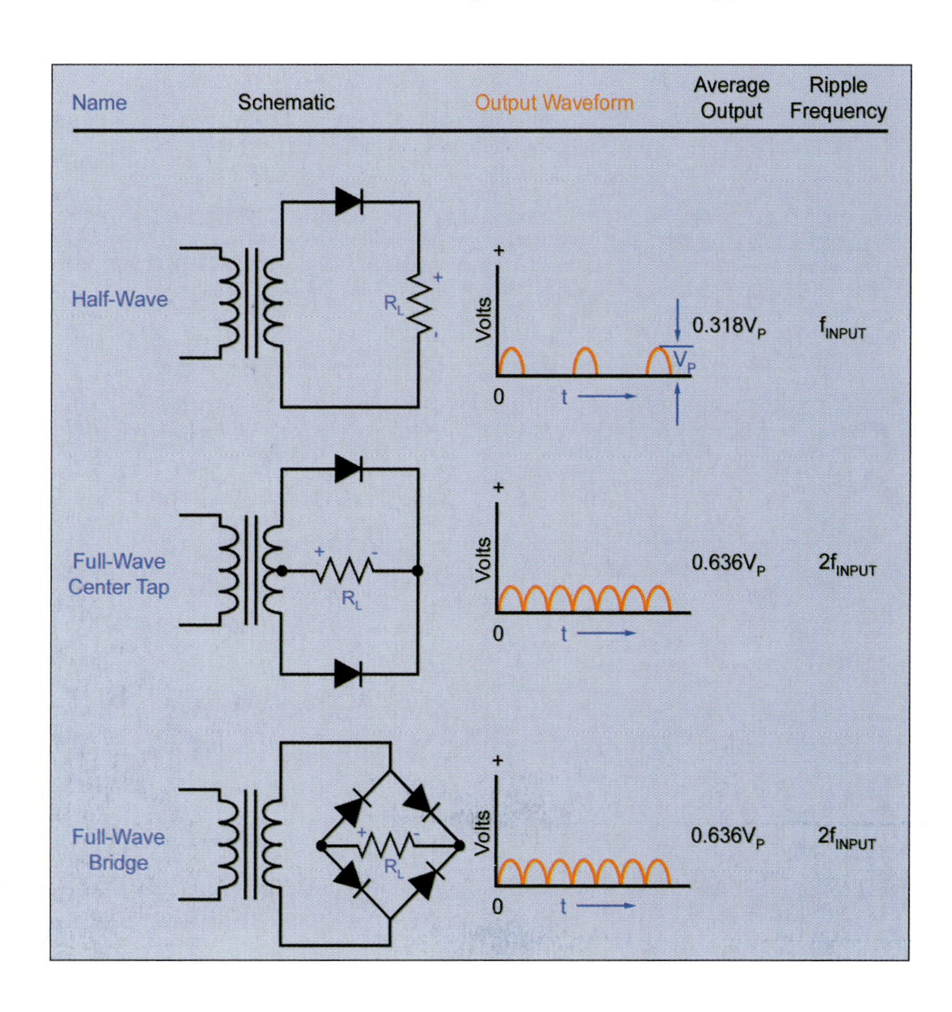

FIGURE 3–15 Rectified AC outputs

of the so-called passive filters use either capacitors or inductors. Although their actual connections vary, the basic principle of both is the same. The capacitor or inductor absorbs energy during the peak of the voltage pulses and then releases it as the voltage falls off. Thus, the filter supports the voltage during its discharge cycle and the pulsations are decreased.

3.11 Capacitors

In earlier lessons in AC theory, you learned that capacitors perform a variety of jobs such as storing an electrical charge to produce a large current pulse. Two conductors, usually plates, separated by some type of insulating material, called the dielectric, make a capacitor. The most common type of capacitor that you will use in power supply work is the electrolytic capacitor. See Figure 3–16 for pictures of actual capacitors.

Voltage Ratings

The voltage rating of the capacitor is actually the working rating of the dielectric. The voltage rating is extremely important, and the life of the capacitor will be greatly reduced if it is exceeded. The voltage rating indicates the maximum amount of voltage the dielectric is intended to withstand without breaking down. If the voltage becomes too great, the

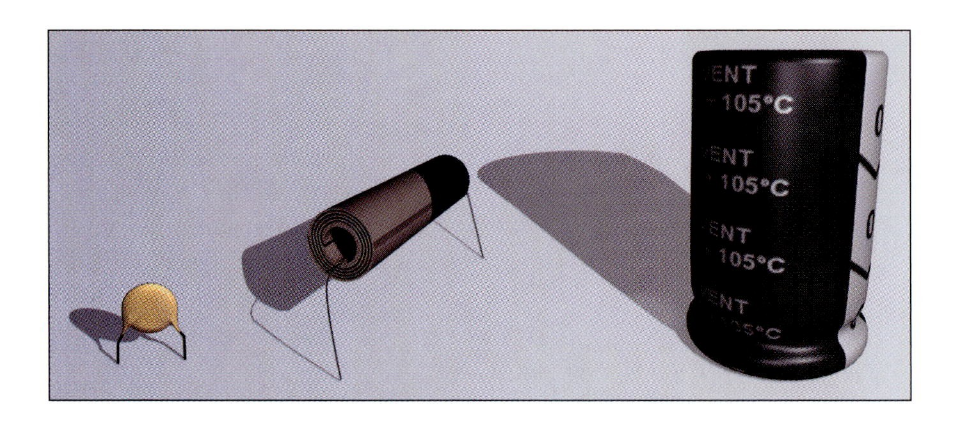

FIGURE 3–16 Types of capacitors

dielectric will break down, allowing current to flow between the plates. At this point, the capacitor is shorted.

For electrolytic capacitors, the voltage rating (also called the DC working voltage) is the DC or average voltage that is applied to the capacitor.

Capacitors as Filters

Capacitors oppose a change in voltage. When connected to alternating current, current will appear to flow through the capacitor. The reason is that in an AC circuit, the polarity is continually changing, causing the current to change direction. This flow of current in and out of the capacitor constitutes current flow through a load connected in the circuit. As noted, the change in voltage in the circuit is opposed by the charge stored in the capacitor. As the voltage potential goes negative from its peak value, the charge that was stored in the capacitor from the peak voltage now discharges into the circuit. This discharging voltage helps prevent the peak voltage from going negative and thus reduces the ripple in the rectified AC output.

Figure 3–17 shows the voltage opposition effect of a "filter" capacitor in a rectifier circuit. The term filter refers to the capacitor's ability to reduce the pulsating DC output ripple. Recall that a capacitor requires five (5) time constants to charge or discharge. To effectively reduce the output ripple, the capacitor should not be allowed to completely discharge between half-cycles. The next section of this lesson will explain this concept in detail.

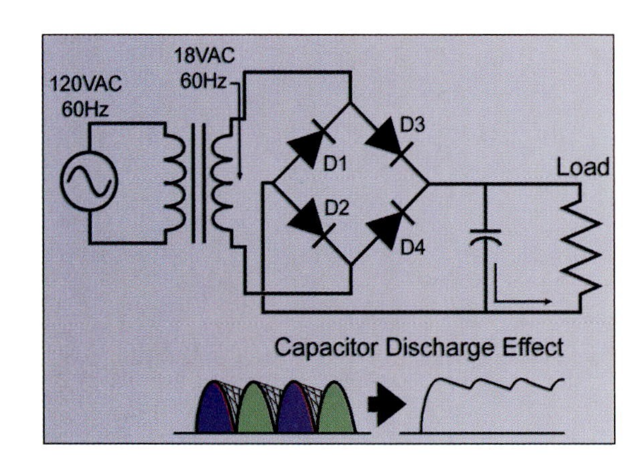

FIGURE 3–17 Bridge rectifier with filter capacitor

3.12 RC Time Constants

Theory

Capacitors charge and discharge at an exponential rate. The curve is divided into five time constants, and each time constant is equal to 63.2 percent of the remaining value. This pattern continues until the capacitor is fully charged. See Figures 3–18a and 3–18b for charge and

FIGURE 3–18 Capacitor time curves; a) charging, b) discharging

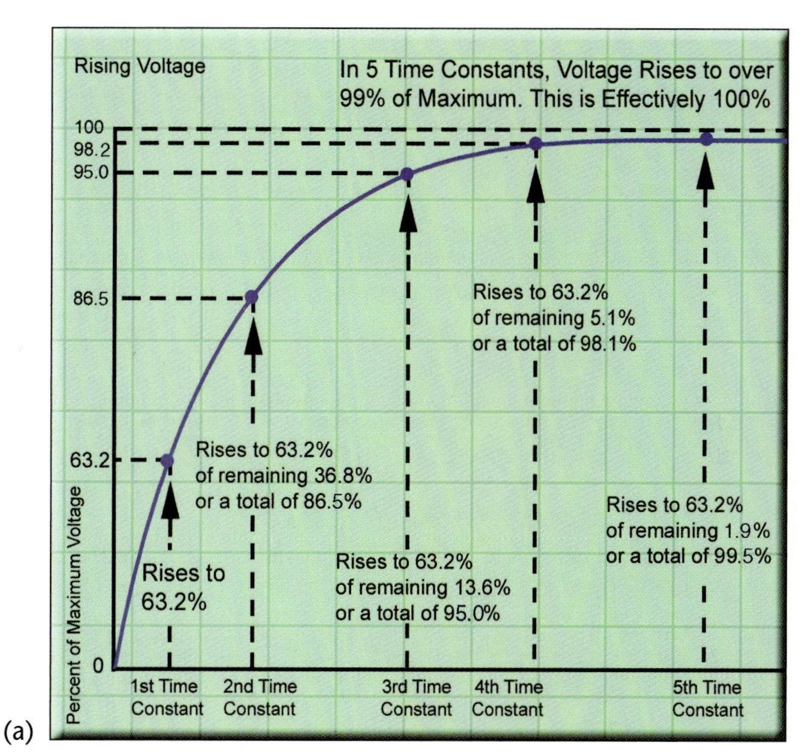

(a)

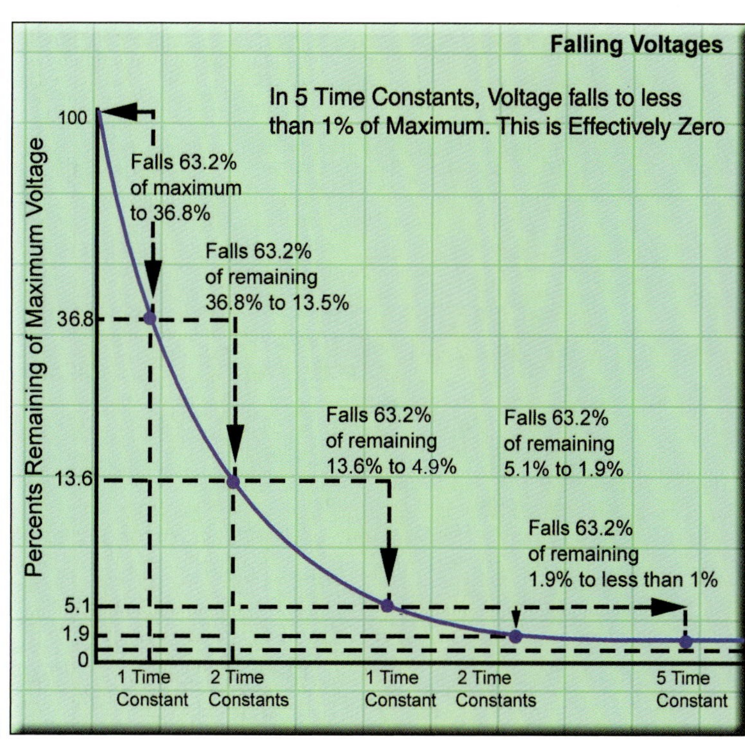

(b)

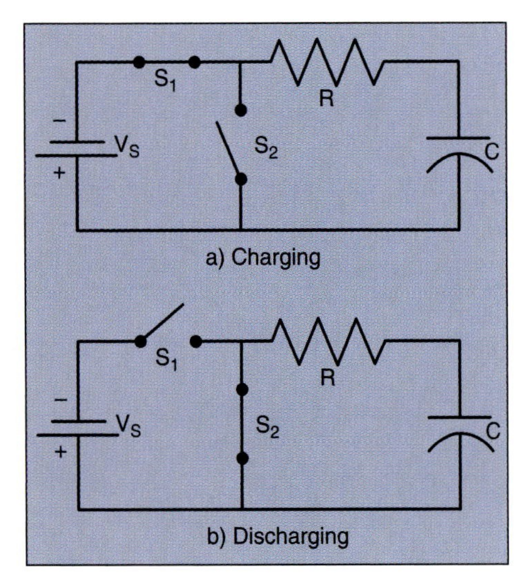

FIGURE 3–19 Capacitor circuits; a) charging, b) discharging

discharge times. This rate can be changed by changing the resistance placed in parallel with the capacitor. Together they form what is called an RC time constant. R stands for resistance and C stands for capacitance. The formula for calculating the required time constant is:

$$T = RC$$

Where:

T = time in seconds
R = resistance in ohms
C = capacitance in farads

The actual formulas that govern the charge and discharge of the capacitor through a resistor can be analyzed by looking at the circuits shown in Figure 3–19.

Charging When switch S_1 is closed, the capacitor will start to charge. The voltage across the capacitor at any time after the switch closes is given by $V_C = V_S\left(1 - e^{-\left(\frac{t}{RC}\right)}\right)$. Although this formula may seem formidable, it is easily calculated using a scientific calculator, electronic spreadsheet, or other such tool.

Table 3–2 shows the values for V_C as calculated from the above formula for t = 0, 1, 2, 3, 4, and 5 seconds, assuming the $T = RC$ = 1 second.

Notice that the values calculated from the formula agree with the values shown in Figure 3–18a. In fact, that is where the values came from.

Discharging If you open S_1 and immediately close S_2, the capacitor will discharge through the resistor. The values of voltage on the capacitor at t = 0, 1, 2, 3, 4, and 5 seconds are shown in the third row of Table 3–2. These values are calculated from the discharge formula, which is $V_C = V_S \times e^{-\left(\frac{t}{RC}\right)}$

Table 3–2 Capacitor Voltages for Charge and Discharge (Figure 3–19)

	Time (t) in seconds					
Condition	**0**	**1**	**2**	**3**	**4**	**5**
V_C—Charging S_1 closed S_2 open	0	0.632 V_S	0.865 V_S	0.950 V_S	0.982 V_S	0.993 V_S
V_C—Discharging S_1 open S_2 closed	V_S	0.368 V_S	0.135 V_S	0.0498 V_S	0.0183 V_S	0.0067 V_S

EXAMPLE 1

You can calculate the effect of different sizes of filter capacitors by using the $T = RC$ formula.

Assume a load of 200 ohms and a filter capacitor of 50 μF (50×10^{-6} farads). The time required for 1 time constant = .01 seconds. ($T = RC = (200) \times (50 \times 10^{-6}) = .01$ second)

This is about the same as a half-cycle for a 60 Hz line voltage. The capacitor will discharge about 63 percent of its charge, causing quite a bit of ripple.

Increasing the capacitor's size will increase the time constant proportionally. With an increase in time constant, the length of discharge decreases and the amount of ripple will decrease.

Using the same formula ($T = RC$), assume a load of 200 ohms and increase the capacitor to 500 μF. The required time for one time constant would be .1 second. This means there is a long time required for capacitor discharge compared to the time allowed by the circuit. Little charge will be lost by the capacitor, and the DC output ripple will be greatly reduced. See Figure 3–20 for an example of ripple outputs using these different filter capacitors.

Notice that the overall DC voltage output of a filtered rectifier will be higher than a nonfiltered one. This is because the charge on the capacitor keeps the voltage across the load near peak value.

As you look at the time constant formula, $T = RC$, you will notice that if resistance decreases, capacitance has to increase for time to remain the same. A practical application of this is when the load starts to draw a lot of current (resistance decreases), the capacitor must have a higher value to maintain the same filtering (ripple reduction) capabilities. This is one of the reasons why capacitors for DC machinery usually have very large values.

3.13 Chokes

Coils (inductors) that are used as rectifier filtering devices are called **chokes**. Chokes get their name from the function they perform in the rectifier circuit. They "choke," or reduce, the current ripple to the load

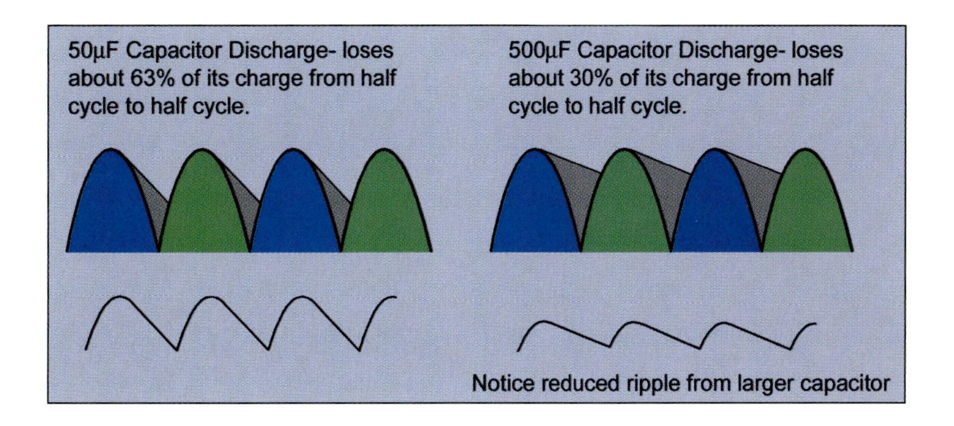

FIGURE 3–20 Filter capacitor effect on ripple

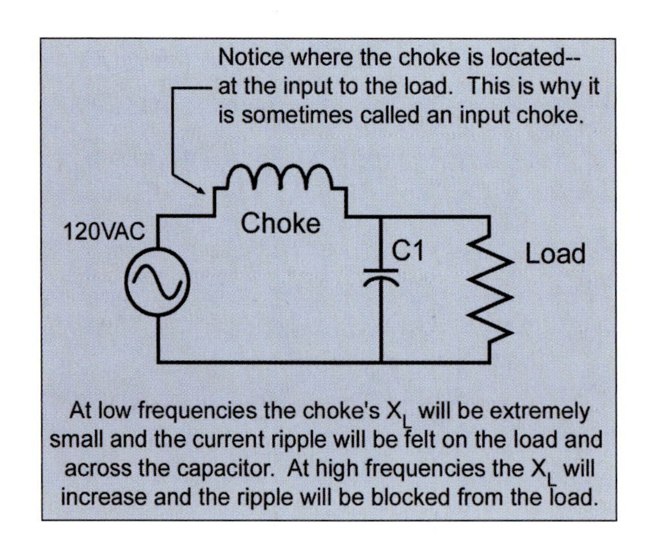

FIGURE 3–21 Choke used as a filter

and thus help reduce the voltage ripple. In earlier AC theory lessons, you learned that inductors (chokes) oppose a change in circuit current. The opposition to change in current by a choke is at the same rate as the opposition to voltage change for a capacitor. It takes five time constants for an inductor to reach maximum current induction. The time constant for an inductor is $T = \frac{L}{R}$. Each time constant is at 63.2 percent of the total current value remaining. Figure 3–21 shows an example of a choke used as an additional filter with a capacitor.

Another common type of choke filter is the power-line filter. Power lines are normally only 60 Hz and carry our common household currents and voltages. The lines do, however, provide a conduction source for rf (radio frequency) currents. These current signals cause interference with motors, fluorescent lighting controls, radios, computers, and other equipment. To filter out or trap these rf currents, an input choke is used. Figure 3–22 shows a balanced low-pass filter. The name low-pass comes from the choke's ability to allow low 60 Hz power fre-

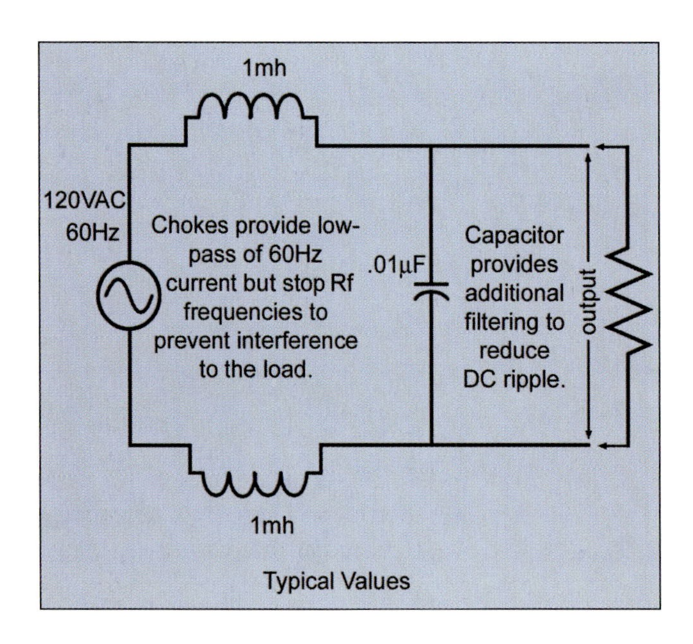

FIGURE 3–22 Balanced choke, low-pass filter

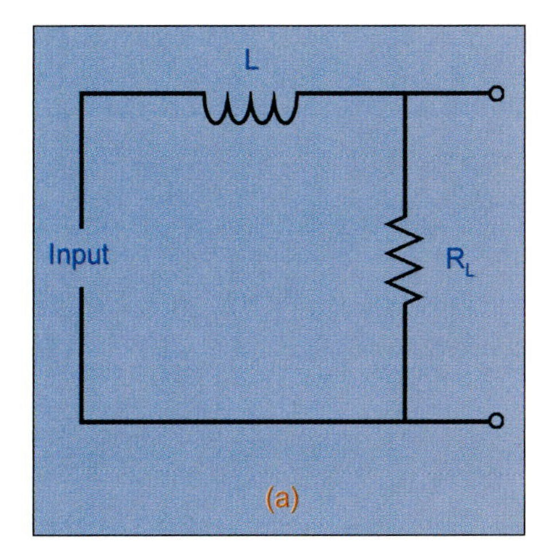

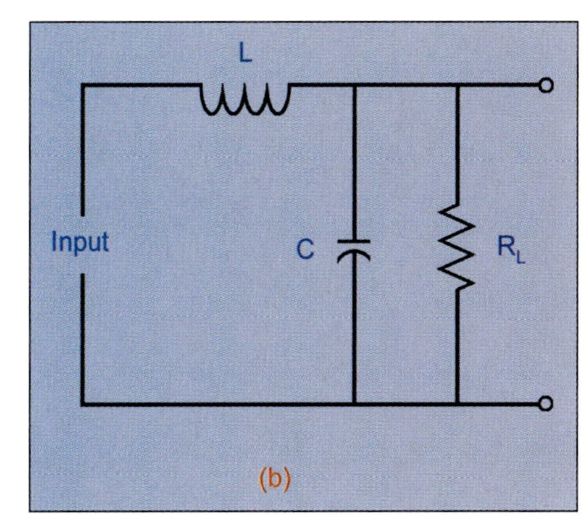

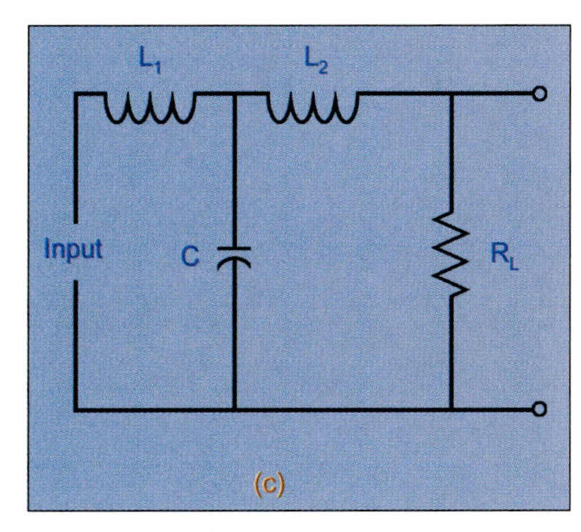

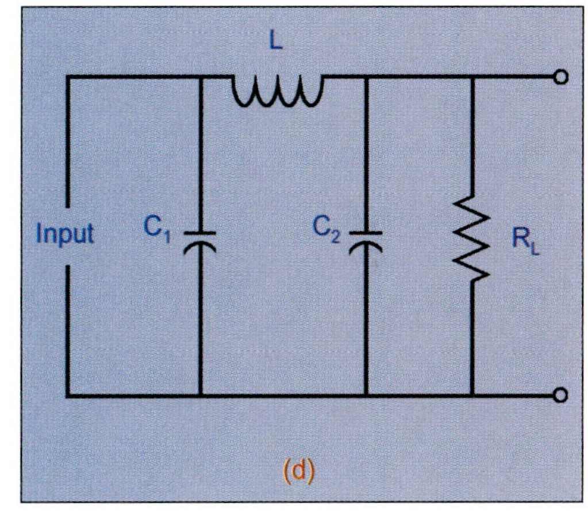

FIGURE 3–23 Typical low-pass filters for power supplies; a) choke acts as a voltage divider with the load resistor, b) adding a bypass capacitor aids filtering by providing sharper cutoff at low frequencies, c) this T-type filter uses an additional choke (L_2) to further reduce ripple, d) Pi (π)-type filters use an initial capacitor (C_1) to provide additional filtering

quencies to pass to the load. At high frequencies, the choke's high impedance drops most of the high-frequency current ripple across the choke. Two chokes are used to balance the line's alternating current with respect to ground. Chokes are best used with unregulated power supplies that have high current demands. For smaller semiconductor power supplies, voltage regulation is used. This has two advantages over the choke, cost and size.

Figure 3–23a, b, c, and d show a number of common low-pass filter configurations for power supplies. In the AC chapter on filters, we covered these and others as low-pass and high-pass filters.

VOLTAGE REGULATION AND OUTPUT

So far in this chapter, you have learned about the different types of rectifiers—half-wave, full-wave, and bridge. You have seen how each of these takes an incoming AC signal and rectifies it to an output pulsating DC (DC with ripple). Finally, you have also learned how capacitors and chokes are used to filter or reduce the DC ripple. The last major sections of a power supply are voltage regulation and output.

3.14 Voltage Regulation

Figure 3–24 shows an example of a regulated power supply. The Zener diode is rated at 9 volts, so the output across the load will remain relatively constant. The capacitor will provide ripple reduction from the pulsating DC output of the bridge. Reducing the ripple in the DC will ensure that the Zener stays in reverse bias condition and operates properly even under varying load conditions.

3.15 Power Supply Output

An alternative to Zener diode regulation for a power supply is to use a variable resister in series with the output. Figure 3–25 shows the alternative schematic for a regulated power supply. Note that the outputs can be combined or used individually. This type of variable output uses a voltage divider network. For example, the voltage output between points A and B is 6 V, between B and C it is 12 V, between C and D it is 18 V, between D and E it is 24 V, and between A and E it is 60 V.

Note that such a power supply is not normally used for circuits with widely varying current requirements. The reason is that the resistor has to be adjusted manually to keep the voltage output constant. The use of a 60 V Zener diode in parallel with the output resistors would be a much more efficient circuit.

3.16 Voltage Doublers

Voltage doublers and other multiplier circuits provide a DC output that can be double, triple, and quadruple the peak input signal. Because the voltage doubler and multiplier circuits are not capable of generating voltage, the increased input voltage is offset by the loss of circuit current.

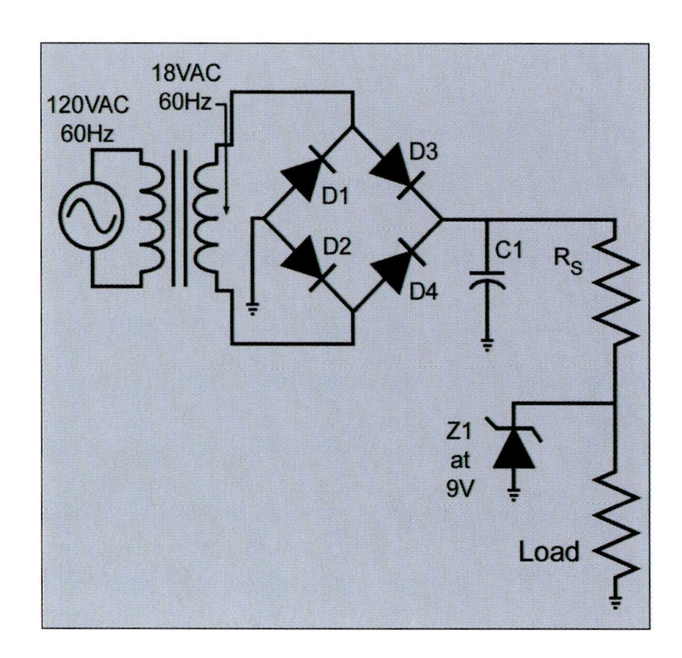

FIGURE 3–24 Regulated power supply

FIGURE 3–25 Power supply with voltage regulation and multiple output voltages

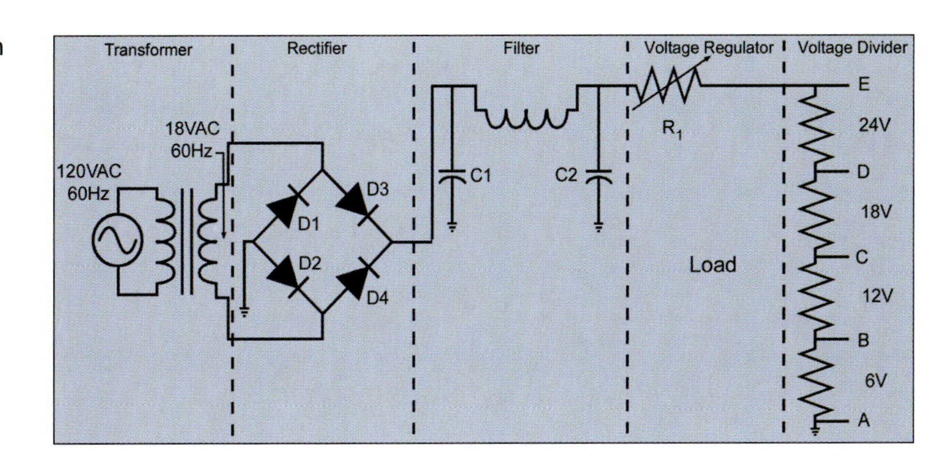

For example, in a voltage doubler circuit, 10 V at 50 mA input would yield slightly less than 20 V at 25 mA output. This is necessary to maintain power balance in the circuit (i.e., power in = power out + losses).

Figure 3–26 shows an example of a voltage doubler.

There are two basic configurations for the voltage doubler. The half-wave is shown in Figure 3–27 and the full-wave is shown in Figure 3–28.

Half-Wave Voltage Doubler (Figure 3–27)

During the negative alternation of the input AC signal (Figure 3–27a), D_1 is forward biased and acts as a short, D_2 is reverse biased and acts as an open. The equivalent circuit shows that C_1 charges to the value of the source voltage and that C_2 discharges through the load resistor.

As the input AC signal goes positive (Figure 3–27b), the diodes switch operation. D_2 becomes forward biased and D_1 is reverse biased. C_1 now discharges and acts as a series, aiding voltage to the AC signal source.

FIGURE 3–26 Voltage doubler

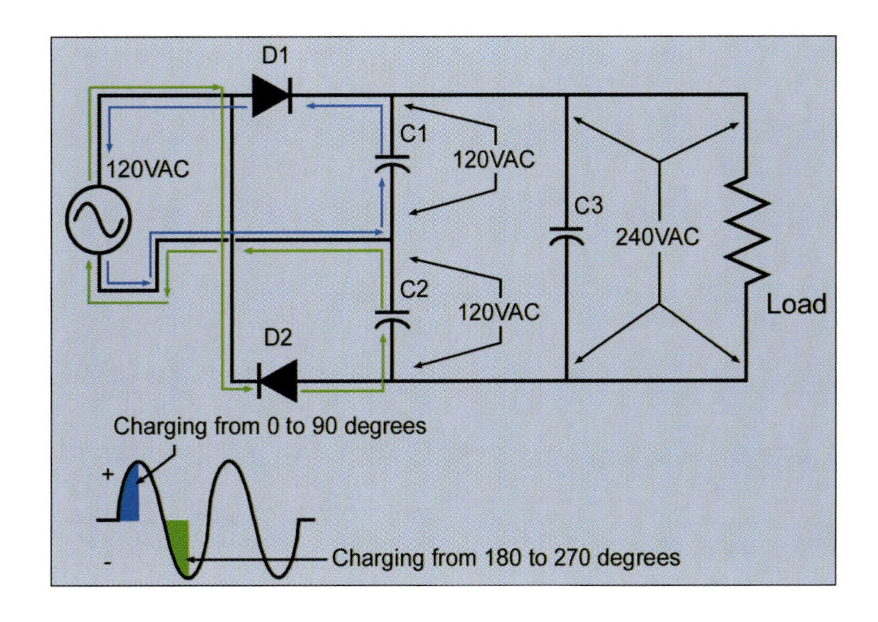

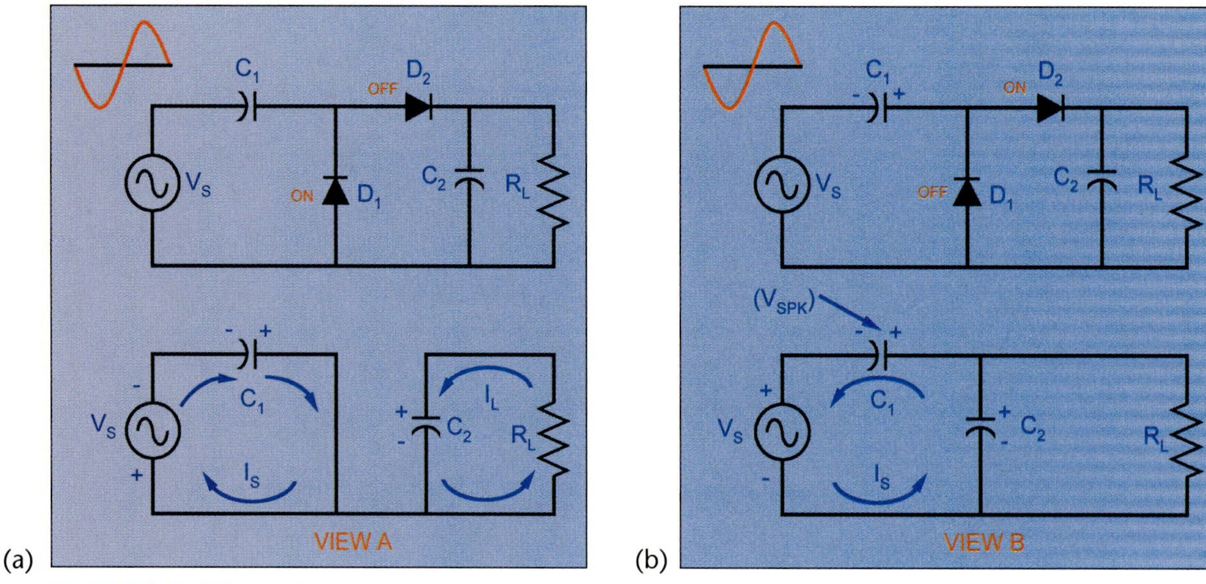

FIGURE 3–27 Half-wave voltage doubler; a) negative half-cycle, b) positive half-cycle

Full-Wave Voltage Doubler (Figure 3–28)

The full-wave voltage doubler circuit is shown in Figure 3–28. During the positive alternation of the input (Figure 3–28b) signal, D_1 is forward biased and D_2 is reverse biased. This allows C_1 to charge to V_{spk}. As the input signal goes negative (Figure 3–28c), the diodes reverse their condition, and C_2 now charges to $V_{s\text{-peak}}$. The two voltages in series provide 2 $V_{s\text{-peak}}$ across C_3 and the load. C_3 does not add to the charge, but has a value designed to reduce the ripple output from the charge–discharge cycle of the input signal and C_1 and C_2.

Principles of Operation

Look at Figure 3–26. C_3 is an additional filtering capacitor that is used to reduce the ripple of the DC output from C_1 and C_2. The circuit operates in the following way. During the first positive half-cycle (from 0° to 180° on the sine wave), D_1 is forward biased and D_2 is reverse biased. This allows a charging current to flow from the AC source to C_1 and from C_1 through D_1 back to the source (see the green line and arrow in Figure 3–26).

On the first negative half-cycle (from 180° to 270° on the sine wave), the current path is reversed. D_2 is forward biased and D_1 is reverse biased. This allows a charging current to flow from the AC source to C_2 and from C_2 through D_2 back to the source (see the blue line and arrow in Figure 3–26).

Now that C_1 and C_2 are charged and because they are in series, their total voltages are added and applied across the parallel branches of C_3 and the load. The combined charges of C_1 and C_2 equal twice the voltage source, hence the name, voltage doubler.

(a)

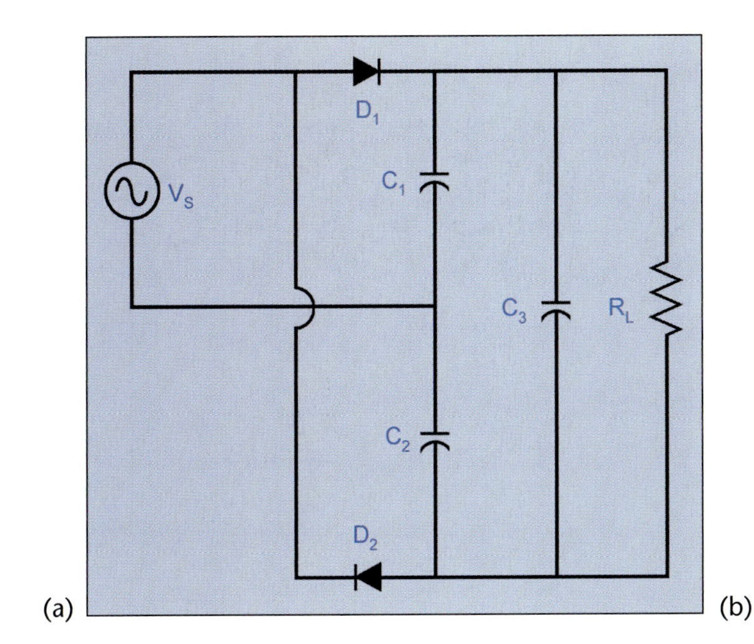

(b)

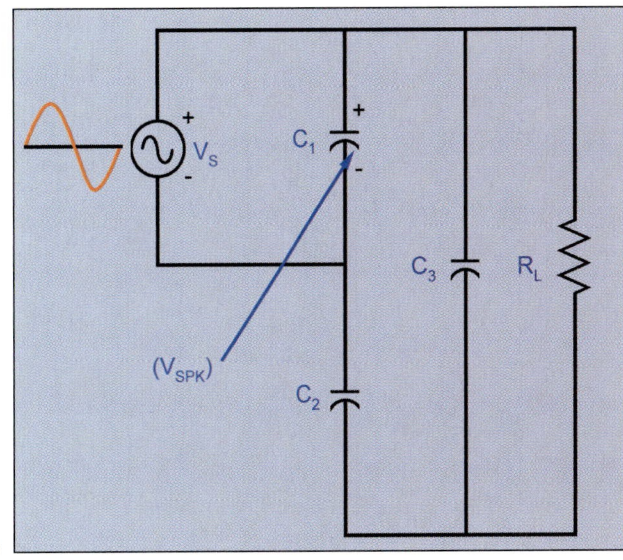

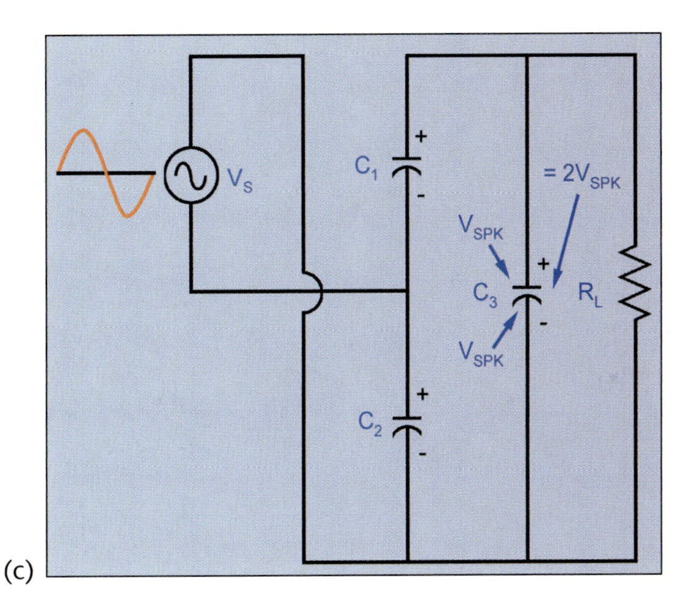

(c)

FIGURE 3–28 Full-wave voltage doubler; a) circuit, b) positive half-cycle, c) negative half-cycle

SUMMARY

Power supplies have the following main components:

- Transformer
- Rectifier
- Filter
- Voltage regulator
- Voltage output or variable outputs

The transformer usually provides a stepped-up (increased) or stepped-down (decreased) AC voltage to the rectifier.

The rectifier converts the AC signal into pulsating DC, or DC with ripple.

The filter removes much of the DC ripple through the combined actions of capacitors and inductors or chokes.

The voltage regulator is made of either a variable resistor connected in series with the load or a Zener diode connected in parallel with the load.

Output voltages can be controlled through voltage divider resisters connected in series. These resistors offer different "tap" points for varying output voltages.

REVIEW QUESTIONS

1. Explain in your own words the purpose of a DC power supply.
2. Refer to Figures 3–18 and 3–19. Explain how the circuits of Figure 3–19 generate the waveforms shown in Figure 3–18.
3. How do capacitors and chokes work to reduce the AC ripple in the output of a power supply?
4. Refer to Figure 3–14. What would happen to the circuit if diode D_3 were reversed?
5. A certain power supply has a capacitor filter. What would likely happen to the output ripple if the capacitor is reduced to half its size?
6. Look at the half-wave and full-wave voltage doublers in Figures 3–27 and 3–28 respectively. Read and memorize the operation description given in the text until you can explain their operation without the book.
7. What is the purpose of the Zener diode in Figure 3–24?
8. In Figure 3–25, what changes would you make to obtain four equal voltages from the output—V_{AB}, V_{BC}, V_{CD}, and V_{DE}?
9. Discuss the advantages you might expect by using a full-wave voltage doubler as opposed to a half-wave voltage doubler.
10. Using what you have learned so far, describe the operation of a capacitor input π filter.

PRACTICE PROBLEMS

1. Look carefully at Figure 3–24 and determine the most probable cause of the following problems:
   a. When the power supply is turned on, the DC output voltage is about one-half of its rated value and shows a large amount of ripple in the output signal.
   b. When turned on, the power supply operates normally for a few minutes and then the output voltage drops to almost zero and the transformer fuse blows.
2. Look at Figure 3–18. A certain RC circuit is charged to 100 volts DC. If the circuit is discharged through a short circuit, what will the capacitor voltage be in 1.5 time constants?
3. A certain half-wave rectifier is being fed a 240 V_{RMS} signal. The load is 500 ohms. What is the load voltage and current (rms, average, peak)?
4. A certain half-wave rectifier is being fed a 240 V_{RMS} signal. The load is 500 ohms. What is the load voltage and current (rms, average, peak)?

Transistors

OUTLINE

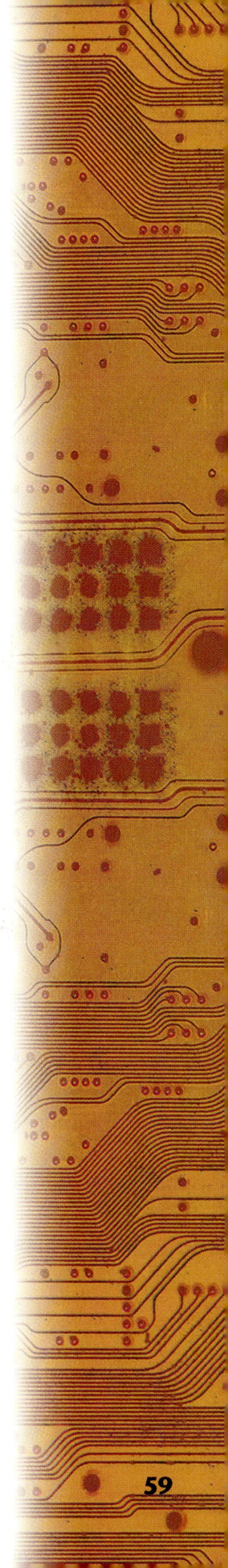

OVERVIEW

The basic component of all modern electronics is the **bipolar junction transistor (BJT)**. Integrated circuits are often nothing more than huge collections of bipolar transistors all on the same piece of semiconductor material. The transistor is very similar to a PN junction diode, but it has an additional PN junction. The output of a transistor provides amplification by taking the input signal and increasing its power.

The fundamental principle underlying the BJT was discovered in 1951 by the Bell Labs research team of John Bardeen and Walter Brattain. These two researchers noticed that when a signal was applied to contacts on a germanium chip, the output taken from one contact was amplified compared to the input.

Dr. William Shockley, the supervisor of the team, took this information and, within a few weeks, developed the theory of the BJT. Using these brilliant theories, Shockley's team developed the first BJT by the end of 1951. The first commercial use of the BJT was in 1952 when it was used in electronic switching for telephones by Bell Telephone in Englewood, New Jersey.

OBJECTIVES

After completing this chapter, the student should be able to:

1. Draw and correctly label schematic symbols for NPN and PNP transistors.
2. Show biasing to establish and control current flow through a transistor.
3. Define the terms *amplification* and *power gain.*
4. Establish transistor characteristic curves.
5. Calculate gain and develop load lines for transistors in a given circuit.
6. Describe the steps involved in testing transistors.

GLOSSARY

Amplify To make larger or more powerful.

Bipolar junction transistor (BJT) A three-terminal semiconductor device that is used for the control and amplification of signals in electronic circuitry. The most common of all types of semiconductors.

The BJT is defined as a three-terminal device constructed of three semiconductor materials forming two PN junctions, whose output current, voltage, and/or power are controlled by its input current.

BJT OPERATING CHARACTERISTICS

4.1 BJT Construction and Symbols

Figure 4–1 shows the construction and schematic symbols of a BJT, and Table 4–1 lists the names and purposes for each of the elements.

The collector and emitter are always the same material (P or N), and the base material is always the opposite (N or P).

For instance, an NPN BJT has a collector and emitter made of N-type material and a base of P-type. The PNP has a collector and emitter made of P-type material and a base of N-type. Note that the physical construction may change (NPN or PNP), but there are always two PN junctions (see Figure 4–1). An easy way to remember whether the transistor is an NPN or PNP is by looking at the arrow on the symbol—the NPN transistor's arrow is **N**ot **P**ointed i**N.**

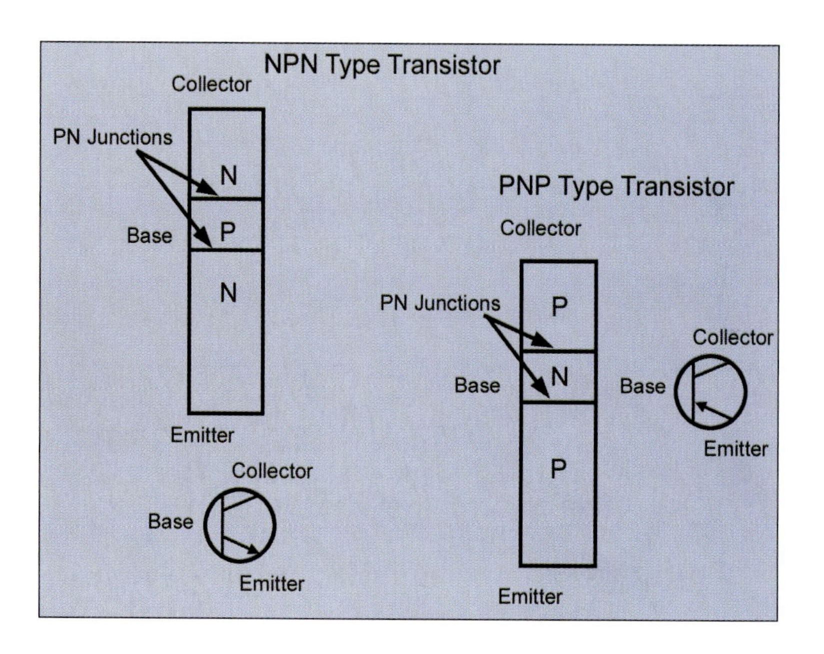

FIGURE 4–1 BJT construction and symbols

Table 4–1 BJT Elements and Their Purposes

Element	Purpose	Comments
Emitter	Serves as a source and "emits" free electrons or holes to the base and collector	The first transistor was of NPN construction. Thus, the emitter was a source of electrons. It is the largest and most heavily doped.
Collector	Receives most of the emitter's electrons	The collector is a "hole" receptor in PNP transistors. It is the second largest and second most heavily doped.
Base	Used as a control and a biasing source for the PN junctions	Small changes between the base-emitter junction create large changes in current between the emitter and collector. The base bias can also be adjusted to block conduction completely. It is the smallest and most lightly doped element.

4.2 Transistor Currents and Voltages

In normal operation, the base-emitter (B-E) PN junction is forward biased, and the base-collector (B-C) PN junction is reverse biased, as shown in Figure 4–2. Because the base is so lightly doped, not much of the emitter current flows through the base region, but passes on to the collector region.

The standard labels for the different currents are shown in Table 4–2.

Remember that emitter current (electron flow) is always against the arrow when the B-E junction is forward biased. Figure 4–3 illustrates the relationship between the three currents. Notice that the emitter current splits between the base and collector current. The base current is quite small compared to the collector current. The ratio of the emitter to collector current is called alpha (α).

The formula for calculating alpha is:

$$\alpha = \frac{I_C}{I_E} = \frac{I_C}{I_B + I_C}$$

Alpha will always be less than 1 because the emitter current splits between the collector and the base. Using the same type of NPN transistor, voltages for the base, collector, and emitter can be identified. Figure 4–4 shows the voltages for each. Note that supply voltages use a double letter to indicate the voltage source for collector, base, or

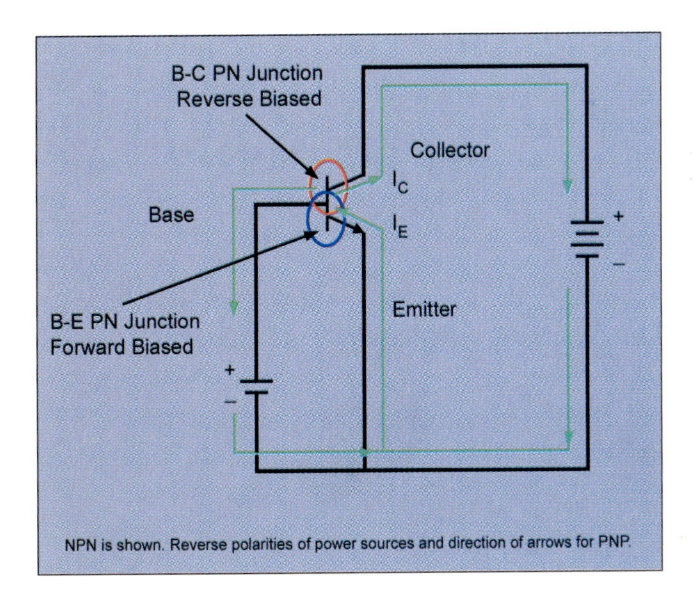

FIGURE 4–2 BJT normal biasing conditions

Table 4–2 Transistor Currents

Name	Symbol
Base current	I_B
Collector current	I_C
Emitter current	I_E

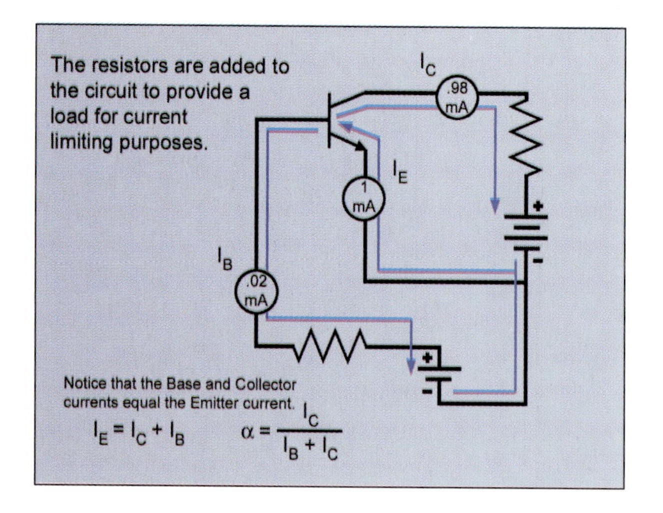

FIGURE 4–3 NPN transistor currents

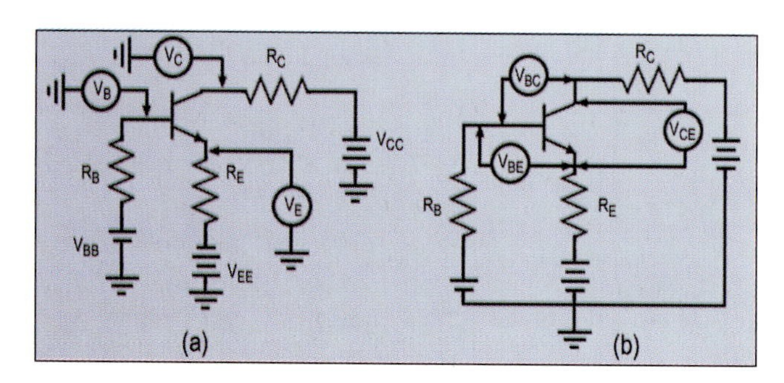

FIGURE 4–4 NPN transistor voltages; a) with respect to Earth, b) directly across transistor

emitter voltages. For example, V_{EE} indicates the emitter supply voltage and V_{BB} indicates the base supply voltage. The biasing voltages may come from separate power supplies (as shown in Figure 4–4) or from a system of resistor voltage dividers. For example, the base and emitter voltages are often supplied by dividing V_{CC} across a set of biasing resistors. Figure 4–4b shows the labeling of the voltage drops across each region of the transistor.

4.3 Transistor Bias and Gain

Transistor Bias

Two additional terms are important to understand the basic operation of a BJT. They are *cutoff* and *saturation*. Cutoff occurs when both PN junctions (B-E and B-C) are reverse biased. This condition yields almost full V_{CC} voltage across the emitter-collector, as shown in Figure 4–5. In other words, the E-C junction is like an open switch. Recall that, under normal operating conditions, the B-E junction is forward biased and the B-C junction is reverse biased.

Saturation is the opposite condition of cutoff. As I_B is increased, I_C increases. Saturation is the maximum current condition of I_C. This value is determined by V_{CC} and the resistance of R_C and R_E. At saturation, any increase in I_B will not affect I_C. Using these relationships, the saturation value is calculated as

$$I_C = \frac{V_C}{R_C + R_E}$$

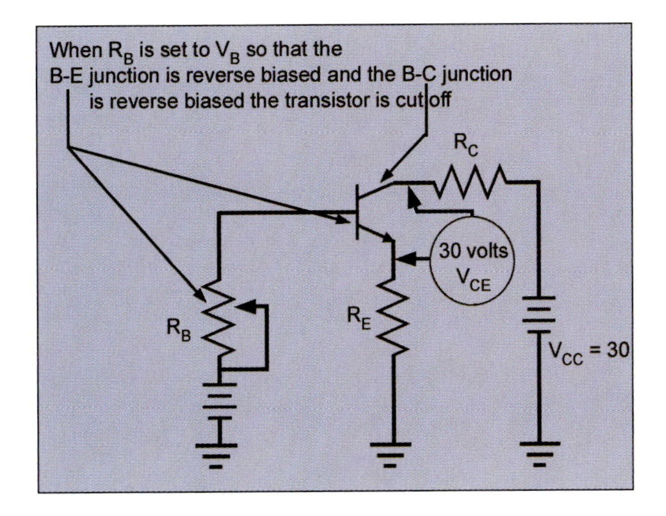

FIGURE 4–5 Transistor biased to cutoff

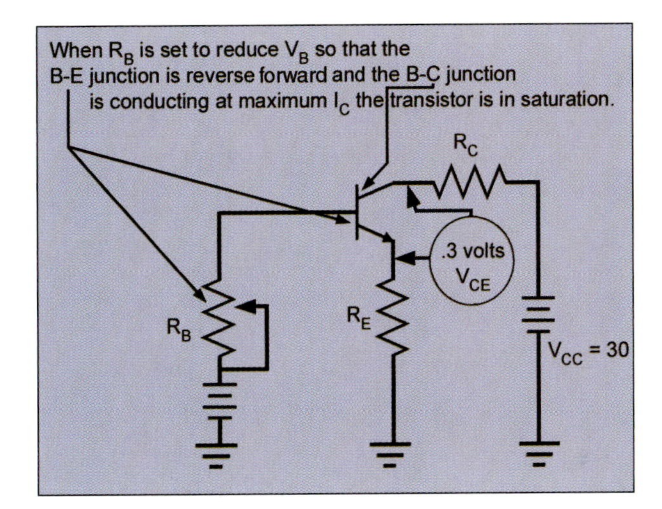

FIGURE 4–6 Transistor biased to saturation

See Figure 4–6.

Figure 4–7 illustrates the biasing and operational state of an NPN transistor when in cutoff, saturation, and normal operation. Table 4–3 identifies key points about each condition. Note especially the biasing differences between cutoff, saturation, and normal operation.

Transistor Gain

A transistor can **amplify** current, voltage, or power, depending on the circuit values that are selected. This amplification is called gain. The amount of DC current gain in a transistor is called beta (β). Beta is the ratio of DC collector current to DC base current. In practical application, a small amount of change in the base current will cause a large change in the emitter and collector currents. The formula for β is

$$\beta = \frac{I_C}{I_B} \Rightarrow I_B = \frac{I_C}{\beta} \quad (1)$$

For example, if the β of a transistor is 75 and $I_B = 100\ \mu A$, then $I_C = 75 \times 100\ \mu A = 7.5\ mA$.

FIGURE 4–7 Transistor biasing configurations; a) cutoff, b) saturation, c) normal

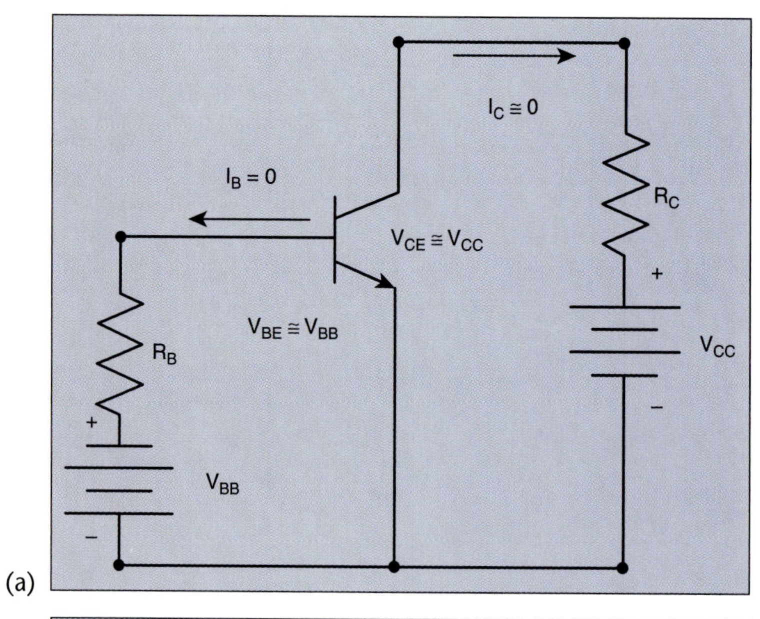

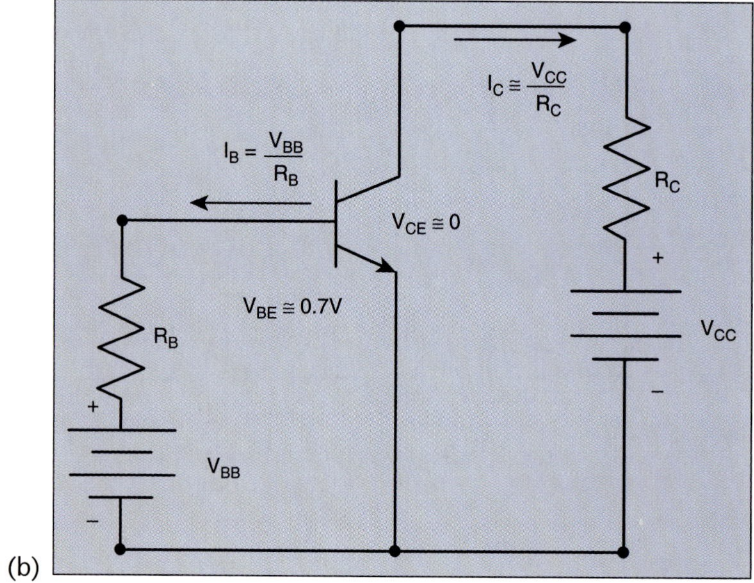

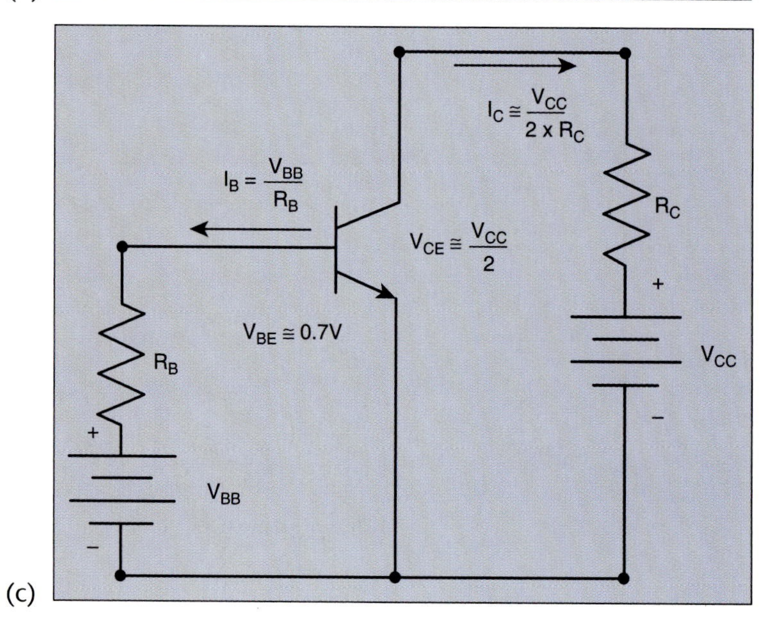

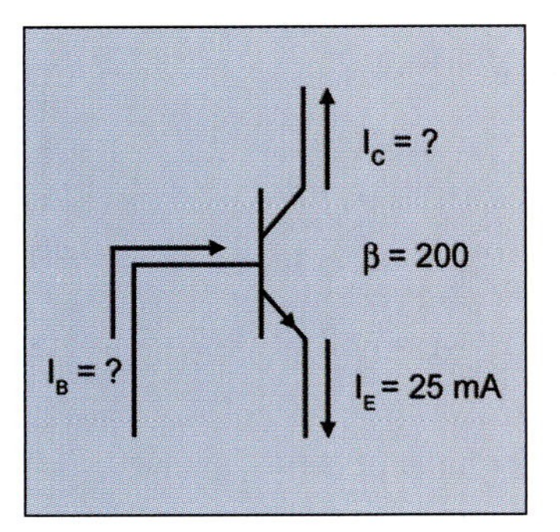

FIGURE 4–8 Transistor gain example

Table 4–3 Bias Conditions for Figure 4–7

Transistor Operation	Cutoff	Saturation	Normal
B-E junction	Reverse biased	Forward biased	Forward biased
C-E junction	Reverse biased	Reverse biased	Reverse biased

EXAMPLE 1

With the information given in Figure 4–8, solve for I_B and I_C.

Solution:
Recall that α of a transistor is the ratio between the collector and emitter current and is expressed as follows:

$$\alpha = \frac{I_C}{I_E} \Rightarrow I_E = \frac{I_C}{\alpha} \tag{2}$$

Also recall that

$$I_E = I_C + I_B \tag{3}$$

Substituting the second parts of Equations 4–1 and 4–2 into Equation 4–3 yields

$$\frac{I_C}{\alpha} = I_C + \frac{I_C}{\beta} \tag{4}$$

Dividing Equation 4–4 by I_C and rearranging terms yields

$$\alpha = \frac{\beta}{1+\beta} \text{ and } \beta = \frac{\alpha}{1-\alpha} \tag{5}$$

And also

$$I_B = I_B\,(1 - \alpha) \tag{6}$$

The first step is to find α.

$$\alpha = \frac{200}{200 + 1} = 0.995 \tag{7}$$

Next

$$I_B = I_E\,(1 - \alpha) = 25 \text{ mA} \times (1 - 0.995) = 125\ \mu\text{A} \tag{8}$$

Now you can solve for I_C:

$$I_C = I_E - I_B = 25 \text{ mA} - 125\ \mu\text{A} = 24.875 \text{ mA}$$

Note: Beta is also equal to the change in the value of I_B to I_C; therefore, $\beta = \frac{\text{Change in } I_C}{\text{Change in } I_B} = \frac{\Delta I_C}{\Delta I_B} = \frac{I_C}{I_B}$.

Remember that beta has no units.

TRANSISTOR CIRCUIT CONFIGURATIONS

There are three ways a transistor can be wired (configured) to produce different types of gain. The following sections will discuss each of these unique configurations—common emitter, common base, and common collector (see Figure 4–9).

Note the location of the input and output connections for all three configurations. The name of each configuration is derived from the element that is common to both the input and the output.

4.4 Common Emitter

Recall that the flow of the current from the emitter, through the base, to the collector is controlled by the amount of forward bias on the base-emitter junction. A slight change in I_B can cause a great change in I_E. The current flow in the base is a function of the base-emitter junction internal resistance (Rb) and the external resistance in the base circuit. The external resistance is either a series-limiting resistor (Rs) or a voltage divider network. The common emitter has a voltage phase reversal due to the collector voltage being 180° out of phase with the input base voltage. The common emitter is the most common of the three configurations. Figure 4–10 shows a simple common-emitter circuit.

EXAMPLE 1

Solve for the voltage gain of the common-emitter circuit in Figure 4–10.

Solution:

$$V_{gain} = \frac{V_{out}}{V_{in}}$$

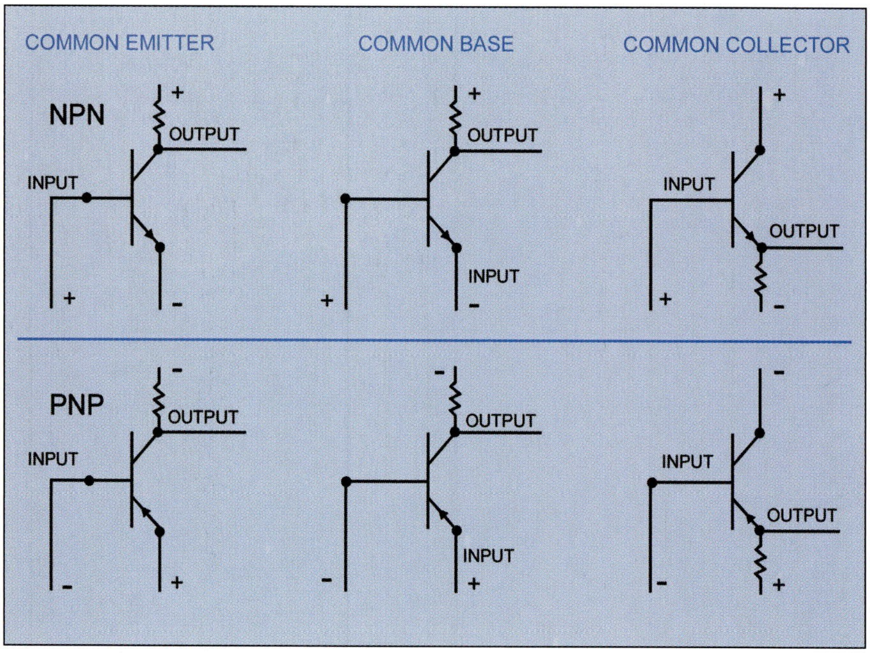

FIGURE 4–9 Three types of transistor circuits

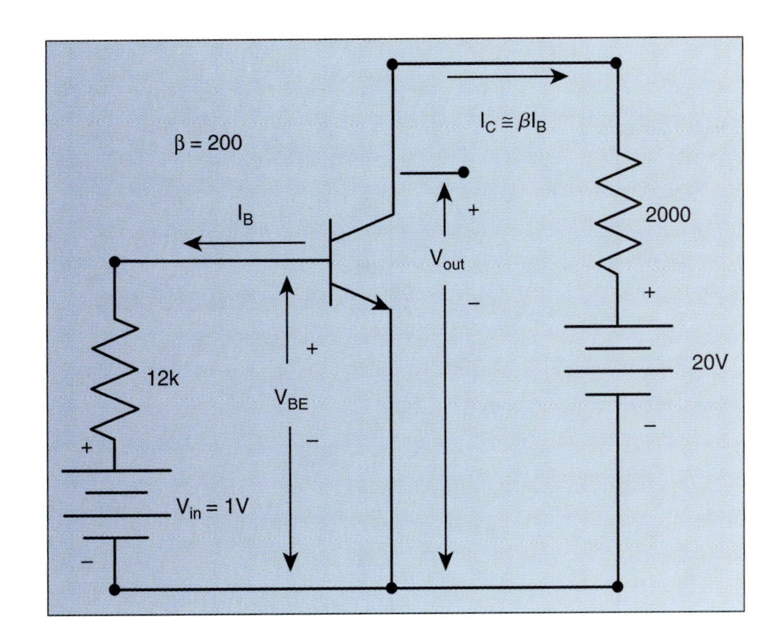

FIGURE 4–10 Simple common-emitter circuit

Writing a KVL around the B-E circuit gives

$$-1\text{ V} + 12{,}000\, I_B + V_{BE} = 0$$

Noting that $V_{BE} = 0.7$ V (for silicon) yields

$$-0.3\text{ V} = -12\text{ k}\Omega \times I_B \Rightarrow I_B = \frac{-.3}{-12{,}000} = 25\ \mu\text{A}$$

From the definition of β,

$$I_C = \beta \times I_B = 200 \times 25\ \mu\text{A} = 5\text{ mA}$$

A KVL around the C-E circuit gives

$$20\text{ V} - V_{out} + I_C R_C = 0 \Rightarrow V_{out} = 20 - 5\text{ mA} \times 2{,}000 = 10\text{ V}$$

$$V_{gain} = \frac{V_{out}}{V_{in}} = \frac{10}{1} = 10$$

Note also that the common-emitter circuit creates power gain. The input power is $P_{in} = V_{in} \times I_B = 1 \times 25\ \mu\text{A} = 25\ \mu\text{W}$, and the output power is $P_{out} = V_{out} \times I_C = 10 \times 5\text{ mA} = 50\text{ mW}$. The power gain then is

$$P_{gain} = \frac{P_{out}}{P_{in}} = \frac{50\text{ mW}}{25\ \mu\text{W}} = 2{,}000$$

Note the interesting result if V_{in} is decreased to 0.9 V.

$$-0.2\text{ V} = -12\text{ k}\Omega \times I_B \Rightarrow I_B = \frac{-.2}{-12{,}000} = 16.7\ \mu\text{A}$$

And

$$I_C = \beta \times I_B = 200 \times 16.7\ \mu\text{A} = 3.33\text{ mA}$$

$$20\text{ V} - V_{out} + I_C R_C = 0 \Rightarrow V_{out} = 20 - 3.33\text{ mA} \times 2{,}000 = 13.34\text{ V}$$

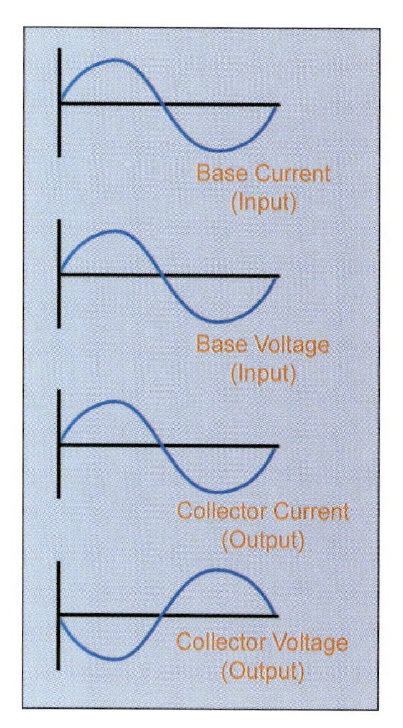

FIGURE 4–11 Common-emitter phase reversal

A negative change in the output voltage created a positive change in the input voltage. A common-emitter amplifier inverts changes in the input signal voltage. This relationship is shown in Figure 4–11.

As seen in Figure 4–11, the common emitter has a voltage phase reversal due to the collector voltage being 180° out of phase with the input base voltage.

4.5 Common Base

The common-base circuit does not exhibit any current gain (Figure 4–12) because the input is into the emitter and the output is from the collector. Remember that typically only 99 percent of the emitter current is seen in the collector. There are two advantages to the common-base circuit:

1. The common base exhibits both voltage gain and power gain.
2. The input impedance of the common base is very useful in certain types of power amplifier circuits.

EXAMPLE 2

Solve for the voltage and power gain of the common-base circuit in Figure 4–12.

Solution:

$$P_{gain} = \frac{P_{out}}{P_{in}} = \frac{I_{out}^2 \times R_{out}}{I_{in}^2 \times R_{in}} = \alpha^2 \times \frac{R_{out}}{R_{in}}$$

This is true because

$$\alpha = \frac{I_C}{I_E} \Rightarrow \alpha^2 = \frac{I_c^2}{I_E^2} = \frac{I_{out}^2}{I_{in}^2}$$

Combining these equations and substituting the values in Figure 4–12 gives

FIGURE 4–12 Simple common-base circuit

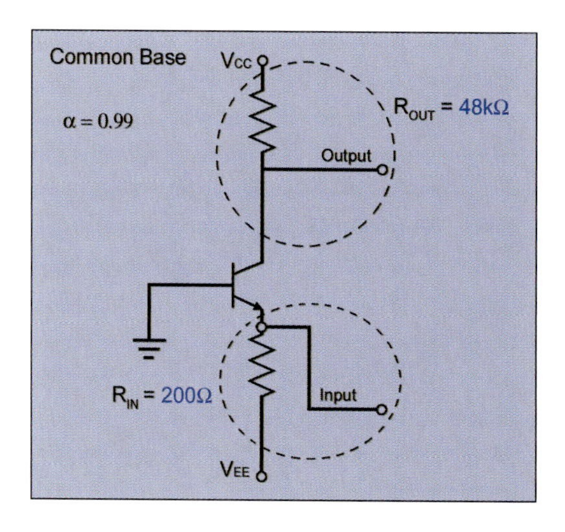

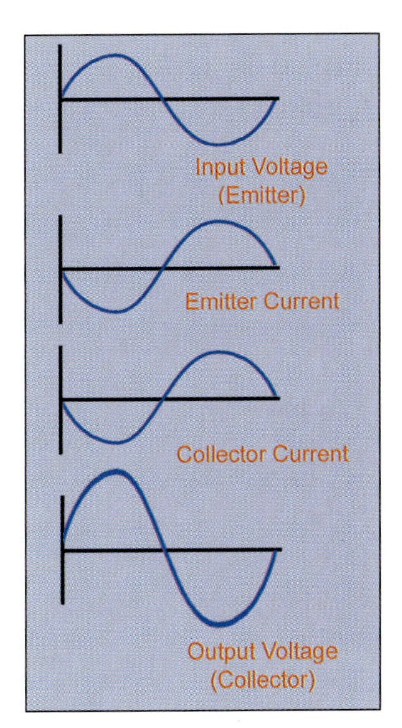

FIGURE 4–13 Phase relationships to the common-base circuit

$$P_{gain} = \frac{P_{out}}{P_{in}} = \alpha^2 \times \frac{R_{out}}{R_{in}} = (.99)^2 \times \frac{48{,}000}{200} = 235.2 \text{ V}_{gain} =$$

$$\frac{I_{out} \times R_{out}}{I_{in} \times R_{in}} = \alpha \times \frac{48{,}000}{200} = 237.6$$

Note that the common-base voltage and power gains are dependent on input versus output resistance because the I in the formulas for voltage or power remains relatively constant for a value of α that is close to unity. In contrast, in a common-emitter circuit, β (the relationship between input and output current) is the determining factor.

The common base is not a current amplifier because the collector current equals α times the emitter current and α is always less then 1. This configuration is sometimes used to match and amplify radio signals being fed from an antenna. The phase relationships for the common-base circuit are shown in Figure 4–13.

4.6 Common Collector

The third transistor configuration is the common collector. In this configuration, a small change in the base current causes a large change in the emitter-collector current. The voltage difference between the emitter and base is very small and almost constant because of the forward bias on the PN junction. The difference in current is then explained because of input versus output resistance. This relationship can be expressed as follows. Assume that 1 mA of current is flowing in the emitter-collector circuit. Using what you know about β, you can calculate the base current to be $\frac{1}{\beta}$. Using Ohm's law, we know that $R = \frac{V}{I}$, and because V is relatively constant, the resistance seen at the base is equal to the emitter resistance times β. This (low) output resistance of the emitter compared to the (high) input resistance on the base is very useful in coupling a high-resistive (impedance) input to a low-resistive (impedance) output without loading the input circuit.

Figure 4–14 shows an example of a simple common-collector circuit. The common collector has a high input resistance and a low

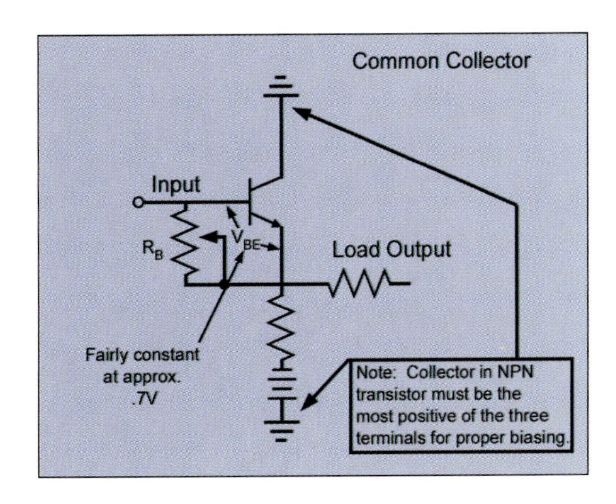

FIGURE 4–14 Simple common-collector circuit

FIGURE 4–15 Summary of circuit characteristics; a) common base, b) common collector, c) common emitter

	Common Base
Input Impedance	Lowest (approx. 50Ω)
Output Impedance	Highest (approx. 1MΩ)
Current Gain	No (less than 1)
Voltage Gain	Yes (500-800)
Power Gain	Yes
Input-Output Phase Voltage	In Phase
Application	RF Amplifier
Configuration Diagram * The configuration is named after the transistor connection that is common to both the input (source) and the output (load) Another explanation is: The transistor connection that does not provide an input or output to the circuit.	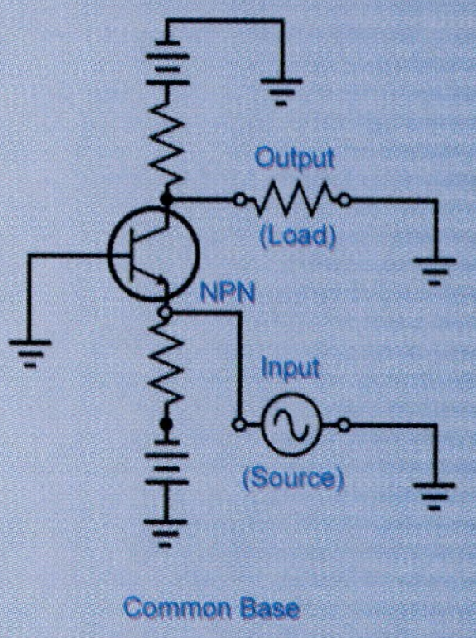 Output (Load) NPN Input (Source) Common Base

(a)

	Common Collector
Input Impedance	Highest (approx. 300kΩ)
Output Impedance	Lowest (approx. 300Ω)
Current Gain	Yes (30-100)
Voltage Gain	No (less than 1)
Power Gain	Yes
Input-Output Phase Voltage	In Phase
Application	Isolation Amplifier Emitter Follower
Configuration Diagram * The configuration is named after the transistor connection that is common to both the input (source) and the output (load) Another explanation is: The transistor connection that does not provide an input or output to the circuit.	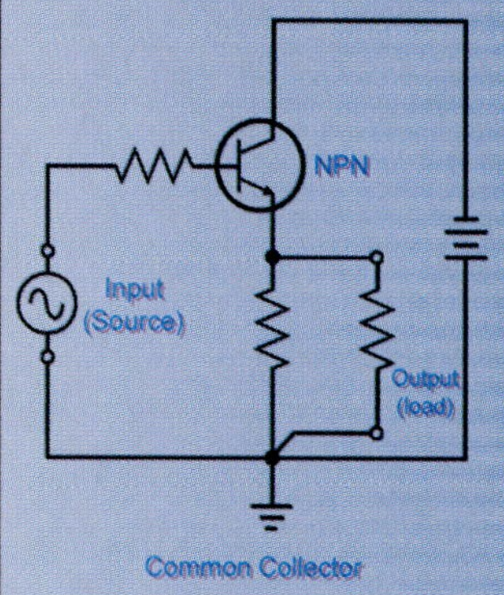 NPN Input (Source) Output (load) Common Collector

(b)

FIGURE 4–15 *(continued)*

	Common Emitter
Input Impedance	Medium (approx. 1kΩ)
Output Impedance	Medium (approx. 50Ω)
Current Gain	Yes (30-100)
Voltage Gain	Yes (300-600)
Power Gain	Yes (Highest)
Input-Output Phase Voltage	180° Out of Phase
Application	Universal
Configuration Diagram * The configuration is named after the transistor connection that is common to both the input (source) and the output (load) Another explanation is: The transistor connection that does not provide an input or output to the circuit.	NPN Input (Source) Output (load) Common Emitter

(c)

output resistance. The output voltage is always less than the input voltage. This configuration is often used as an isolation or buffer amplifier, which explains why it is often called an emitter follower.

4.7 Comparison of the Three Types of Connections

Figures 4–15a, 4–15b, and 4–15c show the features of each of the three types of BJT circuits.

CHARACTERISTIC CURVES

4.8 Introduction

There are many types of characteristic curves that can be used to determine current and voltage relationships in the different types of transistor configurations. Some of these are the collector curves, the base curves, and the beta curves.

Collector curves plot the relationship among I_C, I_B, and V_{CE}. The base curves plot the relationship of I_B and V_{BE}. Beta curves show how the value of β varies with temperature and I_C. Probably the most useful of these characteristic curves are the collector curves, shown in Figure 4–16.

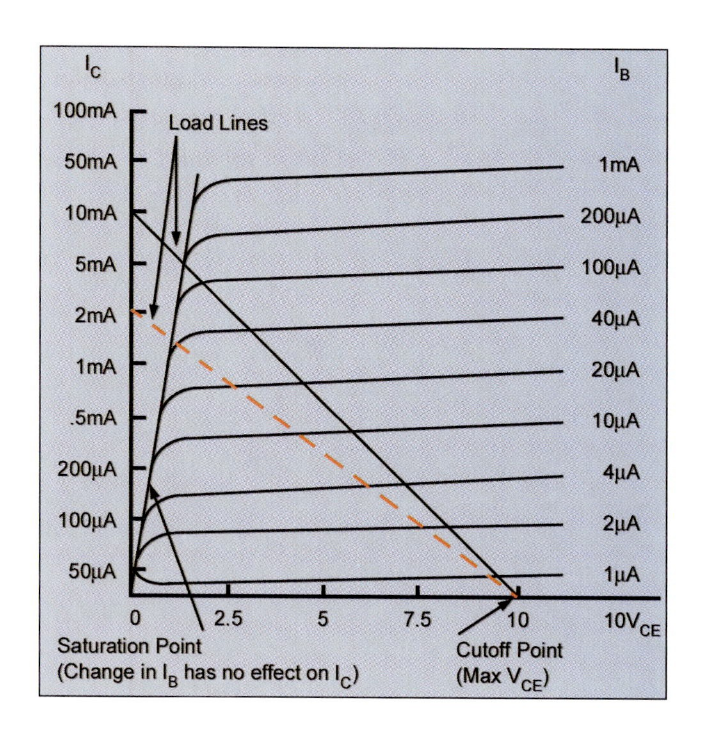

FIGURE 4–16 Typical collector curves with load lines

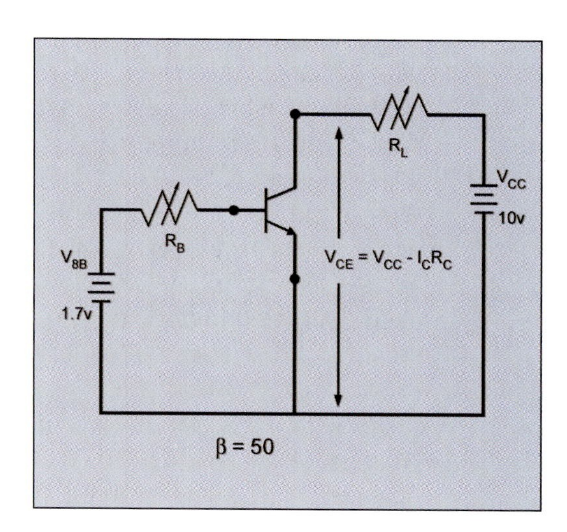

FIGURE 4–17 Typical load line circuit

4.9 Collector Curves

Refer to the circuit in Figure 4–17 for the following discussion. To plot collector curves, the base current must be fixed, and the collector-emitter voltage is varied over a predetermined range. For each value of collector-emitter voltage, I_C is recorded. When the collector current begins to level off (becomes saturated), another base current is set and the process is repeated. Setting the I_B to 0 assumes the transistor is in cut-off. An example set of collector-emitter curves is shown in Figure 4–16. Note that the load line shown in Figure 4–16 has its cutoff at maximum V_{CE}, which is typically V_{CC}. For the same load line, saturation is at 10 mA for the collector current, with 1 mA of base current.

EXAMPLE 1

The load line will shift if the base current is lowered. The dotted red line shows a load line for a base current of 40 mA with an $I_{C\,Sat}$ of 2 mA. Solve for R_B and R_L.

$$R_B = \frac{V_B}{I_B} = \frac{(1.7 - .7)\text{V}}{40\ \mu\text{A}} = 25\ \text{k}\Omega$$

The −.7 V drop comes from the base-collector junction of a silicon transistor. The voltage drop of a germanium transistor is about half that value.

At saturation and cutoff, the load can be calculated by

$$R_L = \frac{V_{CC}}{I_{C\,Sat}} = \frac{10}{.002} = 5\ \text{k}\Omega$$

4.10 Temperature Effects

The bias current of a transistor controls the amount of output the transistor will deliver. Not only do the DC source and biasing resistors set the bias current, but as mentioned above, the temperature also contributes to the bias level. Theoretically, if the collector-base temperature of the transistor is kept constant and I_C is increased, then β will increase. A further increase in I_C beyond the maximum point would cause β to decrease. In reality, when I_C is held constant and the collector-base temperature is varied, the β changes directly with the temperature. Figure 4–18 illustrates how a wide range of temperatures affects β at different I_C for a typical transistor. A transistor data sheet usually includes a temperature-effect curve for that particular transistor. But the temperature effect on several of the same model transistors can vary from transistor to transistor.

When the ambient temperature changes, the junction resistance changes along with the bias and β. This change of bias is called thermal instability. The designer must be aware of the range of ambient

FIGURE 4–18 Temperature effects on operating characteristics

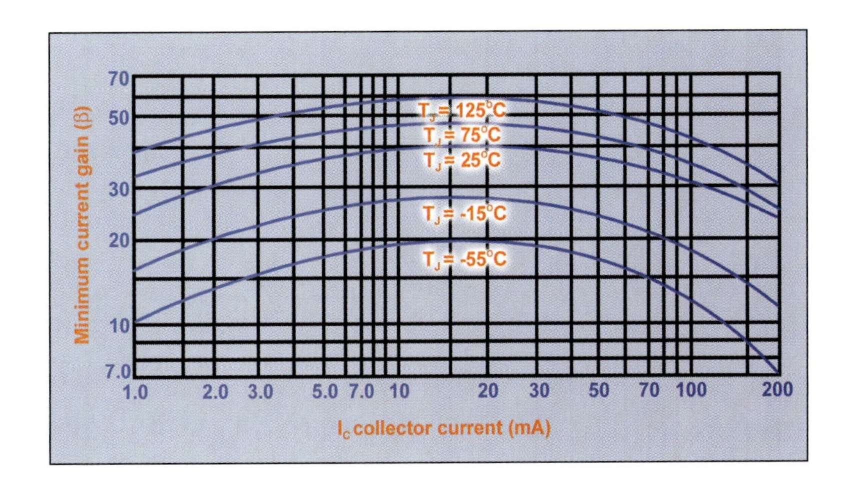

temperatures in which a circuit may be physically located. Several items can be implemented to help stabilize a circuit from thermal instability. The use of heat sinks to pull the heat out of the transistor through conductive heat dissipation will lower the junction temperature. Another approach of thermal protection is within the circuit design itself. Returning a portion of the output back into the input by means of an emitter-feedback, collector-feedback, or combination emitter/collector-feedback circuit will also aid in the stabilization of the biasing.

Finally, the use of the voltage divider with emitter bias circuit can be utilized by including a resistor in the emitter that reacts to the ambient temperature. Resistors also increase in resistance when the temperature is increased; therefore, the voltage on the emitter would decrease and cause an opposition to the base bias. By adding a thermistor or diode in the grounded portion of the voltage divider circuit, additional stabilization can be achieved. The diode or thermistor's resistance value changes as ambient temperature changes and alters the base biasing of the circuit.

NPN VERSUS PNP TRANSISTORS

Both the NPN and PNP transistors are widely used throughout the electronic industry. The two transistor structures are forward biased differently, therefore preventing the replacement of an NPN transistor with a PNP transistor and vice versa. An NPN transistor's collector must be positive with respect to the emitter; conversely, with a PNP transistor, the collector junction must be negative with respect to the emitter.

The PNP transistor is based on hole current and the NPN transistor is based on electron current. Electrons have better mobility than holes and can move faster through the crystal structure; therefore, the NPN transistor is better for high-frequency circuits because it tends to operate faster.

The NPN transistor is more widely used because manufacturers have more NPN types to offer. This fact allows circuit designers better flexibility with respect to the exact parameters their designs require. The NPN is also based on a negative ground design, which is more common that a positive ground system. Because the NPN transistor is implemented in more designs, the cost of the NPN transistor is usually less than the PNP.

An increased performance and more efficient design are often achieved by using both types of transistors in a circuit. Figure 4–19 shows a design utilizing both an NPN and a PNP transistor. In later chapters, you will discover the advantages of this type of design with respect to efficiency and output gain.

TESTING METHODS

4.11 Transistor Checker

There are various methods for testing transistors. A transistor checker is a special piece of equipment used to test most NPN and PNP tran-

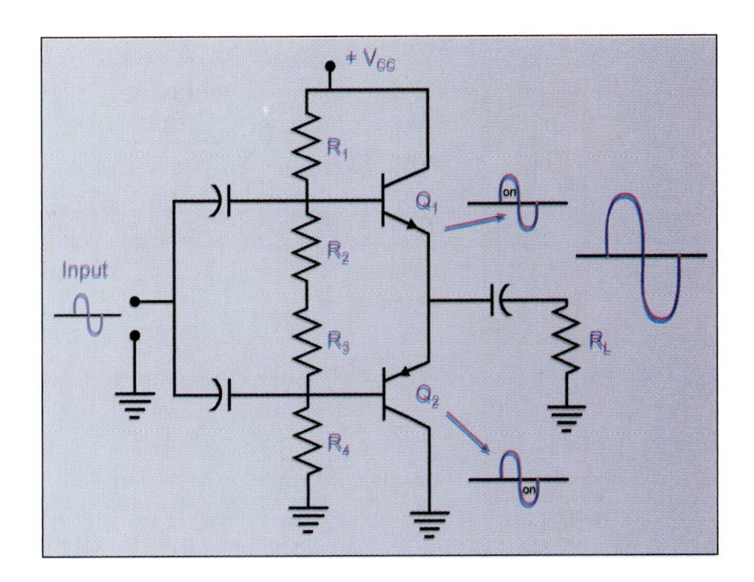

FIGURE 4–19 Complementary class B amplifier

FIGURE 4–20 Typical transistor checker

sistors. It not only tests for proper operation but can also check to determine β. Figure 4–20 shows a photograph of a typical transistor checker.

Other transistor-checking methods are used in test and design laboratories. This type of equipment is called a curve tracer. Some laboratories set up special circuits to test the transistor under actual load and operating conditions. This is called dynamic testing.

4.12 Transistor Testing with an Ohmmeter

The ohmmeter is probably the most common piece of test equipment used to test transistors. Figure 4–21 shows the process for checking an NPN transistor with an ohmmeter. Because there is not a physical PN junction between the collector and the emitter, the resistance will always be very high regardless of the ohmmeter's polarity.

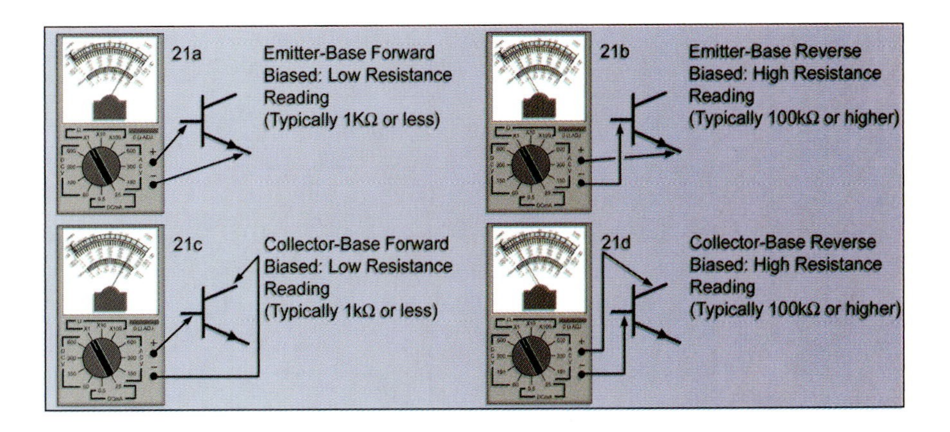

FIGURE 4–21 Ohmmeter and transistor checking; a) E-B forward, b) E-B reverse, c) C-B forward, d) C-B reverse

SUMMARY

In this chapter, you learned about bipolar junction transistors (BJT). The BJT is a three-terminal device constructed of three semiconductor materials forming two PN junctions, whose output current, voltage, and/or power are controlled by its input current. The output of a transistor provides amplification as defined by power = current × voltage.

The BJT has three terminals: the base, emitter, and collector. There are two basic types of BJTs, the NPN and PNP. Under normal operating conditions, a small change in base current produces a large change in collector and emitter current. This change ratio is called beta (β) and is the gain (h_{FE}) of the transistor. Beta is also called h_{FE} in some transistor models. Another important ratio is that of collector current to emitter current. This is called alpha (α).

There are three common configurations for BJTs. They are the common emitter, common base, and common collector. The term common refers to the terminal that is common to both the input (source) and the output (load). For example, the common-emitter circuit has the input at the base and the output from the collector of the transistor.

The load line is another useful characteristic of the BJT. The load line is a function of the collector at saturation and the collector-emitter voltage at cutoff. At saturation, any increase in base current will have little or no effect on the collector current (the collector current is saturated). At cutoff, the collector-to-emitter voltage is effectively that of the applied collector voltage (V_{CC}). This means the transistor is not conducting.

Even though both the NPN and PNP transistors are widely used in the electronic industry, they cannot be interchanged because of their different bias polarities. The NPN transistor is the most common because it is based on a negative ground design, manufacturers offer a larger variety of types, and they cost less. PNP transistors are based on hole current, whereas NPN transistors are based on electron current—usually NPN transistors operate slightly faster, which can make them better in high-speed applications. Also, it is common to find both NPN and PNP transistors utilized in the same circuit to offer better efficiency and output gain.

Temperature affects the bias and β of a transistor circuit and is called thermal instability. Heat sinks can dissipate the heat from the transistor, but usually additional design issues are implemented to help control the increasing β. Feedback circuits such as the emitter-feedback, collector-feedback, or combination emitter/collector-feedback circuit will aid in the stabilization of biasing. The voltage divider with emitter bias circuit also stabilizes the transistor temperature, especially when a diode or thermistor is placed in the circuit.

An ohmmeter can be used to test a BJT for proper operation. If working properly, the forward bias condition of the base-emitter and base-collector junctions should read a low resistance (1 K ohms or less). The reverse bias condition of the same junctions should read a high resistance (100 K ohms or more).

If the transistor base-emitter junction checks good in both directions, then check the base collector in exactly the same way. All readings should be compariable to those shown in Figure 4–21. If the transistor is a PNP, reverse the leads of the ohmmeter and look for the same readings.

REVIEW QUESTIONS

1. What are the two types of transistors? What is the advantage/disadvantage of each?
2. Which of a transistor's four semiconductor layers has the heaviest doping? Which is the largest? Which controls the current between the other two?
3. Describe the differences between the common-emitter, common-base, and common-collector circuits.
4. What is α of a transistor? What is β?
5. You need an amplifier circuit that has a large voltage gain. Which of the three circuit types would give you the best voltage gain? Which one is the worst?
6. Look at the red dashed load line in Figure 4–16. How much collector current will flow when the base current is 4 μA?
7. You are testing a transistor with an ohmmeter. The B-E junction reads a high resistance regard-

less of which polarity lead you attach to the base. What is probably wrong?

8. You are selecting a radio-frequency amplifier for the first stage of a radio receiver. The first stage connects directly to the antenna, which has an impedance of 73 Ω. Which of the three types of circuit connections would probably be the best and why?

9. Explain how current can flow between the emitter and base of a transistor when it has to go backwards across the C-B junction.

10. Describe the conditions you would find in a transistor circuit during cutoff and saturation.

PRACTICE PROBLEMS

1. In Figure 4–10, assume the following values: $\beta = 250$; $R_B = 10{,}000\ \Omega$; $R_C = 2{,}500\ \Omega$. All other values are as shown. Calculate V_{GAIN}, I_{GAIN}, P_{GAIN}.

2. What is the voltage and power gain of the circuit of Figure 4–12 with the following values: $\alpha = .95$; $R_{OUT} = 25{,}000\ \Omega$; $R_{IN} = 400\ \Omega$?

3. A transistor amplifier has the black load line in Figure 4–16. With no input signal, the base current I_B is 10 μA. This point is called the quiescent point. Answer the following questions:
   a. What is the collector current at the quiescent point?
   b. What is the collector voltage at the quiescent point?
   c. If an input signal (I_B) is applied with a peak-to-peak value of 20 μA, what is the peak-to-peak collector current?
   d. What is the peak-to-peak collector voltage for the 20 μA peak-to-peak base current?

4. In the previous example, how much base current is required to
   a. drive the transistor to cutoff?
   b. drive the transistor to saturation?

5. In the example of Figure 4–10, what is the efficiency of this amplifier? Don't forget to include the power being input by the base voltage supply.

JFETs, MOSFETs, and UJTs

OUTLINE

OVERVIEW

In this chapter, you will learn about transistors that do not have bipolar junctions (BJT). These transistors are classified as junction field effect (JFET), metal-oxide semiconductor field effect (MOSFET), and the unijunction (UJT). The phototransistor will be covered in a later lesson with optical devices.

Recall from the past lesson that the BJT is a current-controlled device. In general, field effect transistors are voltage-controlled devices. This means the output characteristics of the FET are controlled by input voltages not input currents.

OBJECTIVES

After completing this chapter, the student should be able to:

1. Draw schematic symbols for JFET, MOSFET, and UJT transistors.
2. List and identify the characteristics of each transistor type.

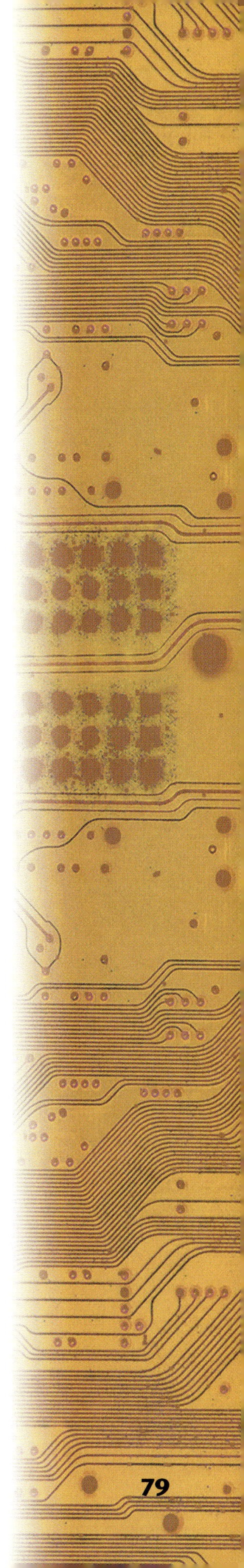

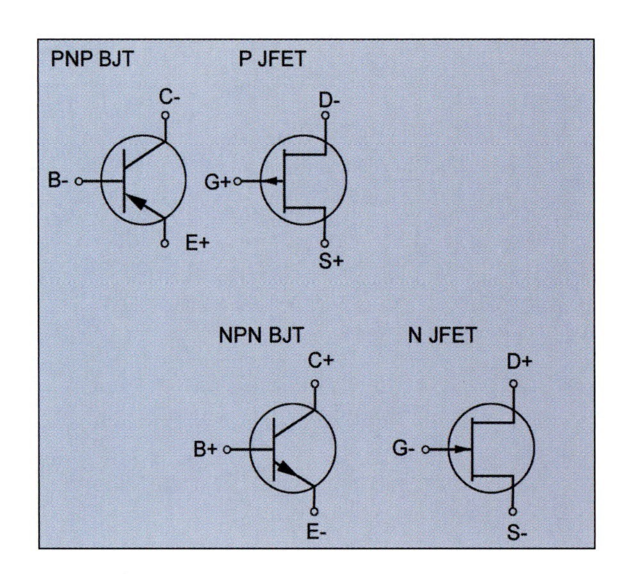

FIGURE 5–1 Comparison of BJT and JFET schematic symbols

JUNCTION FIELD EFFECT TRANSISTORS (JFET)

5.1 Construction

The JFET has three terminals that are similar to the BJT. They are the source, drain, and gate. Figure 5–1 compares the JFET terminals with the equivalent BJT terminals. The BJT counterparts are emitter for the source, collector for the drain, and base for the gate. A fourth JFET component is the channel. The channel is the material that connects the source to the drain.

Note that the gate surrounds the channel (see Figure 5–2). Also note that in the JFET schematic symbol, the arrow points in toward the material for an n-channel JFET.

5.2 JFET Operating Characteristics

Similar to an NPN transistor that requires positive supply voltages (V_{CC} and V_{BB}), the n-channel JFET also requires positive supply voltages. The reverse is true for a PNP transistor and also for a p-channel JFET. Figure 5–3 shows n-channel and p-channel supply voltage configurations. The supply voltage is labeled V_{DD} (similar to the BJT's V_{CC}). Unlike a BJT, which requires holes and electrons for conduction across the collector emitter, the JFET requires only electrons. As an interesting historical note, a patent covering the basic operation of a JFET was issued in the late 1930s. Unfortunately for the patent holder, semiconductor material technology at that time was not sufficient to build one.

Refer to Figure 5–4 for the following discussion. Because the gate surrounds the channel, a change in the width of the gate will control the current flow through the channel at any given voltage. To control the gate width, a reverse bias is applied to the gate-source junction. The reverse bias increases the depletion area (increases the p-material size) and reduces the n-channel. The reduced n-channel restricts the current flow from the drain to the source. The channel width can also be

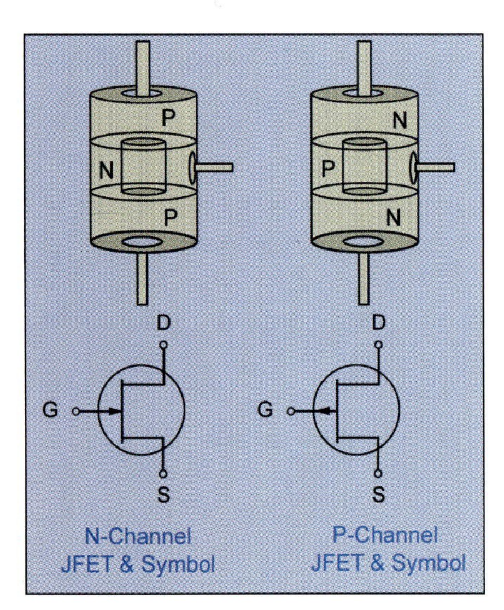

FIGURE 5–2 JFET construction

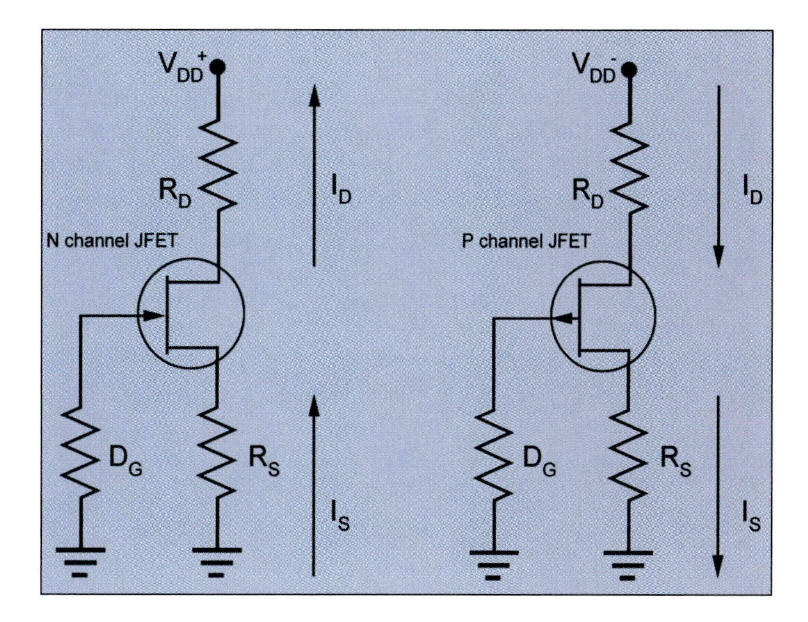

FIGURE 5–3 N- and p-channel JFET circuits

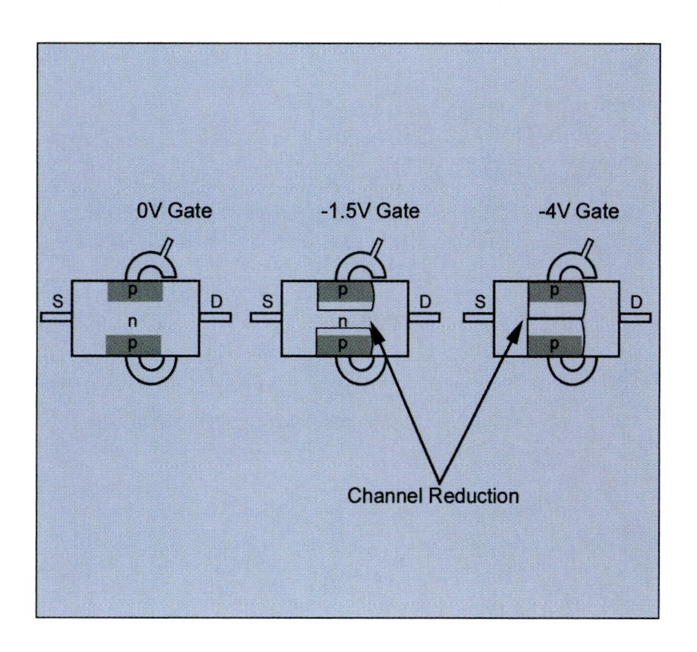

FIGURE 5–4 JFET gate control

changed by varying the drain-source voltage. As the drain-source voltage increases, the difference between the V_{DS} and V_{GS} becomes greater, increasing the depletion area and narrowing the n-channel.

Varying the drain-source voltage, while keeping the gate voltage constant, also changes the channel width. As the drain-source voltage increases, the difference between the V_{DS} and V_{GS} becomes greater; this increased reverse bias causes the channel to narrow as the current increases. Once the pinch-off voltage (V_P is the voltage at which V_{DS} generates the maximum I_D) is reached, an increase in drain-source voltage does not produce a significant drain-source current increase (I_{DS}). Continuing to increase the V_{DS} to and beyond the breakdown voltage (V_{BR}) will damage the JFET.

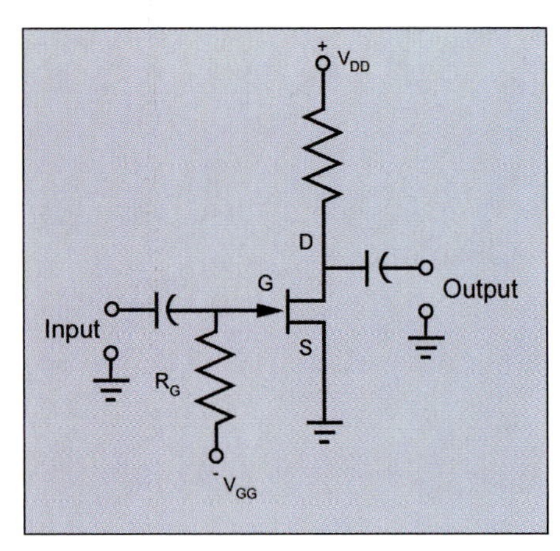

FIGURE 5–5 JFET gate biasing

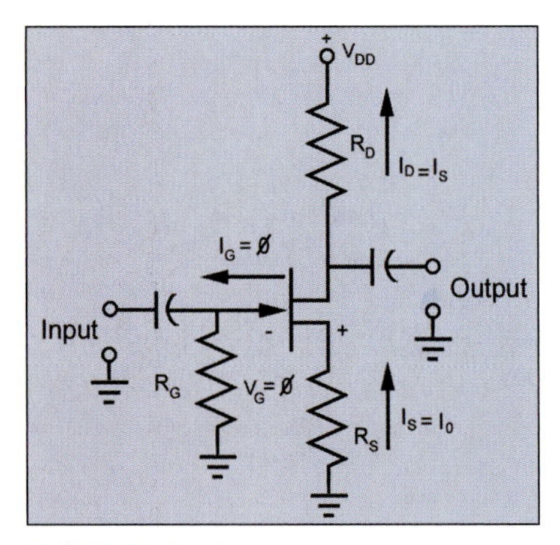

FIGURE 5–6 JFET self-biasing

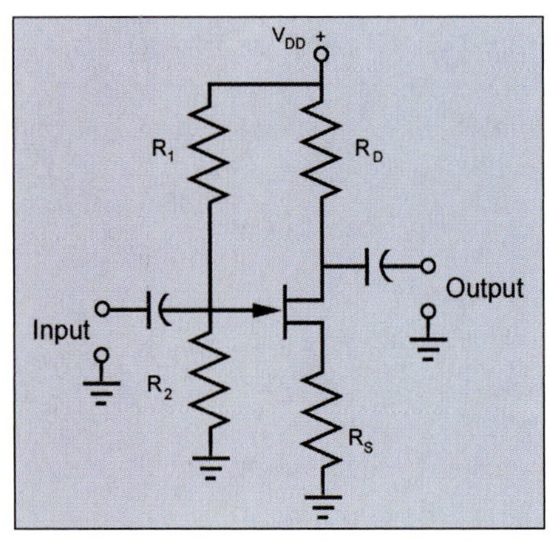

FIGURE 5–7 JFET voltage divider biasing

5.3 JFET Biasing

There are four basic biasing circuits for the JFET. They are: gate biasing (very similar to BJT base biasing), self-biasing, voltage divider biasing (very similar to BJT voltage divider biasing), and current source biasing.

Gate Biasing

Under normal conditions, the gate-source junction of a JFET is reverse biased. In the circuit configuration shown in Figure 5–5, a negative V_{GG} supplies the voltage to the gate to ensure a reverse bias. Because there is no gate current, there is no voltage drop across R_G; therefore, $V_{GS} = V_{GG}$.

Self-Biasing

A simpler biasing configuration is found in the self-bias circuit, as shown in Figure 5–6. In this circuit, the V_{GG} is replaced by a source resistor (R_S). The source resistor provides a voltage on the source that is always positive with respect to the gate. Note that $I_S = I_D$ and that $V_S = I_D R_S$. Because there is no current on the gate, the voltage across R_G is 0 and the corresponding voltage on the gate is 0. If the gate voltage is 0, the voltage drop on V_S will be greater; therefore, the gate will remain reverse biased.

Voltage Divider Biasing

A very stable biasing configuration is the voltage divider. This method is also used in BJTs. The formula is almost the same for finding the gate voltage as for calculating the base voltage. Figure 5–7 shows a voltage divider biasing circuit. This type of biasing is the most commonly employed due to its relative simplicity and stability.

$$V_G = V_{DD}\left(\frac{R_2}{R_1 + R_2}\right)$$

Current Source Biasing

The most stable of all these biasing circuits is current source biasing (see Figure 5–8). This circuit provides the most Q-point stability. In this circuit, the value of $I_D = I_C$. I_C is independent of the JFET characteristics so is the value of I_D. Even though this type of biasing is highly stable, the complexity and added components of the circuit limit its usefulness.

5.4 JFET Application

The high input impedance of the JFET makes it very useful where source loading is not desirable. For instance, the JFET in Figure 5–9 is acting as a "buffer amplifier" to the source amplifier. The JFET's high impedance (typically multiple megOhms) presents no load to the source amplifier and allows nearly all the gain to be passed through to the load.

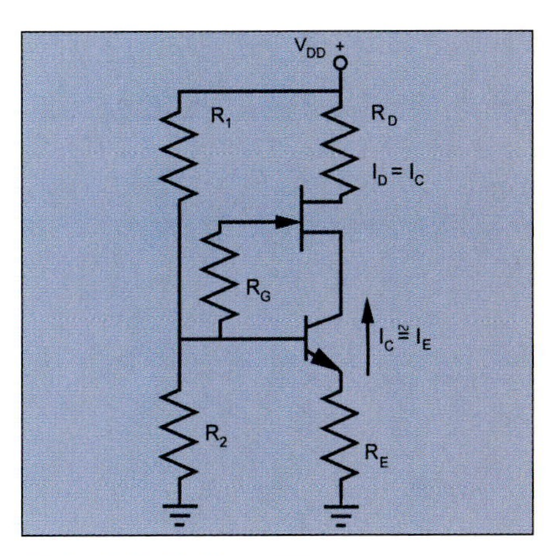

FIGURE 5–8 JFET current source biasing

Unlike the bipolar transistors (BJT), which are current-controlled devices, the JFET is a voltage-controlled device. The BJT is normally off, but the JFET is normally on and draws practically no current from the driving source. Finally, it should be noted that, just as the BJT configurations, the JFET can be designed in a common-gate, common-drain, and common-source circuit configuration.

METAL-OXIDE SEMICONDUCTOR FIELD EFFECT TRANSISTOR (MOSFET)

Metal-oxide semiconductor (MOS) technology is used extensively in most calculators, watches, desktop computers, and other appliances. The main advantage of MOS circuits is their low current requirement. With less current, there is less heat dissipation. Another advantage is that MOS components can be made much smaller than BJT components.

5.5 MOSFET Construction

There are two basic types of MOSFETs—enhancement (E-MOSFET) and depletion (D-MOSFET). The depletion types are more versatile because they can operate in enhancement or depletion mode. The enhancement type can operate only in enhancement mode.

The physical difference between the enhancement and depletion types is the channel between the source and drain. The depletion type has a channel, the enhancement type does not. Figure 5–10 shows how the two types of MOSFETs are constructed. The E-MOSFET is normally an off device, but the proper gate voltage will attract carriers to the gate region and form a conductive channel. Either a negative or positive voltage potential to the gate will allow current to flow between the source and the drain. The D-MOSFET is normally on because of the physical channel, and only when a negative voltage is applied to the gate does the channel resistance increase and current reduce. When a D-MOSFET has a negative gate voltage, it is operating in the depletion mode. Please note that FETs do not require any gate current to operate; therefore, the gate region is not physically connected to the channel.

FIGURE 5–9 JFET amplifier buffer application

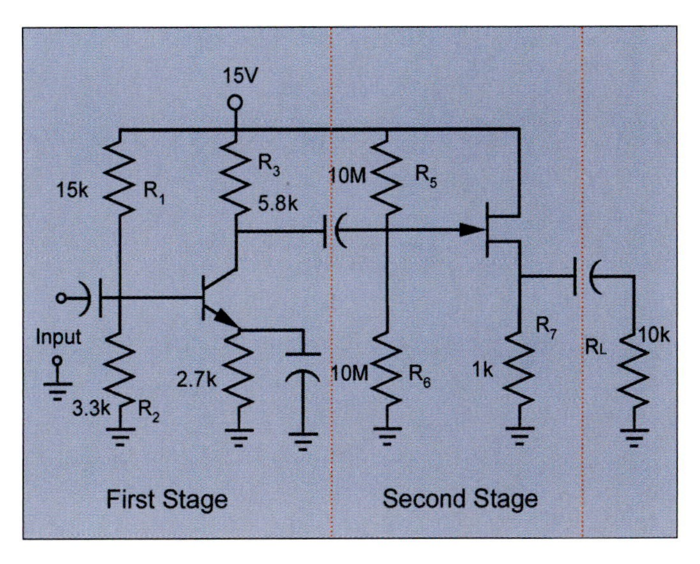

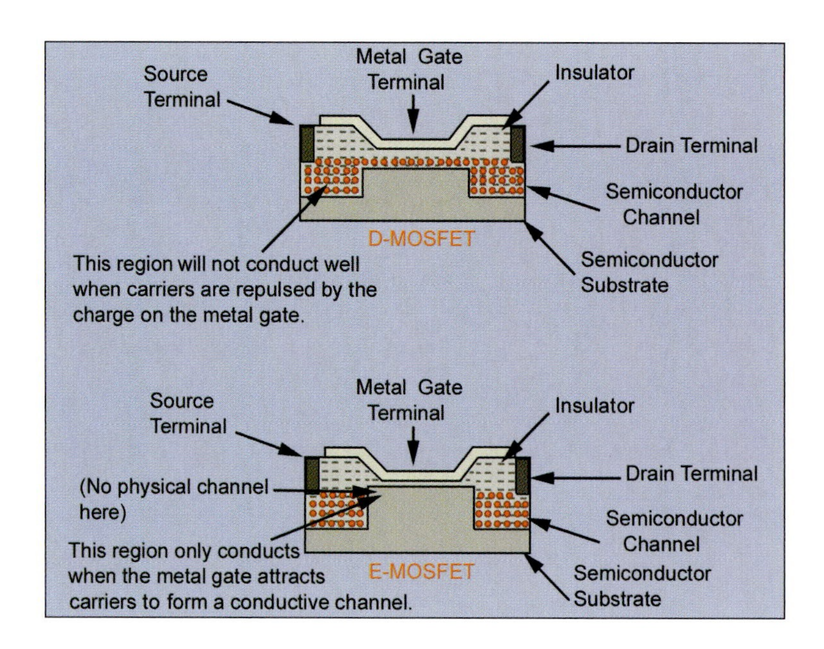

FIGURE 5–10 Depletion and enhancement MOSFET construction

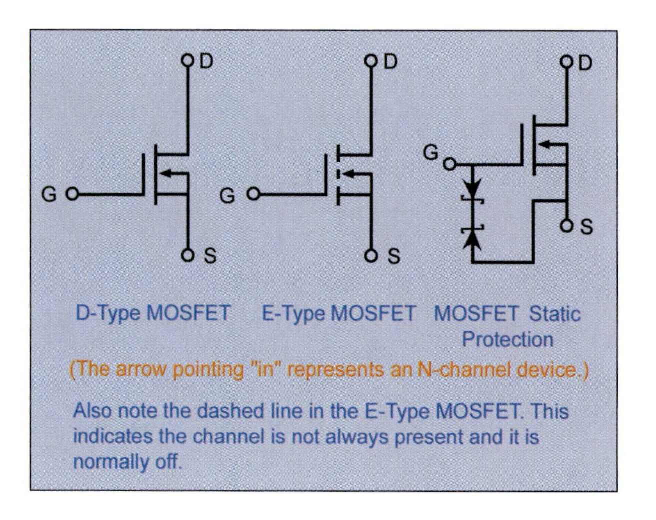

FIGURE 5–11 E- and D-type MOSFET schematic symbols

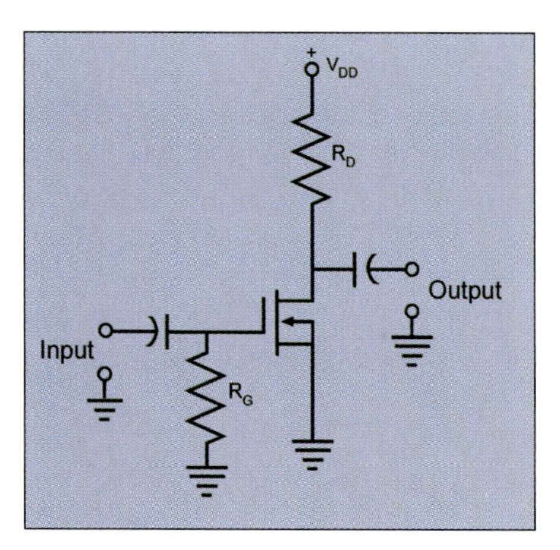

FIGURE 5–12 MOSFET zero bias circuit

Both types have an insulation layer between the gate and the rest of the semiconductor. This layer is made of silicon dioxide (SiO_2), and because the layer is thin, it can easily be damaged by static electricity.

Many of the MOSFETs manufactured today have Zener diodes etched into their construction. The diodes are between the gate and the source and are designed to operate above the rated voltage of the V_{GS}. These diodes protect the gate from damage due to static electricity.

Figure 5–11 shows the schematic symbols for the E- and D-type MOSFET along with a schematic for the diode static electricity protection.

Like the JFET, the MOSFET has very high input impedance. For example, one MOSFET has a maximum gate current rating of 10 pA (picoamps) with a V_{GS} of 35 V. A D-MOSFET has two other advantages over the JFET. It can operate in zero bias (Figure 5–12 shows an example circuit for zero biasing) and it can operate in enhancement mode.

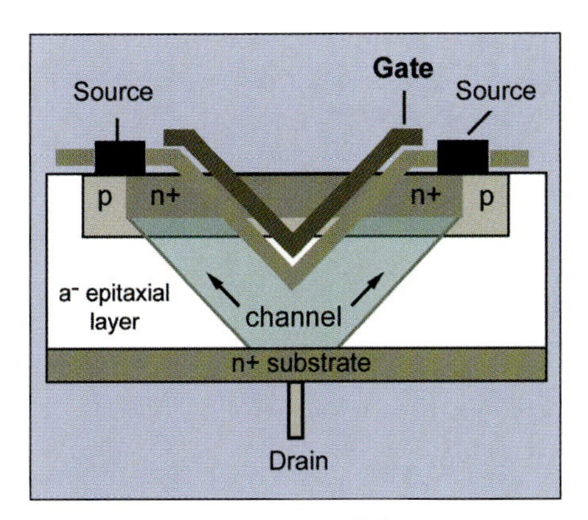

FIGURE 5–13 VMOS construction

5.6 VMOS Power Applications

In modern communications, most signals are transmitted and received in digital form. The receiver converts the incoming signal and information back into its original form. This requires the use of a power amplifier that can produce high-speed, high-current digital outputs (i.e., respond to on-off signals very quickly with no distortion). This is the perfect application for power MOSFET amplifier drivers.

The power MOSFET (Figure 5–13) is known as the VMOS. The VMOS name comes from the way the semiconductor is constructed. The gate is V-shaped and is surrounded by heavier doping of the N-type material. This heavier doping produces a wider channel and allows the VMOS to handle higher drain current levels. The VMOS operates only in enhancement mode.

Notice in Figure 5–13 that the VMOS current flow is vertical. This short and wide channel results in a higher current capacity, greater power dissipation, and improved frequency response. Because the VMOS is an enhancement mode device, it is normally off and has no physical channel. Current flows between the source and the drain when the gate is made positive with respect to the source.

5.7 Other Types of MOSFET Semiconductors

Two other types of MOSFET semiconductors are the dual gate and the LDMOS (lateral double-diffused MOSFET). The dual gate MOSFET is used in applications where the two-gate construction decreases the capacitance between the gate and the channel. This high capacitance is caused by the gate acting as a conductor. Using two gates reduces the plate size and overall capacitance. By lowering the capacitance, a better high-frequency operation can be obtained from the MOSFET.

LDMOS is an enhancement-type MOSFET. Its main advantage is low channel resistance and its ability to handle very high drain currents without generating a lot of heat dissipation. A very small channel and a heavily doped N-type substrate make this possible.

UNIJUNCTION TRANSISTOR (UJT)

5.8 UJT Construction

The UJT has three leads but only two doped regions. The terminals are emitter, base 1 (B1), and base 2 (B2). Figure 5–14 shows the schematic symbol for a UJT. The emitter is heavily doped P-type material, and the base is only slightly doped N-type material. Notice that the construction of the UJT is very similar to the JFET. The only difference is that the P-type gate material of the JFET surrounds the n-channel, whereas the P-type material of the emitter does not surround the base in the UJT.

5.9 UJT Operating Characteristics

The main function of the UJT is switching. Oscillators make frequent use of UJTs. UJTs are not used as amplifiers.

FIGURE 5–14 UJT schematic symbol

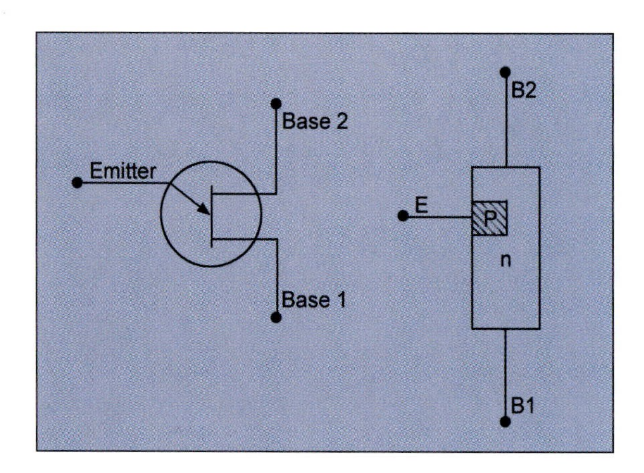

FIGURE 5–15 UJT biasing

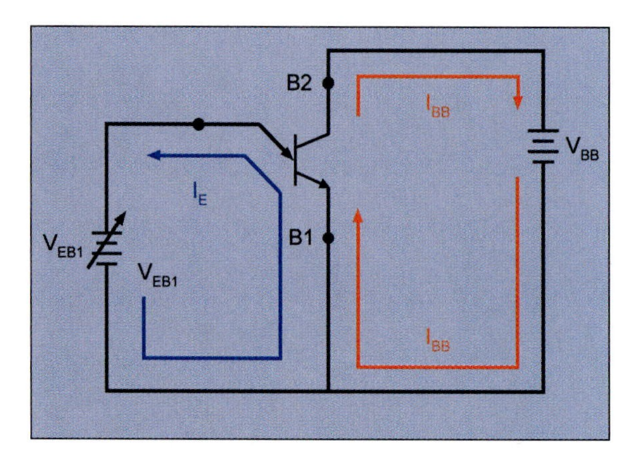

Refer to Figure 5–15 for the following discussion. When a base biasing voltage (V_{BB}) is applied between B1 and B2, the emitter-B1 junction is seen as open until the voltage across V_{EB1} is increased to a specific value. This value is called peak voltage (V_P). When V_P is reached, the E-B1 junction triggers or fires. After firing, the UJT will remain on until the emitter current falls below a specific value. This value is called peak current (I_P). When the emitter current falls below I_P the UJT turns off.

Figure 5–16 shows the UJT characteristic curve. When the emitter current (I_E) increases above I_P, the value of the internal resistance R_{B1} drops dramatically. This results in a reduction of V_{EB1} as I_E increases. The reduction in V_{EB1} continues until I_E reaches the valley current (I_V). Increasing I_E beyond I_V puts the UJT into saturation. Notice that the region where the internal resistance is reduced (between I_P and I_V) is called the negative resistance region. Negative resistance is when any device has current and voltage values that are inversely related to each other.

The relationship between resistance of B1 and B2 in the UJT can be expressed as a voltage divider using an equivalent circuit. Figure 5–17 shows an example of this equivalent circuit.

The actual voltage drops and operating principle are straightforward. The E-B1 diode must be forward biased. This means that a voltage drop of .7 V is required from the anode to the cathode. This forward bias potential (V_K) is calculated using the voltage divider rule.

$$V_K = V_{BB}\left(\frac{R_{B1}}{R_{B1} + R_{B2}}\right)$$

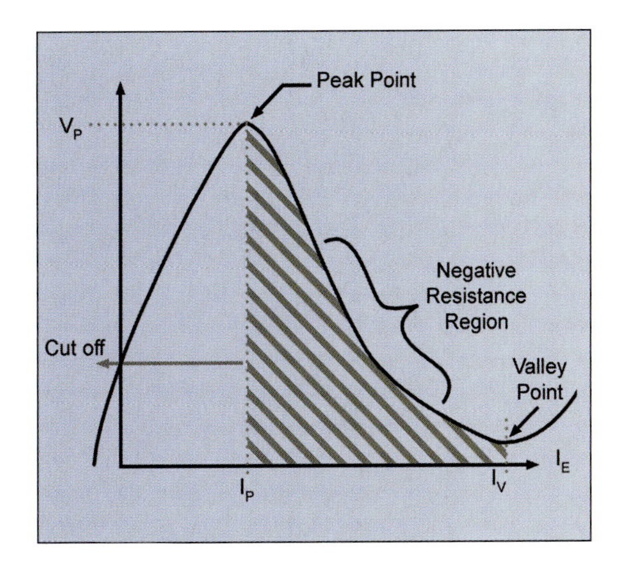

FIGURE 5–16 UJT characteristic curve with negative resistance region

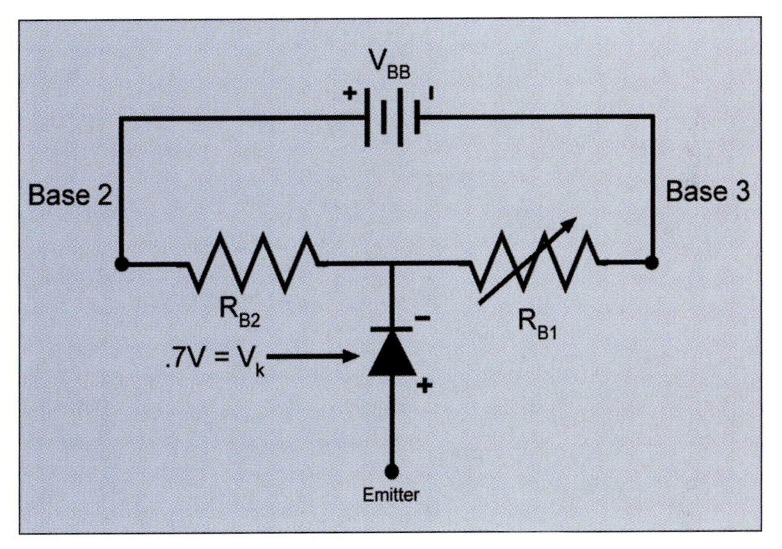

FIGURE 5–17 UJT equivalent circuit

The resistance ratio $\left(\frac{R_{B1}}{R_{B1} + R_{B2}}\right)$ is called the intrinsic standoff ratio (η). This ratio is listed on the specification sheet for any UJT.

$$V_K = \eta\ V_{BB}$$

$$V_p = \eta\ V_{BB} + 0.7\ V$$

5.10 UJT Example Circuit

EXAMPLE 1

Calculate the V_P required to trigger the UJT in the following circuit (see Figure 5–18). Assume the following: η max from the specification sheet for this UJT is .75.

$$V_P = \eta\ V_{BB} + 0.7$$
$$V_P = .75(15\ V) + .7\ V = 11.95\ V$$

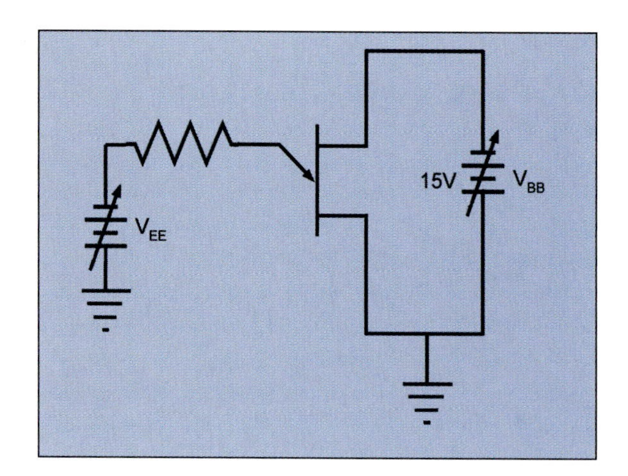

FIGURE 5–18 Example UJT circuit

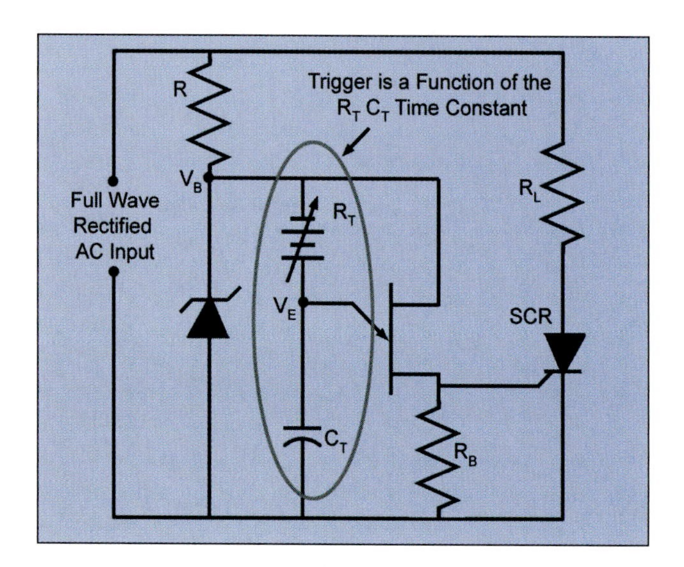

FIGURE 5–19 UJT circuit with RC time constant

Because one of the primary functions of the UJT is to trigger triacs or SCRs into conduction, an RC time constant could be added to the circuit to provide alternating charge and discharge pulses. In the example in Figure 5–18, once the capacitor charges to 11.95 volts, the UJT would "turn on." When the capacitor discharges below V_P, the UJT will "turn off." An example UJT circuit with added RC components is shown in Figure 5–19.

SUMMARY

In this chapter, you have learned about three major types of transistors: the JFET, MOSFET, and UJT. Each has its own advantages and disadvantages. Figure 5–20 summarizes the main characteristics of each.

The JFET has a relatively high input impedance and is used as a buffer against loading down the source amplifier and reducing overall amplifier gain. The JFET comes in two basic types, n and p. JFETs are voltage-controlled rather than current-controlled devices.

The MOSFET also comes in two basic types—depletion and enhancement. The depletion type can operate in either depletion or enhancement mode. The enhancement type also has a high-power version (VMOS). It is used when low heat dissipation and high currents are required. The MOSFET has a higher input impedance than the JFET and is used extensively in integrated circuitry. MOSFETs also require special handling to protect them from possible static electricity damage.

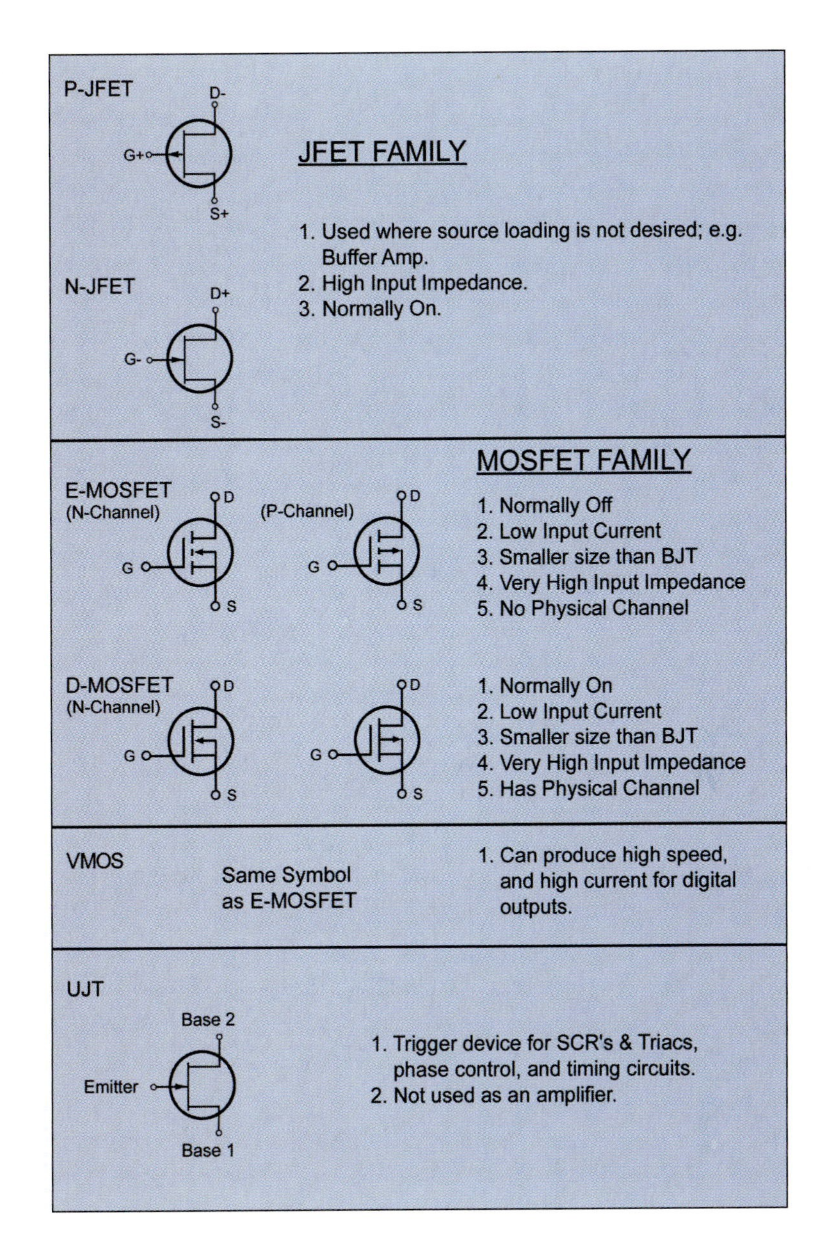

FIGURE 5–20 Summary of JFETs, MOSFETs, and UJTs

The UJT is not used as an amplifier, but as a switching device. Frequently, the UJT is used to trigger other devices such as triacs or SCRs. UJTs are very similar in construction to JFETs but do not have the same lead configuration. Unlike FETs that have a gate, a source, and a drain, the UJTs have an emitter, a base 1, and a base 2.

REVIEW QUESTIONS

1. Which of the devices in this chapter would you probably use to make a triangle-wave signal generator? Why?
2. MOSFETs are used extensively on chips for integrated circuits. Discuss the possible reasons for this.
3. Discuss the advantages of high input impedance for a semiconductor device used in an amplifier circuit.
4. What does the 'V' in VMOS stand for? Discuss the application of the VMOS.
5. Discuss the various families for the semiconductor types discussed in this chapter. Use Figure 5–20 to guide the discussion.

Amplifiers

OUTLINE

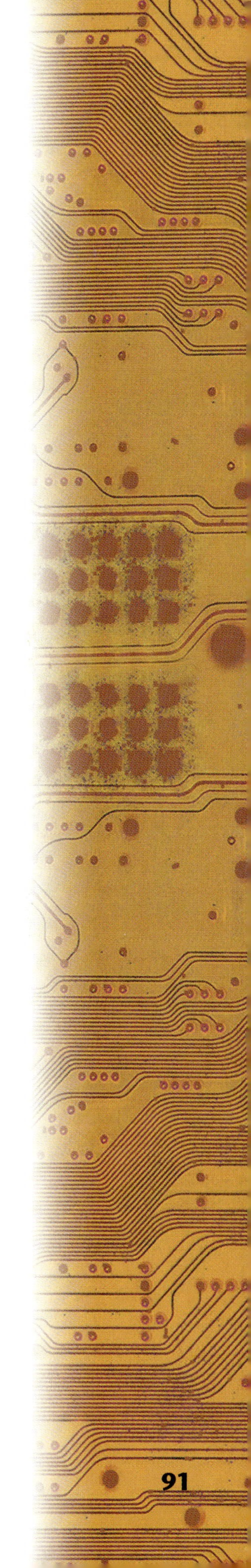

OVERVIEW

In this chapter, you will learn about the basic types of amplifiers and how they operate. Amplifiers are used in most electronic circuits. In audio equipment, the normal input signal to the receiver is too small to power the speaker system, so it must be amplified before it is sent to the speaker. The same is true with video equipment; the input signal must be amplified before it can be presented on a cathode-ray tube (a TV screen).

The electronic circuits that perform this job of increasing the voltage, current, or power level of electrical signals are called **amplifiers**. In the strict definition, power gain must be involved for the circuit to be an amplifier. A transformer, for example, will increase the voltage level of a signal, but it does not increase the power level; consequently, the transformer is not an amplifier.

In electronics, you normally do not worry about such issues. Amplifier circuits may be designed for voltage, current, or power gain depending on the application. In almost every case, an electronic amplifier provides power gain as well.

OBJECTIVES

After completing this chapter, the student should be able to:

1. Define the various terms associated with amplifier circuits.
2. Draw basic circuit configurations for the three types of amplifiers.
3. Describe the unique characteristics for each type of amplifier.

GLOSSARY

A_I Current gain.

A_P Power gain.

A_V Voltage gain.

Active component Components of an electronic circuit that use a power source to process a signal. The processing usually involves amplification or some other change in the signal that requires additional power. Transistors, FETs, and UJTs are examples of active components.

Amplification The process of increasing the voltage, current, or power of a signal.

Amplifier A device that provides gain without much change in the original signal waveform.

Amplitude The size of a signal. Most commonly, the amplitude is expressed in terms of the signal voltage.

Gain The ratio of the output signal to the input signal of an active component. $V_{GAIN} = \frac{V_{OUT}}{V_{IN}}$. Gain is normally thought of as being greater than 1; however, gain is also defined for values that are less than one.

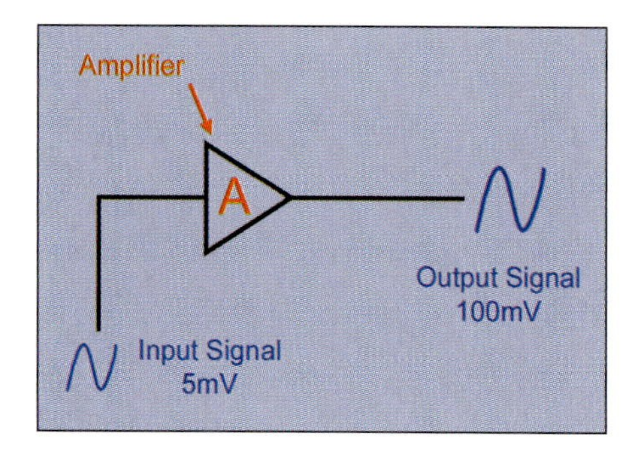

FIGURE 6–1 Amplifier block diagram

AMPLIFIER GAIN

An amplifier provides **gain**—the ratio of the output to the input. An amplifier achieves gain by converting one thing to another. For example, a lever converts force to movement or movement to force; one pound of force on the end of a beam may lift three pounds on the other end, but it will not lift it far. An electronic amplifier uses circuit DC power and divides that power between the output terminal and a load resistor based on the strength of the input signal. Figure 6–1 shows a block diagram of an amplifier circuit.

6.1 Types of Gain

If the objective for an amplifier is to produce gain, then a measure of the gain of the amplifier is a measure of the amplifier's success. Gain is expressed as the ratio of output to input. Gain is shown in formulas using the symbol A with a subscript of P, I, or V to indicate power, current, or voltage respectively.

Take care to write the result correctly; A_P, A_I, and A_V are unitless. This is because A is calculated by dividing power by power, current by current, or voltage by voltage. By not tacking units on the ends of amplification factors, phrases like, "We have six watts times as many watts as Mr. Watts." can be avoided. Normally, gain is expressed as a dimensionless number—"The amplifier has a gain of 10, 20, 50, 110, . . . "

6.2 Measuring Gain

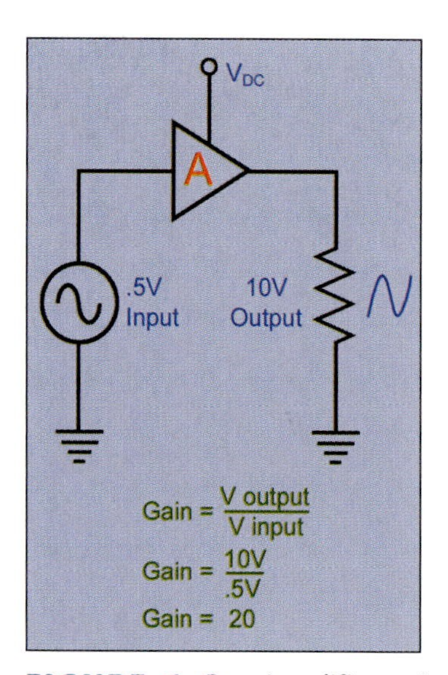

FIGURE 6–2 Amplifier gain

To calculate voltage gain, divide the output voltage by the input voltage levels. Refer to Figure 6–2.

$$\text{Gain} = \frac{\text{Output signal}}{\text{Input signal}}$$

$$A_V = \frac{V_{OUT}}{V_{IN}} = \frac{10\text{ V}}{0.5\text{ V}} = 20$$

Notice that the volts symbols cancel each other and the final result is unitless.

As another example, assume an amplifier has a 1 mW input and an output level of 10 mW, as shown in Figure 6–3.

What is the amplifier's power gain?

$$A_P = \frac{P_{OUT}}{P_{IN}} = \frac{10\text{ mW}}{1\text{ mW}} = 10$$

If another amplifier takes the 10 mW input and yields a 100 mW output level, what is this amplifier's power gain?

$$A_P = \frac{P_{OUT}}{P_{IN}} = \frac{100\text{ mW}}{10\text{ mW}} = 10$$

What is the gain of both stages from input to output?

$$A_P = \frac{P_{OUT}}{P_{IN}} = \frac{100\text{ mW}}{1\text{ mW}} = 100$$

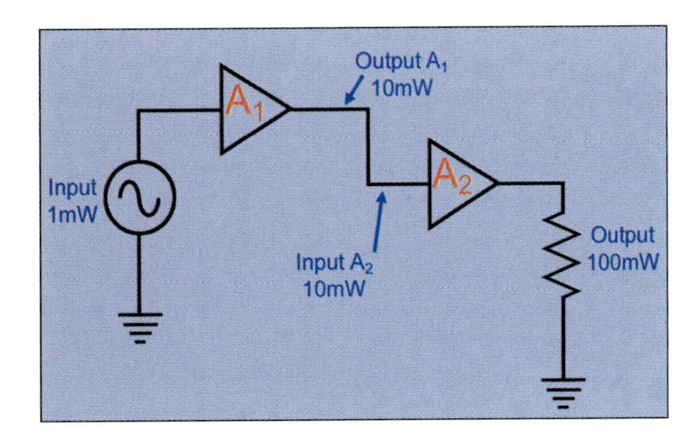

FIGURE 6–3 Power gain

Notice that the total gain of both stages is the product of the individual gains.

6.3 Gain in Decibels

Power Versus Audio Response

If we listen to the sound of a tone that starts at 1 mW and increases to 100 mW in two steps (as in the previous two examples), we might expect to be deafened by the volume of the second step. The gain can be calculated as 100. In fact, we would perceive the sound to be only four times as loud as the original. Furthermore, the 10 mW tone would sound twice as loud as the 1 mW tone, and the 100 mW tone would sound twice as loud as the 10 mW tone.

The human ear has a logarithmic response to sound; this means that we perceive the relative volume of two sounds instead of the absolute volume. We perceive other things in the same way. A poor person will look at the gift of a dollar in a very different way than a millionaire. This is to say that the effect of the sound on our ears is relative instead of absolute (in math calculation terms).

Decibels

Because of the logarithmic nature of human hearing, the decibel system of expressing power levels was developed. It is used to describe the performance of audio systems, but it is also useful for describing any system of electrical amplifiers or relative amplitude levels. The power gain in Bels is calculated as:

$$A_{G(b)} = \log_{10}\left(\frac{P_{OUT}}{P_{IN}}\right)$$

For most "real world" applications, the Bel[1] is simply too large a unit to work with. A more convenient unit is the decibel (dB), which is defined as:

$$A_{G(dB)} = 10 \times A_{G(B)} = 10 \times \log_{10}\left(\frac{P_{OUT}}{P_{IN}}\right)$$

[1]The Bel is named in honor of Alexander Graham Bell.

In the previous example, the difference between 1 mW and 10 mW is 10 decibels or 10 dB. This is calculated by using the formula:

$$A_{P(dB)} = 10 \times \log_{10}\left(\frac{10\text{ mW}}{1\text{ mW}}\right) = 10 \times \log_{10} 10 = 10 \times 1 = 10\text{ dB}$$

Note that the gain of the second stage is also equal to 10 dB. The gain of both stages together is calculated as

$$A_{P(dB)} = 10 \times \log_{10}\left(\frac{P_{OUT}}{P_{IN}}\right) = 10 \times \log_{10}\frac{100}{1} = 10 \times \log_{10} 100 = 10 \times 2 = 20\text{ dB}$$

The conversion is easy with a scientific calculator. With an algebraic entry calculator, the following steps will do the job:

1. Key in the number 10.
2. Press the multiply button.
3. Press the log button. [Be sure to use log to the base 10, not the *ln* button, which calculates logs to the natural (Naperian) base *e*.]
4. Key in the left parenthesis [( ].
5. Key in the output power.
6. Press the divide key.
7. Key in the input power.
8. Key in the right parenthesis [ )].
9. Press the equal sign (=).

The result is the power gain expressed in decibels.

For an RPN-type calculator, such as many Hewlett-Packard models, the procedure is slightly different:

1. Key in the output power.
2. Press the <Enter> key.
3. Key in the input power.
4. Press the divide key (÷).
5. Press the log key. (As before, be sure you use the common logarithm, not the natural logarithm.)
6. Key in the number 10.
7. Press the multiply key (×).

The result is the power gain expressed in decibels.

6.4 Multistage Gain

Figure 6–4 shows the gains of each stage of a multistage amplifier. They are 14 dB, −4 dB, 18 dB, and 22 dB respectively. Note that a gain of −4 dB is actually a reduction in level, as you will see.

In order to calculate the gain using decibels, simply add them together, as follows:

$$A_{P(dB)} = 14\text{ dB} + (-4\text{ dB}) + 18\text{ dB} + 22\text{ dB} = 50\text{ dB}$$

Figuring gain with the same system using ratios involves multiplication instead of addition. First you need to know the gain of each

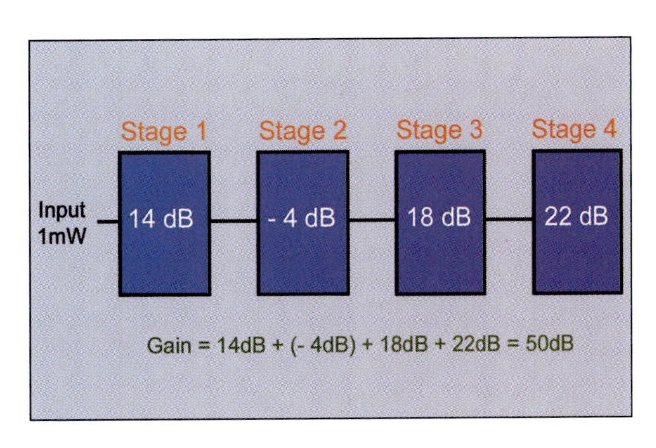

FIGURE 6–4 Multistage gain in decibels

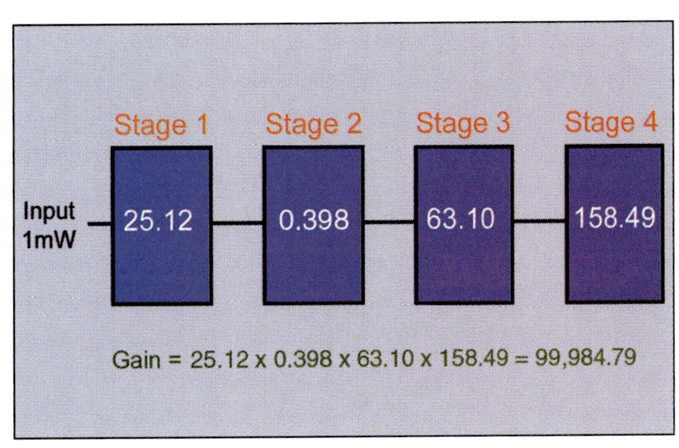

FIGURE 6–5 Multistage gain in linear amplification factors

stage as expressed by a ratio rather than in dB. Start by considering the definition of the decibel:

$$A_{P(dB)} = 10 \log_{10}\left(\frac{P_{OUT}}{P_{IN}}\right)$$

Solving this formula for the ratio $\frac{P_{OUT}}{P_{IN}}$ involves a little algebra.

$$\frac{A_{P(dB)}}{10} = \log_{10}\left(\frac{P_{OUT}}{P_{IN}}\right) \Rightarrow \left(\frac{P_{OUT}}{P_{IN}}\right) = 10^{\left(\frac{A_{P(dB)}}{10}\right)}$$

Calculating the ratio gain of the four stages (see Figure 6–5) gives:

$$A_P\big|_{\text{Stage 1}} = 10^{\frac{14}{10}} = 25.12$$

$$A_P\big|_{\text{Stage 2}} = 10^{\frac{-4}{10}} = 0.398$$

$$A_P\big|_{\text{Stage 3}} = 10^{\frac{18}{10}} = 63.10$$

$$A_P\big|_{\text{Stage 4}} = 10^{\frac{22}{10}} = 158.49$$

$$A_P\big|_{\text{Total}} = 25.12 \times 0.398 \times 63.10 \times 158.49 = 99{,}984.79$$

To check your work, you can put this value into the original formula, as follows:

$$A_{P(dB)} = 10 \times \log_{10}(99{,}984.79) = 49.99 \text{ dB}$$

The slight difference in result is caused by rounding error. If you perform the previous steps using a scientific calculator and do not round, you will get a much closer answer. Perhaps you can see that it is easier to work with decibels than linear amplification factors—using decibels generally causes fewer mistakes.

6.5 Voltage Gain in Decibels

The voltage gain in decibels uses a slightly different formula than the power gain in decibels. The reason for the difference is the logarithmic nature of decibels and the relationship between voltage and power. Recall that $P = \frac{V^2}{R}$

Therefore

$$A_{P(dB)} = 10 \times \log_{10}\left[\frac{\left(\frac{V_{OUT}^2}{R_{OUT}}\right)}{\left(\frac{V_{IN}^2}{R_{IN}}\right)}\right]$$

If R_{OUT} equals R_{IN}, then the resistances will divide out of the equation, leaving only

$$A_{P(dB)} = 10 \times \log_{10}\left(\frac{V_{OUT}^2}{V_{IN}^2}\right)$$

Note that you can divide the resistances out only if they are equal. If they are not, you must include them and use the slightly more complicated formula.

From logarithms, recall the following:

$$\log X^2 = 2 \times \log X$$

This means that

$$A_{P(dB)} = A_{V(dB)} = 2 \times 10 \times \log_{10}\left(\frac{V_{OUT}}{V_{IN}}\right) = 20 \times \log_{10}\left(\frac{V_{OUT}}{V_{IN}}\right)$$

Notice the change from power gain to voltage gain. This occurs because the ratio of the logarithm $\left(\frac{V_{OUT}}{V_{IN}}\right)$ is a voltage ratio. Remember that this formula applies ONLY if the resistances across which the two voltages are dropped are equal.

6.6 Practical Uses of dB Ratios

The advantage of dB ratios is the logarithmic ability to express large numbers with small numbers. Named after Alexander Graham Bell, the Bel is a logarithm ratio of two quantities. The decibel (1⁄10th of a Bel) is normally used in sound systems because the number scaling is more natural. By applying the dB power formula (gain in decibels), you will find that if a 400-watt audio amplifier was replaced with an 800-watt amplifier, there would be only an actual 3 dB gain. The power difference between the two amplifier gains is twofold; therefore, the application of the power formula

$$dB = 10 \times \log_{10} \frac{P_{OUT}}{P_{IN}} = 10 \times \log_{10} 2 = 3.01 \text{ dB}$$

In Table 6–1, the relationship between power values and dB levels is illustrated. For example, an 800-watt amplifier output is equal to 29 dB ($dB = 10 \times \log_{10} 800 = 29$). In this case, we are comparing the 800 watts to 1 watt, and the gain is simply selected from the 800 watt row.

In comparison, if a 400-watt amplifier was replaced with an 800-watt amplifier, the results would be a 3 dB gain. Here, we are comparing the new 800-watt amplifier to the replaced 400-watt amplifier. By subtracting the 26 dB in the 400 watt row from the 29 dB in the 800 watt row, the result is 3 dB.

Table 6–1 Power Values (Watts) Versus Equivalent dB Levels

Power Values (Watts)	Levels in dB (compared to 1 watt)
1.0	0
1.25	1
1.6	2
2.0	3
2.5	4
3.15	5
4.0	6
5.0	7
6.3	8
8.0	9
10.0	10
100	20
200	23
400	26
800	29
1,000	30
2,000	33
4,000	36
8,000	39
10,000	40
20,000	43
40,000	46
80,000	49
100,000	50

Therefore, Table 6–1 illustrates the values in dB compared to 1 watt or, by subtracting, compared to other wattage ratings within the table.

If you turned your stereo up from 10 watts to 100 watts and someone commented, "It's too loud," you could say, "But I only turned it up ?? dB." By subtracting the dB in the 10 watt row from the dB in the 100 watt row, the answer is found to be 10 dB.

The decibel expresses the "logarithmic" ratio of two acoustic quantities (sound levels) and is widely used in audio and communication electronics.

An electrical power level is expressed as dBm, where the *m* stands for milliwatts and is the reference. The formula for dBm is

$$\text{dBm} = 10 \times \log_{10}\frac{P\text{mW}}{1\ \text{mW}}$$

The dBm has no direct relationship to impedance or voltage. Originally, the dBm was devised from a 600-ohm telephone load, and in most cases today the 0 dBm is usually referenced to $0.775\ \text{V}\left(P = \frac{V^2}{R} = \frac{(.775)^2}{600} = 1\ \text{mW}\right)$, but it can be referenced to other voltage levels.

A voltage level is expressed as dBu and is referenced to 0.775 V. It is not dependent on the resistance load. The voltage expressed by dBu and dBm are equal only if the dBm voltage is derived with a 600-ohm load. A sound system console specification could list its maximum output level as +20 dBu into an 8k-ohm (or higher) impedance load.

The dBm and dBu mainly express small wattage and voltage levels. The dBW was derived to express large wattage outputs of large amplifier systems. Zero dBW is equal to 1 watt. Therefore, an 800-watt amplifier is a 29 dBW amplifier. This is found by applying the power formula

$$\text{dBW} = 10 \times \log\left(\frac{800}{1}\right) = 29\ \text{dBW}$$

AMPLIFIER CLASSIFICATION

6.7 Amplifier Operating Characteristics

There are two major operating characteristics that affect the classification of an amplifier—function and frequency response.

1. Function—an amplifier is usually designed to amplify either voltage or power.
2. Frequency response—an amplifier may be designed as:
   a. Audio amplifier—designed to amplify frequencies from 5 Hz to 20 kHz
   b. Radio frequency amplifier—designed to amplify frequencies between 10 kHz and 100,000 MHz
   c. Video amplifier—designed to amplify wide bands of the frequencies between 10 Hz and 6 MHz

6.8 Amplifier Classes

The class of operation of an amplifier is determined by the amount of time the current flows in the output circuit, compared to the input signal. There are four classes of operation: A, AB, B, and C. Each of these classes has certain uses and characteristics, but no one class of operation is considered to be better than another class. The selection of the class to use is determined by how the amplifier is used. The best class of operation for a servo-feedback loop would not be the same as for a radio transmitter.

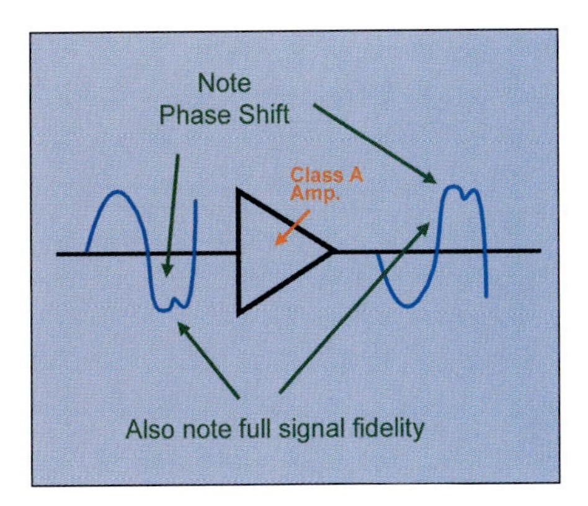

FIGURE 6–6 Class A amplifier

Class A Amplifier

Figure 6–6 illustrates the basic operation of a class A amplifier. In a class A amplifier, current will flow in the output during the entire input signal period. The output is an exact copy of the input except for increased **amplitude**. A class A amplifier is called a Fidelity amplifier because of its ability to recreate a larger scale image of the input signal with little or no distortion. A class A amplifier may change the phase of a signal, but not the shape. Because the class A amplifier operates during the entire time that the input signal is present, it is less efficient than an amplifier that operates only during half the input, but it produces better fidelity.

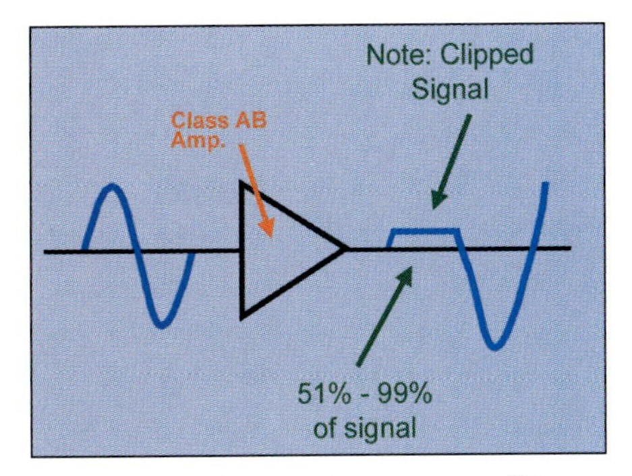

FIGURE 6–7 Class AB amplifier

Class AB Amplifier

The class AB amplifier operates during 51 percent to 99 percent of the input signal. Because it does not operate for the complete input cycle, the output signal cannot be the same shape as the input signal. In this case, the output signal will be distorted. Distortion is any undesired change in signal from the input to the output. Figure 6–7 shows the clipping that occurs during the positive part of the input. Notice that not all of the positive part is clipped. In a class AB amplifier, the transistor going into cutoff causes the clipping; thus, it is more efficient than the class A amplifier because it does not always amplify. Nor does it produce a fidelity signal because the signal is distorted due to clipping.

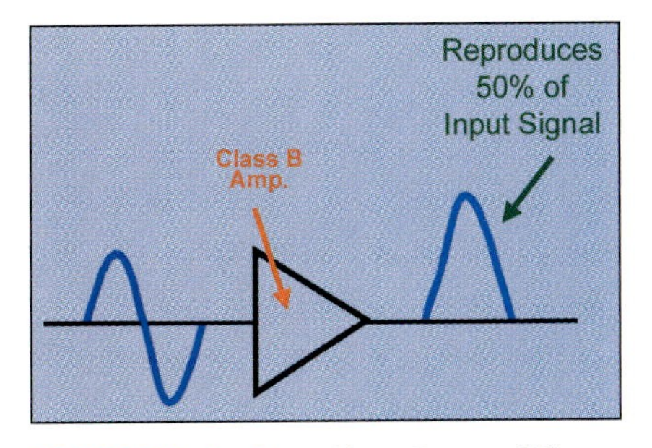

FIGURE 6–8 Class B amplifier

Class B Amplifier

The class B amplifier operates for 50 percent of the input signal. The amplifier will amplify one half of the signal very well, but the other half is completely lost. This amplifier is twice as efficient as the class A amplifier and is used when only half of the signal is needed. Figure 6–8 shows an example of a class B amplifier.

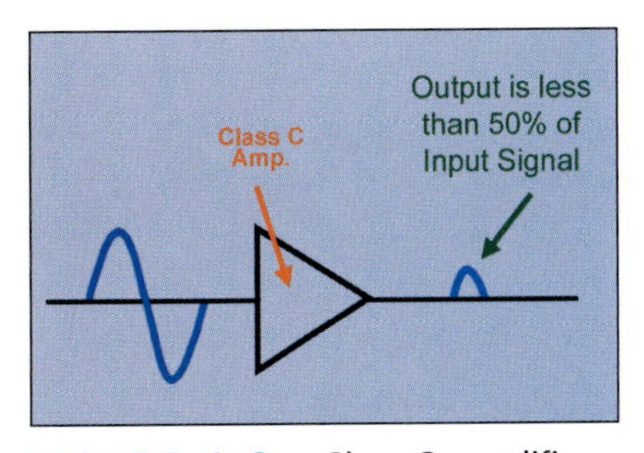

FIGURE 6–9 Class C amplifier

Class C Amplifier

The class C amplifier is the most efficient of the four classes because it operates only during a small portion of the input. It also produces the most distortion of all the amplifiers. It is useful when only a very small part of a signal must be amplified. Figure 6–9 shows an example of a class C amplifier.

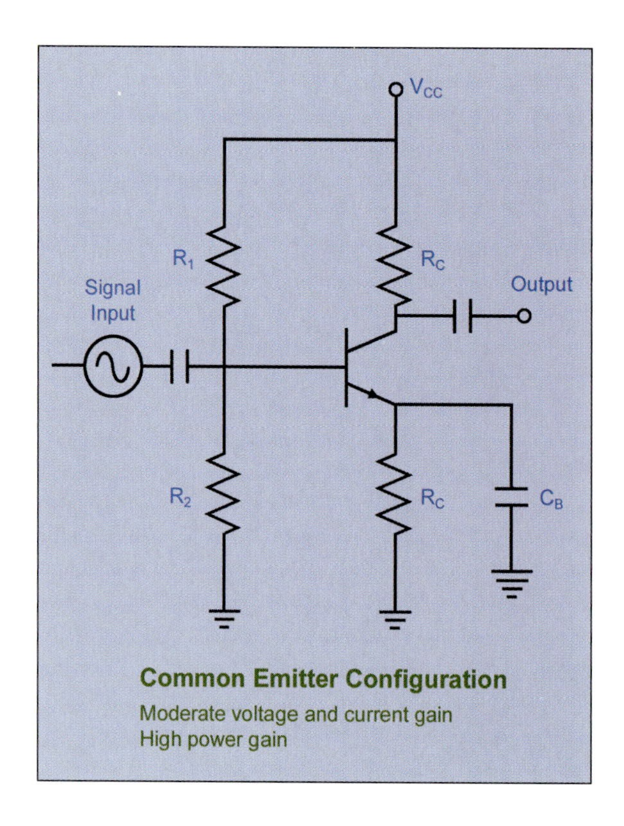

FIGURE 6–10 Common-emitter amplifier

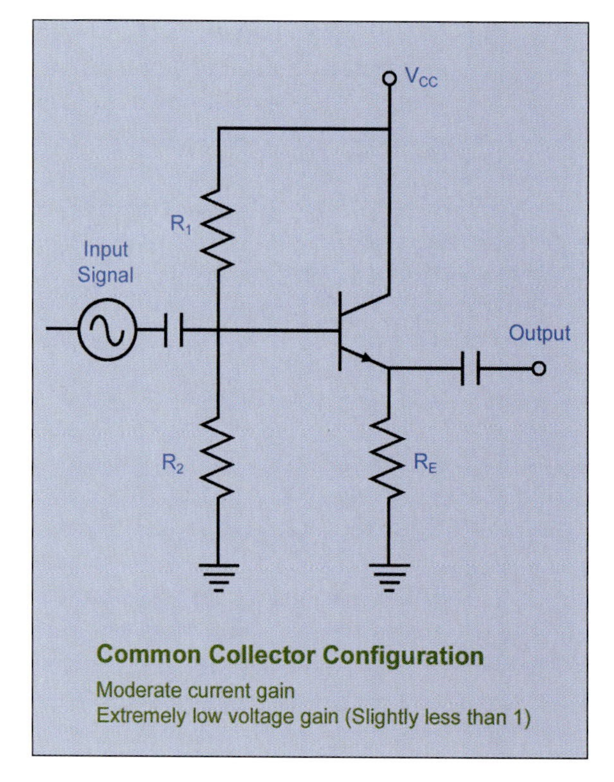

FIGURE 6–11 Common-collector amplifier

6.9 Amplifier Configurations

Figures 6–10, 6–11, and 6–12 show the three types of BJT amplifiers that were discussed in an earlier chapter. They are the common emitter, common collector, and common base. Each of these configurations can function as an amplifier.

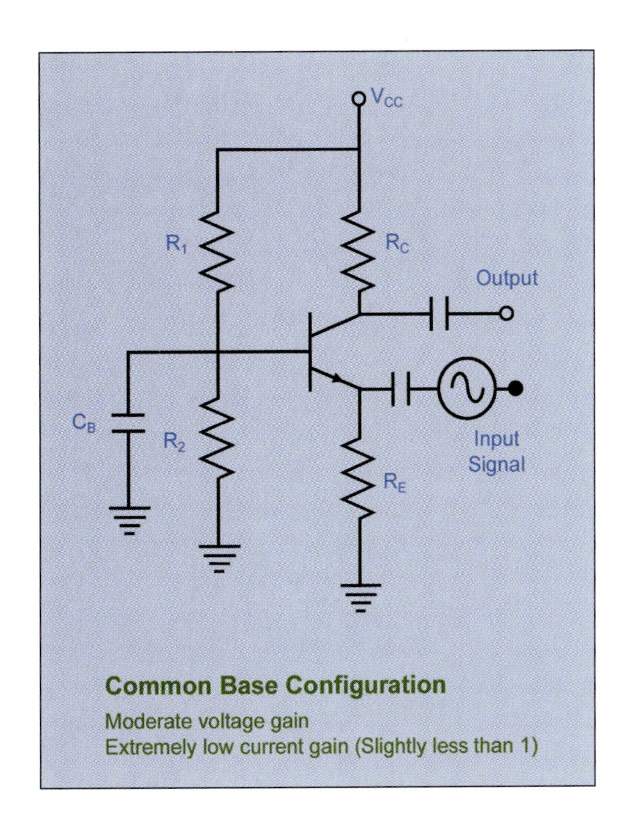

FIGURE 6–12 Common-base amplifier

The characteristics of a common-base amplifier's gain are very different from those of a common emitter or common collector. Table 6–2 shows the relationship of the gain characteristics among the three configuration types.

6.10 Load Line and the Operating (Q) Point

Load Line Development

Transistor amplifiers are current-controlled devices. The amount of current in the base of a transistor controls the amount of current in the collector. The secret to understanding amplifiers is to remember the fact that current controls the gain. If the current is controlled in the amplifier, then simply decreasing or increasing the current that flows through the collector can control the output voltage. If only 5 percent of the total current flows through the base, then this small portion of the current will control the larger collector current flow. The capability to control the larger collector current flow with the smaller base

Table 6–2 Amplifier Gain Characteristics

Configuration Types	Power Gain	Voltage Gain	Current Gain
Common Emitter	High	Moderate	Moderate
Common Collector	Moderate	Very low, less than 1	Moderate
Common Base	Moderate	Moderate	Very low, less than 1

Gains: Low = less than 100 Moderate = 100 to 1,000
High = greater than 1,000

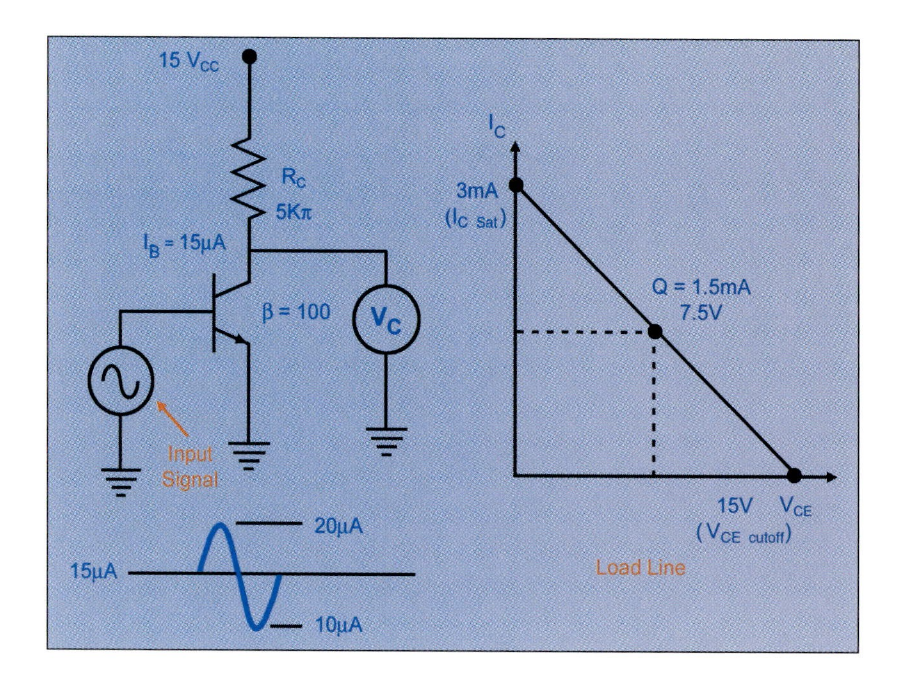

FIGURE 6–13 Load line for amplifier

flow is the basic concept of a transistor amplifier. This is why the load line characteristics of a transistor amplifier are so important. The load line is developed from two assumptions; one is the transistor in saturation (use $I_{C\ Sat}$) and the other is the transistor in cutoff (use $V_{CE\ cutoff}$).

$$I_{C\ Sat} = \frac{V_{CC}}{R_C + R_E}$$

$$V_{CE\ cutoff} = V_{CC}$$

Use the values of Figure 6–13 to calculate $I_{C\ Sat}$: 15 V / (5,000 Ω + 0 Ω) = 3 mA and $V_{CE\ cutoff}$: 15 V.

$$I_{C\ Sat} = \frac{15\ \text{V}}{5{,}000\ \Omega + 0\ \Omega} = 3\ \text{mA}$$

$$V_{CE\ cutoff} = 15\ \text{V}$$

On the right-hand side of Figure 6–13, these two values have been plotted and connected with a straight line. The straight line is the load line and can be used to determine the various operating values for the amplifier.

Q-Point (Quiescent Point)

The Q-point is the point on the load line that indicates the values of V_{CE} and I_C when there is no active (AC) input signal. Figure 6–14 shows an example of how the input AC signal causes the amplifier output of Figure 6–13 to vary above and below the load line Q-point. Note the Q-point of the amplifier in Figure 6–13 is V_{CE} = 7.5 and I_C = 1.5 mA. This shows a class A amplifier because the whole signal is amplified and that the biasing is at midpoint on the load line.

If the input signal becomes too large, the amplifier is overdriven and the collector current will reach saturation ($I_{C\ Sat}$). This is what

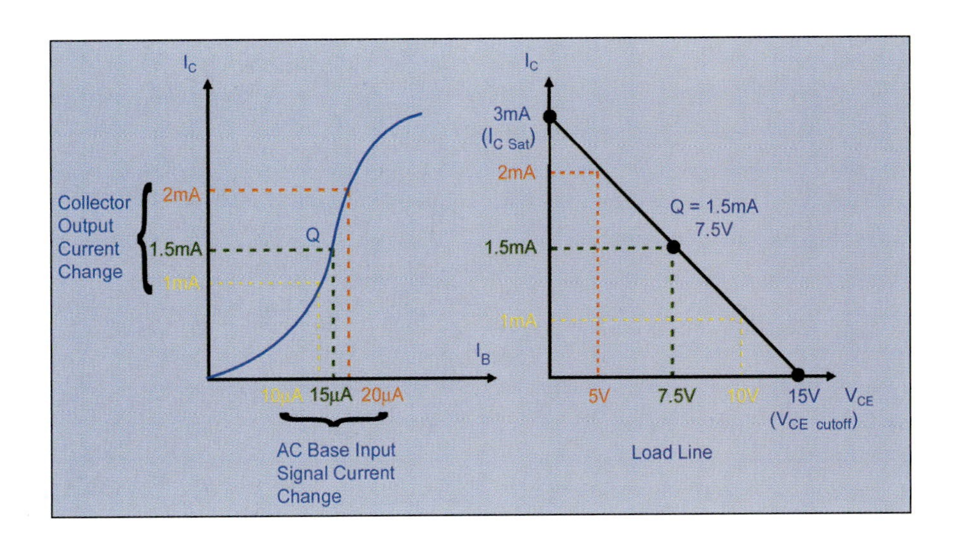

FIGURE 6–14 AC signal effect on operating (Q) point

causes the clipping in audio signals, and the result is poor quality sound.

An example of this can be seen in Figure 6–13. If the input signal rises to 45 mA, the resulting I_C would be

$$I_C = I_B \times \beta = 45\ \mu\text{A} \times 100 = 4.5\ \text{mA}$$

Thus, 4.5 mA would produce an output signal across R_C of (4.5 mA × 5,000) = 22.5 V. This is greater than V_{CC}; therefore, the amplifier would be driven into saturation, and the last 7.5 V of signal would be clipped or distorted.

EXAMPLE 1

Use the common-emitter circuit in Figure 6–15 for the following example. Assume a gain (β) of 50 and solve for the following:

- $I_{C\ Sat}$
- $V_{CE\ cutoff}$
- I_B

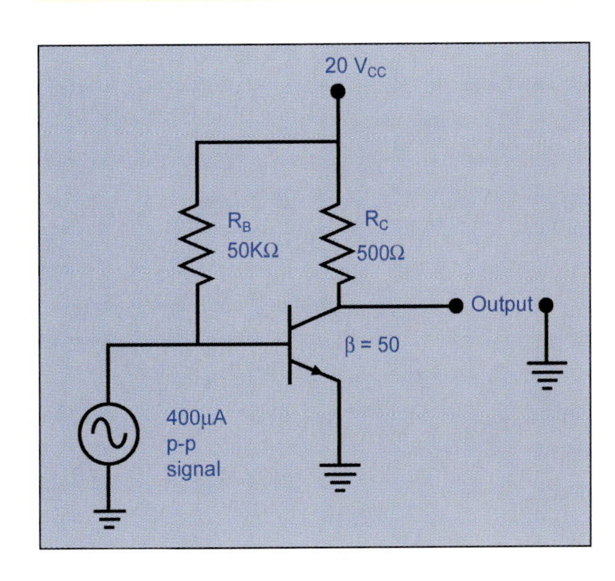

FIGURE 6–15 Common-emitter example circuit

- I_{CQ} (operating current with static AC signal)
- V_{CQ} (operating voltage with static AC signal)
- Minimum and maximum base current due to the dynamic swing of the AC signal input
- Minimum and maximum collector current due to the dynamic swing of the AC signal input
- Minimum and maximum voltage output due to the dynamic swing of the AC signal input

Step 1: Determine $I_{C\ Sat}$.

$$I_{C\ Sat} = \frac{V_{CC}}{R_C + R_E} = \frac{20\ V}{500\ \Omega + 0\ \Omega} = 40\ mA$$

Step 2: Determine $V_{CE\ cutoff}$.

$$V_{CE\ cutoff} = V_{CC} = 20\ V$$

Step 3: Determine the quiescent base current (I_B). (Quiescent means with no AC signal.)

$$I_B = \frac{V_{CC} - V_{BE}}{R_B} = \frac{20\ V - 0.7\ V}{50{,}000\ \Omega} = 386\ \mu A$$

Step 4: Determine the quiescent collector current (I_{CQ}).

$$I_{CQ} = \beta \times I_B = 50 \times 386\ \mu A = 19.3\ mA$$

Step 5: Determine the quiescent output voltage.

$$V_{CQ} = V_{CC} - V_C = 20\ V - (500\ \Omega \times 19.3\ mA) = 10.35\ V$$

Step 6: Determine the minimum and maximum base current due to the dynamic swing of the AC signal input.

$$I_{B\ Min} = I_{BQ} - \Delta I_B = 386\ \mu A - 200\ \mu A = 186\ \mu A$$
$$I_{B\ Max} = I_{BQ} + \Delta I_B = 386\ \mu A + 200\ \mu A = 586\ \mu A$$

Step 7: Determine the minimum and maximum collector current due to the dynamic swing of the AC signal input.

$$I_{C\ Min} = \beta \times I_{B\ Min} = 50 \times 186\ \mu A = 9.3\ mA$$
$$I_{C\ Max} = \beta \times I_{B\ Max} = 50 \times 586\ \mu A = 29.3\ mA$$

Step 8: Determine the minimum and maximum voltage output due to the dynamic swing of the AC signal input.

$$V_{CE} = V_{CC} - V_C$$
$$V_{CE\ Min} = 20 - (29.3\ mA \times 500\ \Omega) = 5.35\ V$$
$$V_{CE\ Max} = 20 - (9.3\ mA \times 500\ \Omega) = 15.35\ V$$

SUMMARY

In this chapter, you learned about the different classifications of amplifiers: class A, AB, B, and C. The class A amplifier has the most fidelity in reproducing the input signal, but it is the least efficient. The class AB operates over a range of 51 percent to 99 percent of the input signal. The class B operates over exactly 50 percent of the input signal, and the class C operates for less than 50 percent of the input signal. The class C is the most efficient of the classes.

Amplifier configurations were also reviewed. There are three basic types: common emitter, common collector, and common base. As presented earlier, the input and output to the amplifier determine which part of the transistor is considered "in common." For example, if the signal input is at the base and the output is taken from the collector, then the emitter is common to both.

Amplifiers produce gain. The gain can be in current, voltage, or power. The different configurations are designed for one or two of these gain types.

Decibels are used as a measurement of audio signal strength and gain. Multistage amplifiers' gain (in decibels) can be simply added to obtain the total gain for the circuit configuration. Decibels are not linear but are logarithmic.

A load line was developed for an example amplifier. The two key points that determine the load line are $I_{C\,Sat}$ and $V_{CE\,cutoff}$. The Q-point (or operating point) on the load line is approximately centered for a class A amplifier. The operating limits of the circuit are at $I_{C\,Sat}$ and $V_{CE\,cutoff}$. Attempted operation beyond these points will cause the amplifier to have a clipped and distorted signal.

REVIEW QUESTIONS

1. Define and discuss gain. What is it? How is it calculated? Does a transformer have gain?
2. A certain amplifier has an input voltage of .25 V and an output voltage of 15 V. What is its gain?
3. What is the relationship between gain expressed as a ratio and decibels?
4. How do you calculate the gain of a two-stage amplifier if you know the gain of each stage expressed in dB?
5. A certain amplifier has a gain of −3 dB. What does this mean about the amplifier?
6. Why does a decibel value always have to have a reference?
7. Discuss the advantages and disadvantages of the various classes of amplifier.
8. Which of the classes of amplifier has the best fidelity? Which has the worst?
9. What happens to the output of an amplifier that is driven so hard that its collector voltage reaches the cutoff value?
10. Name some applications for each of the four classes of amplifier.

PRACTICE PROBLEMS

1. A certain amplifier has a power gain of 15. What is its gain in decibels?
2. A certain amplifier has a power gain of 0 dB. What is its gain expressed as a ratio?
3. An amplifier has a voltage (V_{IN}) of 5 V and an input resistance (R_{IN}) of 1,000 Ω. The output voltage (V_{OUT}) is 10 V, and the output resistance (R_{OUT}) is 100 Ω. What is the amplifier's power gain in decibels?
4. Redo the common-emitter problem with the following changes:
   - **a.** $R_B = 100\text{ k}\Omega$
   - **b.** $V_{CC} = 25\text{ V}$
   - **c.** $R_C = 750\text{ k}\Omega$
   - **d.** $\beta = 110$
   - **e.** Input signal $= 250\ \mu A$ p-p
5. Looking at your results from question 4, answer the following:
   - **a.** Does this amplifier have a high fidelity?
   - **b.** What class amplifier is this?

More on Amplifiers

OUTLINE

OVERVIEW

This chapter builds on the last chapter's subject of amplifiers. The common-emitter circuit configuration is used to show the analysis of an amplifier circuit and how a DC-biased circuit is used to further demonstrate the usefulness of coupling transistors together to achieve a specific response. This lesson expects you to "add to" previous knowledge to further your abilities to interpret and predict the operation of transistors.

Coupling amplifiers together is used in practically every modern solid-state device. Variable-speed drives, measuring instrumentation, photosensitive lighting, process control, security systems, and voice, video, and data systems are just a few of the components that you may be required to troubleshoot over the years of your electrical career. Understanding the operation of amplifiers used in these devices will provide you with the tools that you will need.

OBJECTIVES

After completing this chapter, the student should be able to:

1. Understand the voltage gain of cascaded amplifiers.
2. Draw the load line for a common-emitter circuit.
3. Describe the unique characteristics for each method of amplifier coupling.

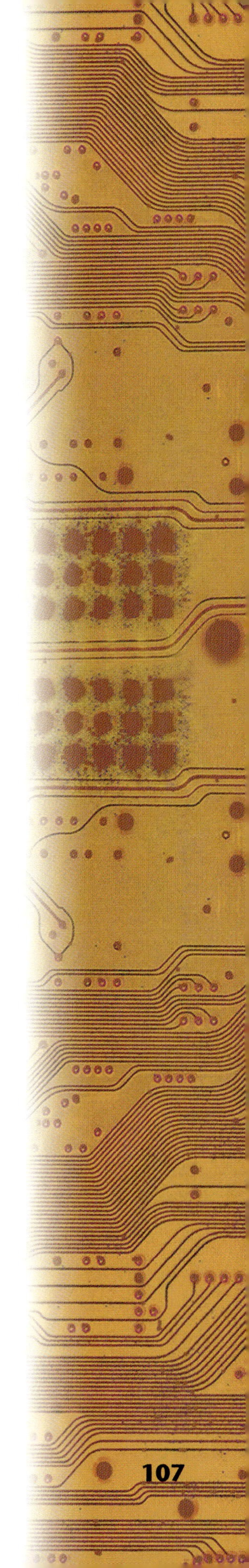

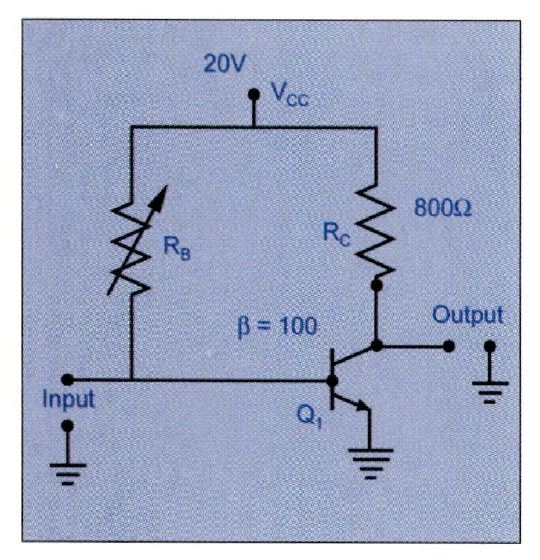

FIGURE 7–1 Common-emitter analysis

COMMON-EMITTER ANALYSIS

Solve for the missing values for transistor Q_1 in Table 7–1. Use Figure 7–1.

7.1 Row 1 (Cutoff)

When Q_1 is in cutoff, $V_{CE} = V_{CC}$ and $I_C = 0$ mA.

This means that there is 0 V across R_C, that there is 0 μA for I_B, and that R_B is seen as ∞.

7.2 Row 2 (Normal Operation)

Given the value of I_C as 15 mA, the rest of the table values can be solved using the following steps:

Step 1:

$$I_B = \frac{I_C}{\beta} = \frac{15 \text{ mA}}{100} = 150\ \mu\text{A}$$

Step 2:

$$R_B = \frac{V_{CC} - V_{BE}}{I_B} = \frac{(20 - 0.7)}{150\ \mu\text{A}} = 128.667\ \text{k}\Omega$$

Step 3:

$$V_{RC} = R_C \times I_C = 800 \times 15 \text{ mA} = 12 \text{ V}$$

Step 4:

$$V_{CE} = V_{CC} - V_{RC} = 20 \text{ V} - 12 \text{ V} = 8 \text{ V}$$

7.3 Row 3 (Saturation)

When Q_1 is in saturation, $V_{CE} = 0$ and $I_C = 25$ mA. Recall this can be calculated by using the formula

$$I_{C\,Sat} = \frac{V_{CC}}{(R_C + R_B)} = \frac{20}{800} = 25 \text{ mA}$$

This means that

$$I_B = \frac{25 \text{ mA}}{\beta} = \frac{25 \text{ mA}}{100} = 250\ \mu\text{A}$$

Table 7–1 Common-Emitter Analysis Data

	R_C (kΩ)	I_C (mA)	V_{RC}	V_{CE}	I_B (μA)	R_B (kΩ)
Q_1 in Cutoff	.8	0	0	20	0	∞
Q_1 in Normal Operation	.8	15				
Q_1 in Saturation	.8	25	20	0	250	77.2

With an I_B of 250 mA, then

$$R_B = \frac{(V_{CC} - V_{BE})}{I_B} = \frac{20\text{ V} - 0.7\text{ V}}{250\ \mu\text{A}} = 77.2\text{ k}\Omega$$

Now that you have reviewed the basic analysis for a common-emitter amplifier, the next step is to look at amplifier coupling.

AMPLIFIER COUPLING

7.4 Multiple Stages

Many times, it is either impractical or impossible to amplify a small signal to a usable level with a simple one-transistor amplifier. This could be compared to trying to enlarge a slide of a single-celled organism so that an entire audience could see it straight from the microscope, Figure 7–2. To project the image so that the class could see it would require a lot of light. If we tried to illuminate the specimen with enough light that we could project the image straight through the microscope to the wall, we would incinerate the little bug. The only reasonable answer to this problem is to amplify in stages.

Amplifying in stages presents a problem. We have already talked about measuring the gain of multiple stages of amplification and discovered that the easiest way to deal with it is to use the decibel system of adding and subtracting gain factors. We have discussed how to produce voltage, current, and power gain with single-transistor amplifiers. Each of these amplifiers may be used as one stage of a multistage amplifier system. Now we tackle the problem of coupling the amplifier stages together.

Coupling refers to the method of transferring a signal from one stage to the next. There are three methods for coupling amplifiers together:

- Direct coupling—connecting the output of one stage directly to the input of the next stage. The only allowed element between the two stages is a resistor.

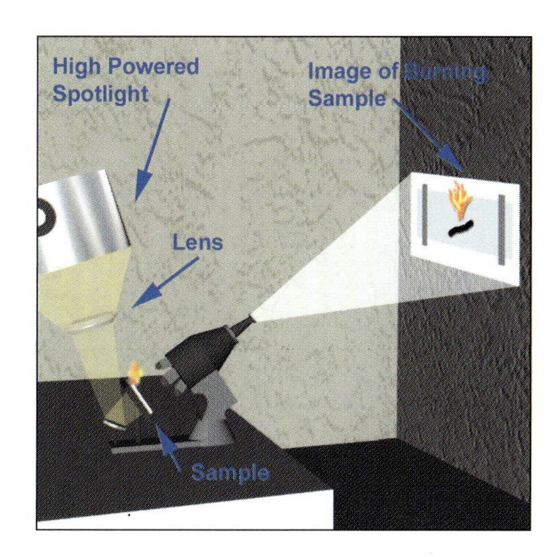

FIGURE 7–2 From microscopic image to wall projection in one step

- Capacitive coupling—using a capacitor to couple two stages together.
- Transformer coupling—using a transformer to couple two stages together.

7.5 Impedance Matching

Regardless of the coupling type used, a characteristic of the amplifier that must be taken into consideration is its impedance. The input impedance of the amplifier will directly affect how it loads the source. For example, an input signal from cable or an antenna could have an input impedance of 75 Ω, or the input could be from a microphone with over 150 kΩ. Either way, to get the best power transfer (from input to amplifier), the impedances must be matched (or as closely matched as possible). The three block circuits in Figures 7–3a, 7–3b, and 7–3c illustrate this principle.

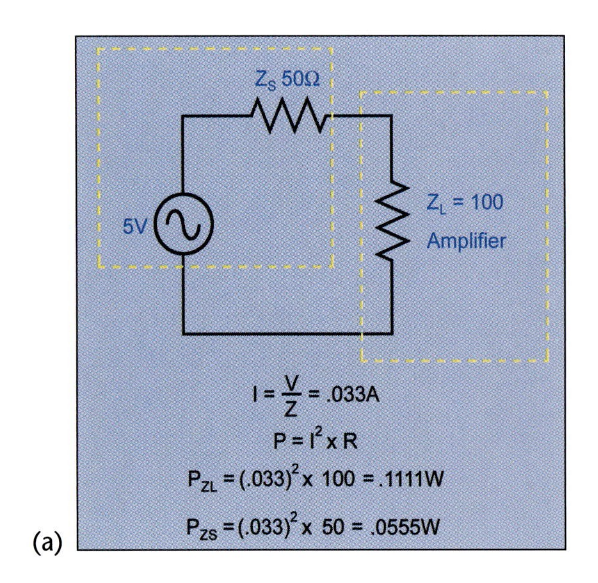

(a)

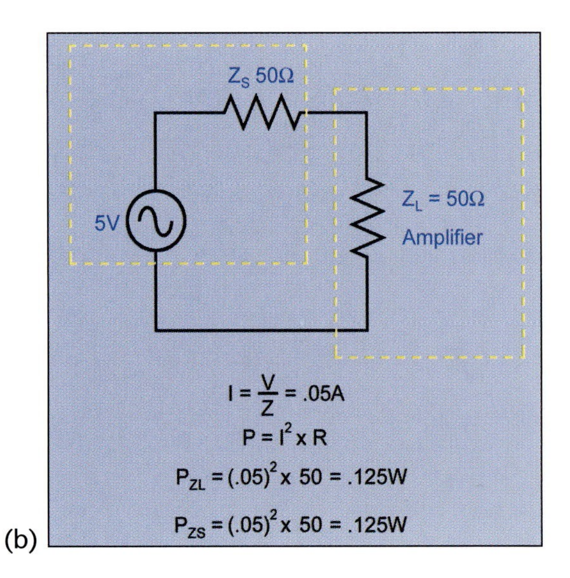

(b)

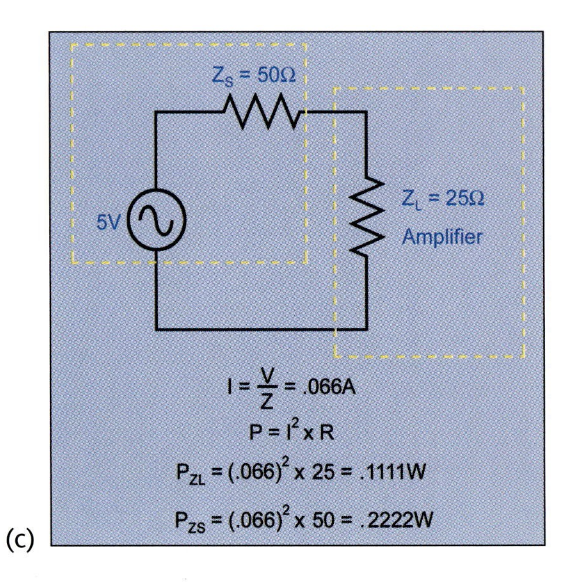

(c)

FIGURE 7–3 Impedance matching; a) $Z_L > Z_S$, b) $Z_L = Z_S$, c) $Z_L < Z_S$

Table 7–2 summarizes the results of the three different loads.

Notice that the power delivered to the load is greatest when the impedance of the signal and the load are matched; consequently, for maximum power transfer, the impedance of the load must be equal to the impedance of the amplifier. In other words, the output impedance of one stage of an amplifier must be equal to the input impedance of the next stage.

7.6 Direct Coupling

The simplest method of coupling is called direct coupling. As the name implies, it is passing current straight from one stage to another over a wire or some other conductor. Both AC and DC signals are passed with direct coupling. Because neither capacitors nor transformers are capable of passing a DC signal, they cannot be used in direct coupling. Direct coupling finds wide use in medical equipment, voltmeters, and other equipment that must amplify DC or very low frequency inputs. Figure 7–4 shows a pair of common-emitter amplifier stages directly connected.

The output voltage of the first stage becomes the input voltage of the second stage. The arrangement itself is simple enough, but there are limitations. In most applications, the base voltage of Q_2 would become excessive and cause Q_2 to burn out. A resistor tied to the emitter of Q_2 can reduce the danger of Q_2 burnout, but there is a gain sacrifice

Table 7–2 Summary of Power Transfers—Figure 7–3

Figure	R_S (Ω)	R_L (Ω)	P_S (Watts)	P_L (Watts)
7–3a	50	100	0.055	0.111
7–3b	50	50	0.125	0.125
7–3c	50	25	0.222	0.111

FIGURE 7–4 Directly coupled common-emitter amplifier stages

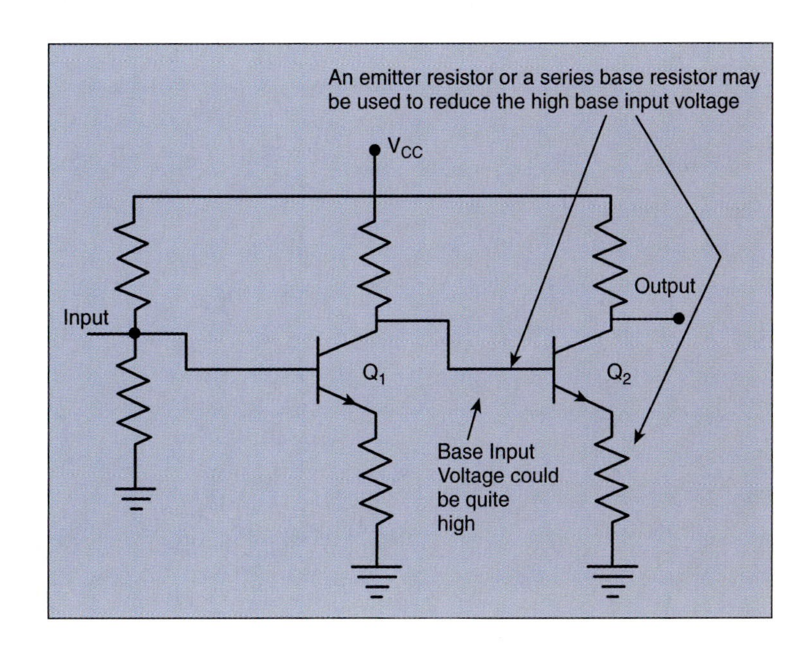

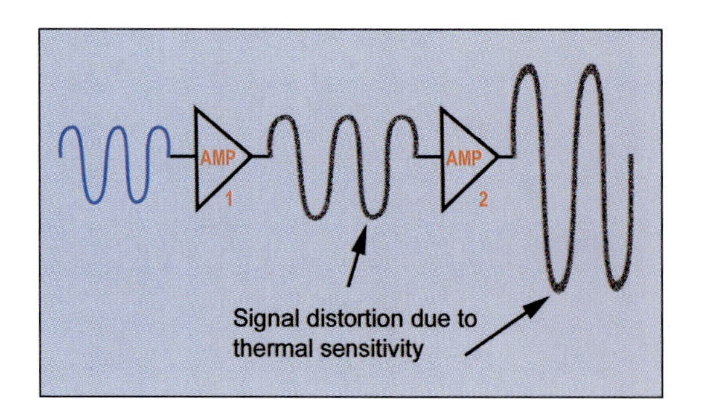

FIGURE 7–5 Effect of temperature sensitivity

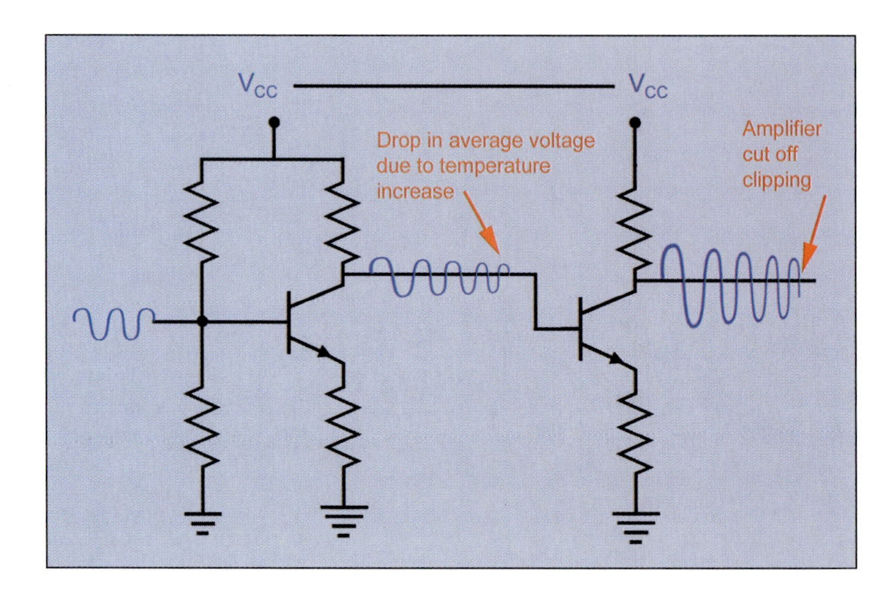

FIGURE 7–6 Clipping due to thermal drift

through Q_2 as a result. A resistor can also be placed in series with the base of Q_2, but this also causes a loss of overall gain.

Another problem with direct-coupled amplifiers is temperature sensitivity. A slight rise in temperature in one of the early stages of amplification can result in a big error by the time the little bit of noise gets amplified several times. Figure 7–5 shows the effects of amplifying temperature-induced noise.

Temperature also induces thermal drift. This is the rise in internal resistance of transistors, such as BJTs, with the rise in temperature. The drift shows up as both voltage and current changes in various parts of the amplifier. Depending on the configuration, a little drift early on can result in clipping or saturation in later stages of amplification, as shown in Figure 7–6.

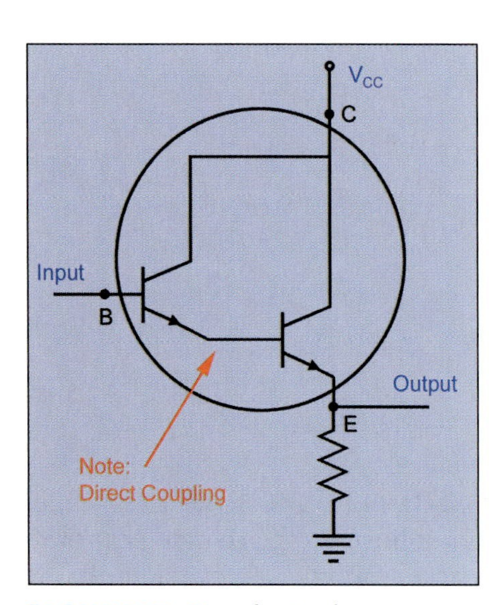

FIGURE 7–7 The Darlington pair

One of the most common direct-coupled devices is the Darlington pair, shown in Figure 7–7. The advantages of this direct-coupled amplifier are simple circuitry, high input impedance, and an overall current gain that is slightly more than the mathematical product ($T_1 \times T_2$) of the current gain of the two transistors ($I_{total} > I_{Q1} \times I_{Q2}$). The Darlington pair is often sold packaged as a single transistor and used for small signal inputs for low noise applications.

Direct coupling is great for DC or very low-frequency signals that do not respond well to transformers and capacitors, but its strength with these signals is its weakness with higher frequency signals. It does

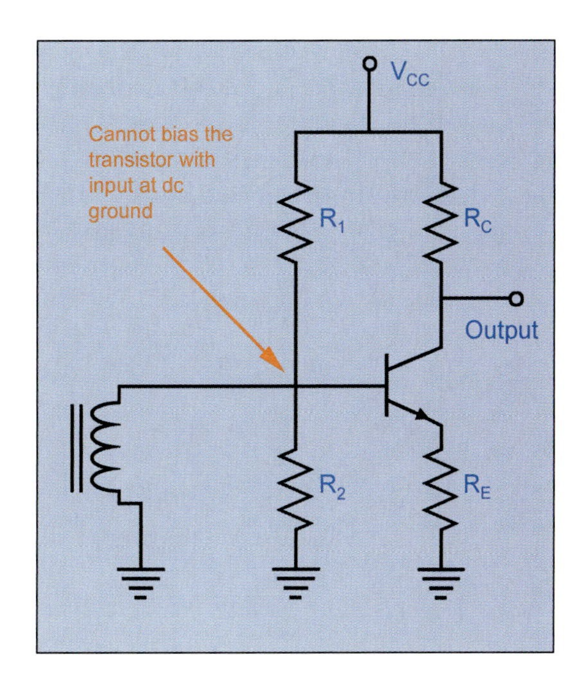

FIGURE 7–8 When direct coupling does not work

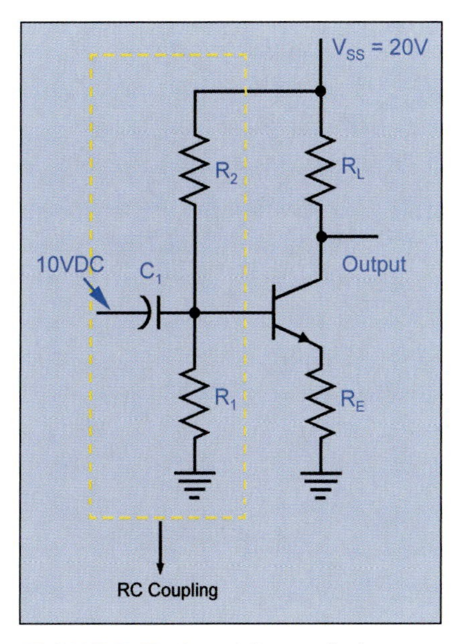

FIGURE 7–9 RC-coupled amplifier

not isolate them and may not pass them. If you were to use direct coupling of a speaker or microphone, the signal would be shunted to ground; consequently, some other type of coupling should be used for such applications (see Figure 7–8).

7.7 Capacitive Coupling

Overview

The most common form of coupling for analog signals is capacitive coupling, as shown in Figure 7–9. It is also known as RC (resistive-capacitive) coupling. RC coupling offers the advantages of being cheap and easy with good frequency response over a wide range of frequencies. Capacitive coupling also ensures fixed bias by blocking DC. Its main drawback is that it does nothing to match impedances; this sometimes requires the designer to include an extra stage of amplification to improve the poor power gain due to impedance mismatching.

Coupling capacitors are used for two main reasons:

1. To pass an AC signal from one amplifier to another and to provide DC isolation between the two stages
2. To short-circuit the AC signal to ground and not affect the DC parameters

An RC coupling network is shown in Figure 7–9. Notice that the dotted box includes the biasing resistors as part of the coupling. This is because the biasing resistors reduce the amount of voltage supplied to the base of the second transistor and are, therefore, part of the coupling problem.

Also note that the capacitor is electrolytic. This is common in RC coupling, and care must be taken not to install replacement capacitors with the wrong polarity. Installing an electrolytic capacitor backward may ruin the capacitor and make a loud pop followed by a burning smell.

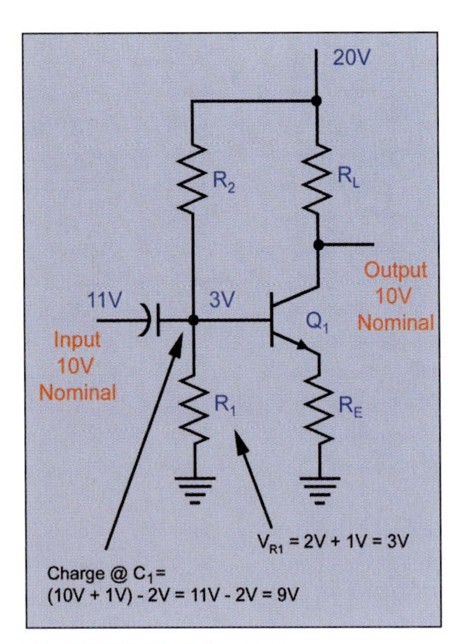

FIGURE 7–10 RC coupling during positive half-cycle

Refer to Figure 7–9 for the following discussion. The capacitive reactance of C_1 must be low in the frequency range of the desired signal, and the resistance of R_1 must be at least 10 times the capacitive reactance of C_1 to prevent signal loss across the coupling. Assume Q_1 is in the static state (Q-point quiescent—no signal applied) and that the biasing of Q_1 is set at 2 V by the voltage divider and R_E. C_1 will develop a charge equal to the static voltage of the input signal minus the bias (2 V) of Q_1. Assume that the previous stage of amplification had an output level of 10 V; the resulting level of charge on C_1 is 10 V − 2 V = 8 V.

As the input signal swings higher, C_1 charges through the base of Q_1 and R_1. (11 V − 2 V = 9 V). Notice that the base current and the voltage across R_1 both increase during the charging of C_1 (see Figure 7–10).

When the input signal swings lower, C_1 discharges through R_2. As the current through R_2 increases, the voltage drop across it increases, lowering the current through Q_1 and decreasing the voltage drop across R_1. See Figure 7–11.

AC and DC Equivalent Circuits

The circuit effect of capacitors in amplifier stages can be seen in an AC equivalent circuit. Figure 7–12a shows the full amplifier circuit.

To construct the AC equivalent circuit of this amplifier, do the following:

- Short-circuit all capacitors by adding a jumper wire across them. This step assumes that the impedance of the capacitor is low in the frequency range of the amplifier. If it is not, you must substitute an impedance of the capacitor that is suitable for the range of frequencies. This, of course, will make the analysis more complex.
- Replace all DC sources with a ground symbol. This is effectively the same thing as placing a short circuit across all DC sources.

See Figure 7–12b for the following discussion. The reasoning behind replacing all capacitors with a wire (straight line) is that the ca-

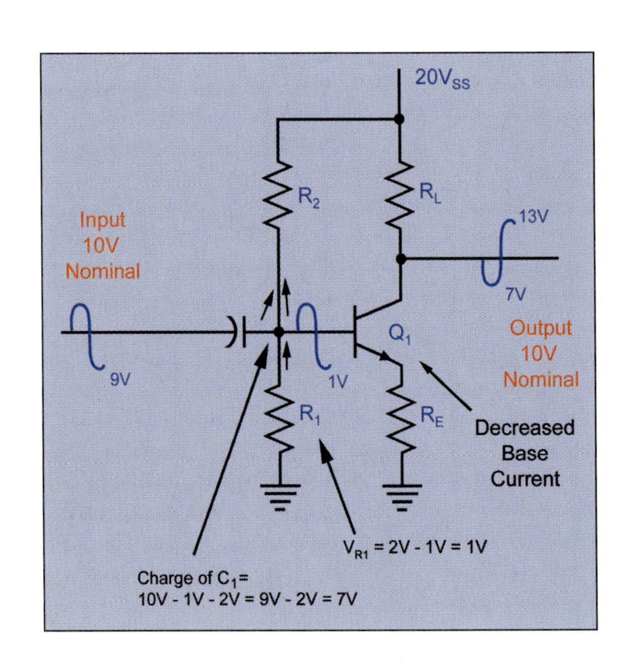

FIGURE 7–11 RC coupling during negative half-cycle

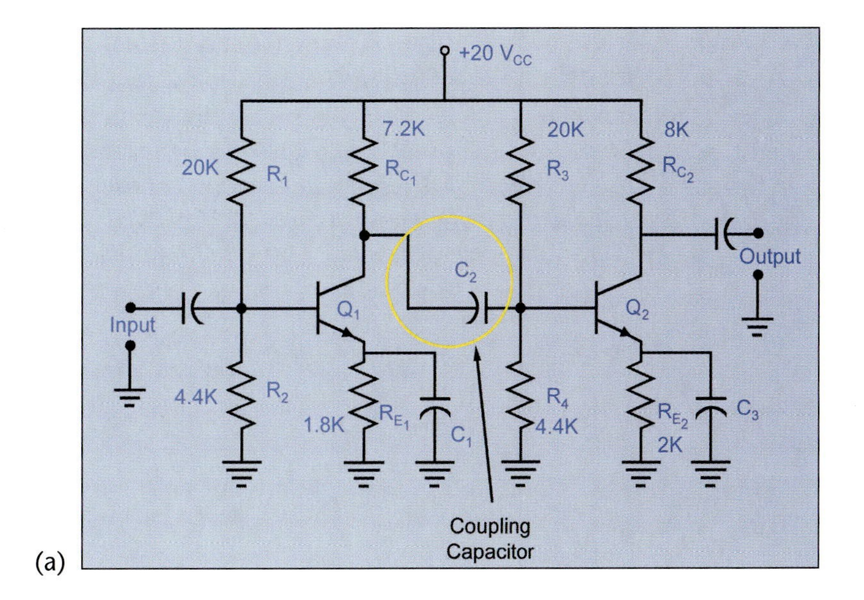

(a)

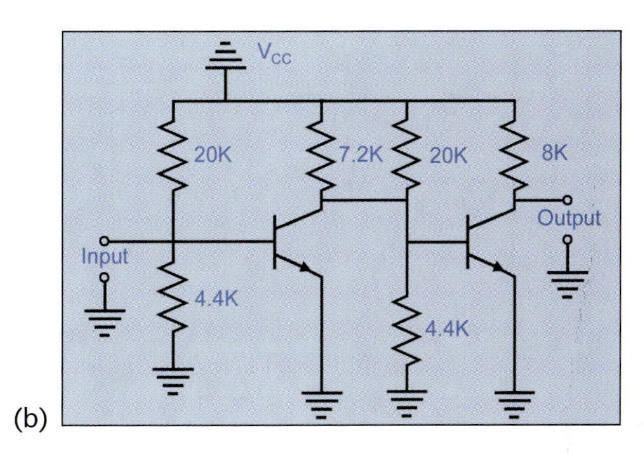

(b)

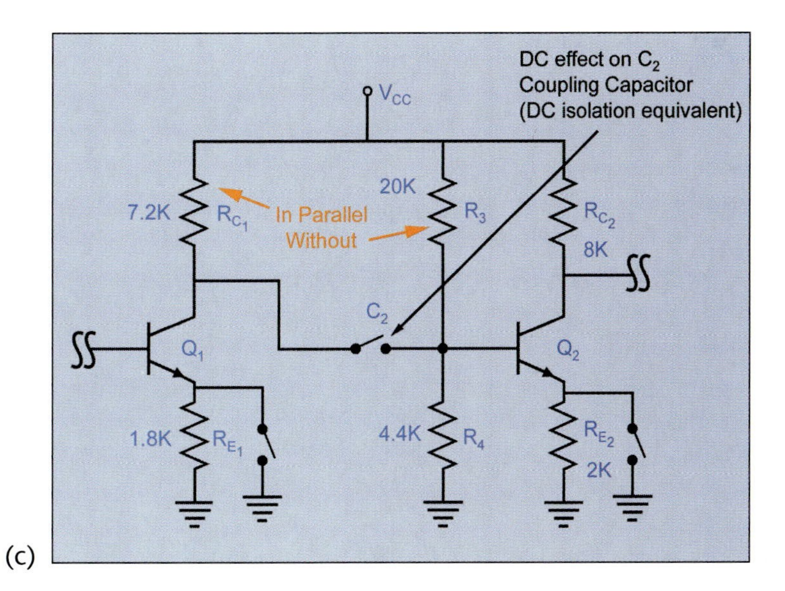

(c)

FIGURE 7–12 RC-coupled amplifier; a) full circuit, b) AC equivalent, c) DC equivalent

pacitor is built to offer little or no resistance to the AC signal frequency (recall that X_C is low in the frequency range of the signal). In practice, it acts as a short circuit. Replacing DC sources with a ground is based on similar logic. Recall from your study on batteries that the internal resistance of a DC source is extremely low. In theory, it is 0 Ω. This causes the wire (straight line) DC source replacement to be at ground potential.

Note that C_2 is considered an open for the DC calculations of the circuit (see Figure 7–12c). If C_2 did not exist, the impact of R_{C1} on the base of Q_2 would be considerable. R_{C1} would be in parallel with R_3 and the resulting R_{EQ} would impact the voltage divider value used for Q_2. This is shown in the equation for R_{EQ}.

V_B without the Coupling Capacitor C_2

$$R_{EQ} = \frac{(R_{C1} \times R_3)}{(R_{C1} + R_3)} = \frac{(7.2\ \text{k}\Omega \times 20\ \text{k}\Omega)}{(7.2\ \text{k}\Omega + 20\ \text{k}\Omega)} = 5.298\ \text{k}\Omega$$

Using R_{EQ} to solve for V_B

$$V_B = V_{CC}\left(\frac{R_4}{R_{EQ} + R_4}\right) = 20\ \text{V} \times \frac{4.4\ \text{k}\Omega}{5.294\ \text{k}\Omega + 4.4\ \text{k}\Omega} = 9.08\ \text{V}$$

V_B with the Coupling Capacitor C_2

$$V_B = V_{CC}\left(\frac{R_4}{R_3 + R_4}\right) = 20\ \text{V} \times \frac{4.4\ \text{k}\Omega}{20\ \text{k}\Omega + 4.4\ \text{k}\Omega} = 3.6\ \text{V}$$

7.8 Transformer Coupling

As the name implies, a transformer-coupled amplifier, Figure 7–13, uses a transformer between the collector of the transistor and the load. The load can be the final output, or it could be another stage of amplification. Transformer coupling is mainly used for its ability to match impedance and provide maximum power transfer.

Consider Figure 7–14. As you learned previously, the maximum power transfer will occur between the two amplifiers if the output impedance of A_1 is equal to the input impedance of A_2.

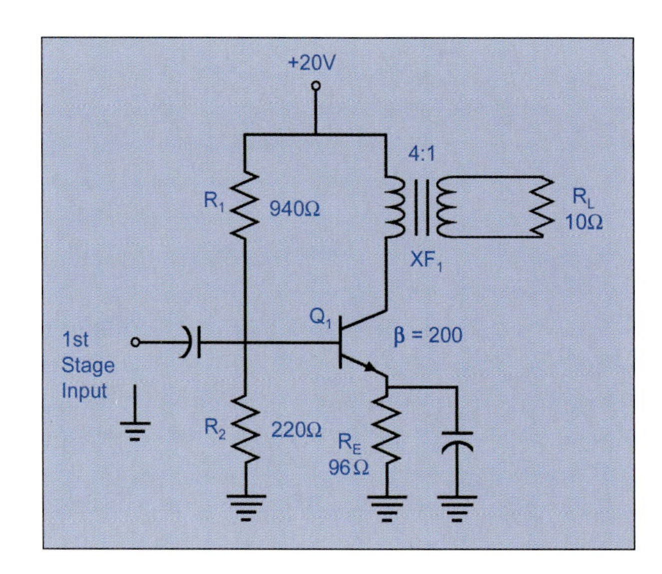

FIGURE 7–13 Transformer-coupled amplifier

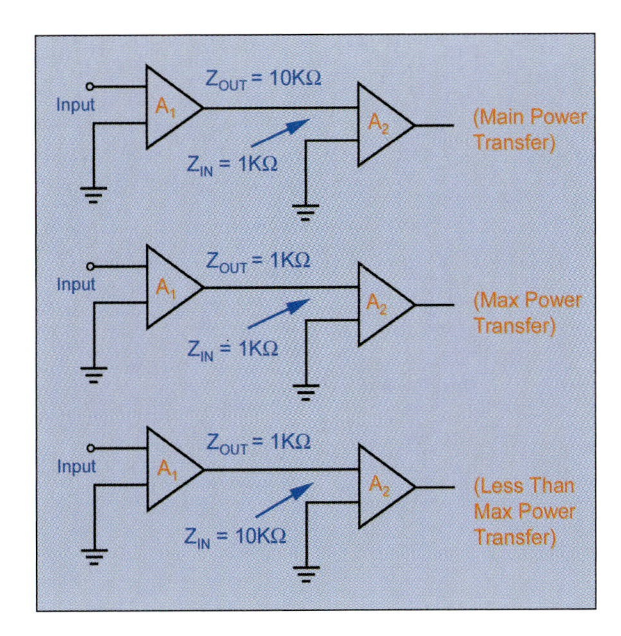

FIGURE 7–14 Importance of amplifier impedance matching

Transformer coupling provides DC isolation from the primary to secondary windings, and it can provide the necessary impedance matching. Recall from earlier lessons that the relationships in the transformer are:

$$\frac{N_P}{N_S} = \frac{V_P}{V_S} = \frac{I_S}{I_P}$$

and

$$\left(\frac{N_P}{N_S}\right)^2 = \frac{Z_P}{Z_S}$$

The DC biasing of a transformer-coupled amplifier and the analysis on the collector curves (Figure 7–15) are slightly different than a direct-coupled amplifier. The analysis must take into account that the transistor sees two different values for R_{CE}. The DC current does not see the load resistance. This is because the transformer does not pass DC current, so the load resistance is effectively disconnected from the transistor for DC values.

The AC current, on the other hand, does see the load resistance that is reflected back by the transformer. The analysis starts with the DC circuit.

DC Analysis

Refer to Figure 7–13.

To promote linear operation, the quiescent voltage is usually selected as one-half (1/2) V_{CC}. Assume a primary winding resistance of 4 Ω.

$$V_{CEQ} = V_{CC} - V_{XF} - V_{RE} = 20\text{ V} - (100\text{ mA}) \times (96\ \Omega + 4\ \Omega) = 10\text{ V}$$

Note, however, that the Q-point on the curves (Figure 7–15) falls at $V_{CE} = 10$ V and $I_C = 105$ mA. This is because there is a slight resistance

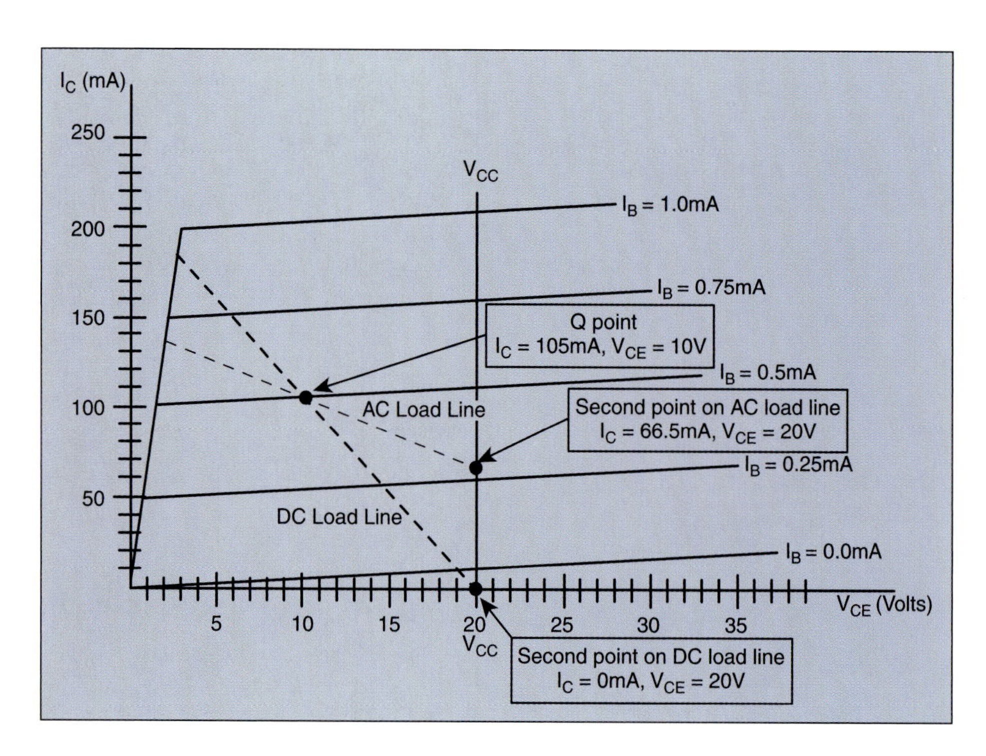

FIGURE 7–15 Load lines of transformer-coupled amplifier

between the collector and the emitter. So, it takes a little more current than we had previously supposed. If the collector curves were perfectly horizontal, this problem would not exist. However, that would also imply that the transistor is perfect, which it cannot be.

In any event, the cutoff voltage for the transistor is 20 V and 0 mA. This makes up another point on the load line. You can now draw the load line by connecting the Q-point and 20 V.

AC Analysis

The AC load line is a little more difficult to understand. Recall that any alternating current that is applied will pass through the reflected load impedance.

Step 1: Calculate the AC collector impedance (Z_{CE})

$$Z_{CE} = \left(\frac{N_P}{N_S}\right)^2 \times Z_S = 4^2 \times 10\ \Omega = 160\ \Omega$$

Step 2: Determine the Q-point for AC operation.

The Q-point is, by definition, the 'at rest' point for the amplifier. That is, it is the point at which the AC input is equal to zero. Clearly, this means that the Q-point is the same for both AC and DC operation.

Step 3: Calculate a second point on the AC load line.

Note that the total resistance seen by the AC signal is 160 Ω + 96 Ω + 4 Ω = 260 Ω. The 4 Ω is included because the transformer winding still is in the circuit. Because we are dealing with an *alternating* current value, we must work with the changes in voltage and current. Consider the following:

$$I_C\,(R_E + R_L + 4\ \Omega) + V_{CE} = V_{CC}$$

This is just Kirchoff's voltage law for the collector circuit. Note that we have to include the 4 Ω resistance of the transformer primary as well as the reflected load resistance (R_L). Now consider if we look at the change in current versus the change in voltage for an AC signal.

$$\Delta I_C (R_E + R_L + 4\ \Omega) + \Delta V_{CE} = V_{CC} - V_{CC} = 0$$

Note the '0' on the right-hand side. This is because V_{CC} is the same at both points. Therefore, any AC quantity will not be reflected across the voltage supply. This can be further rearranged as

$$(R_E + R_L + 4\ \Omega) = -\frac{\Delta V_{CE}}{\Delta I_C}$$

which can be further reduced to

$$260\ \Omega = -\frac{V_{CE1} - V_{CE2}}{I_{C1} - I_{C2}}$$

Where the values V_{CE1} and I_{C1} are one point on the load line, and V_{CE2} and I_{C2} are another point on the load line. Take point 1 to be the quiescent point, so that $V_{CE1} = 10$ V and $I_{C1} = 105$ mA. For the second point, we can take the cutoff value, so that $V_{CE2} = 20$ V and $I_{C2} = ??$

Substituting these values gives

$$260\ \Omega = -\frac{V_{CE1} - V_{CE2}}{I_{C1} - I_{C2}} = -\frac{10\text{ V} - 20\text{ V}}{105\text{ mA} - I_{C2}}$$

rearranging terms

$$260\ \Omega \times (105\text{ mA} - I_{C2}) = -(10\text{ V} - 20\text{ V})$$

which simplifies to

$$27.3\text{ V} - (260\ \Omega \times I_{C2}) = 10\text{ V} \Rightarrow I_{C2} = \frac{10\text{ V} - 27.3\text{ V}}{-260} = 66.5\text{ mA}$$

The AC load line can now be drawn by connecting the quiescent point with the second point just calculated.

Recalling geometry, you may notice that the slope $\left(\frac{\text{rise}}{\text{run}}\right)$ of the DC load line is $-\frac{1}{100}$ and the slope of the AC load line is $-\frac{1}{260}$.

7.9 Tuned Transformers

A tuned transformer-coupled amplifier uses an LC (inductive-capacitive) parallel "tank" circuit. By varying the L and C values in the tank, you can control the total output impedance to the load. Figure 7–16 shows an example tuned transformer circuit.

Figure 7–17 shows how the impedance (AC resistance) of a parallel LC circuit varies with frequency. Notice that for frequencies well below the resonant frequency $\left(f_R = \frac{1}{2\pi\sqrt{LC}}\right)$, the impedance is very low. As the frequency increases, the impedance rises until at f_R the impedance is very, very high. Theoretically, the impedance is infinite at

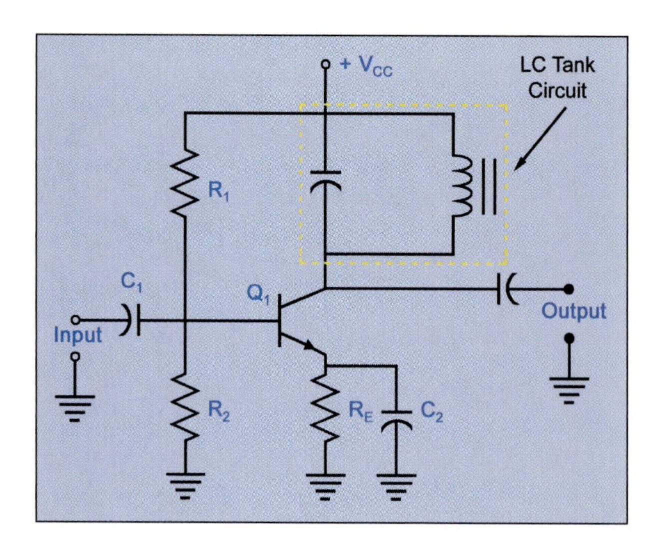

FIGURE 7–16 Tuned transformer-coupled amplifier

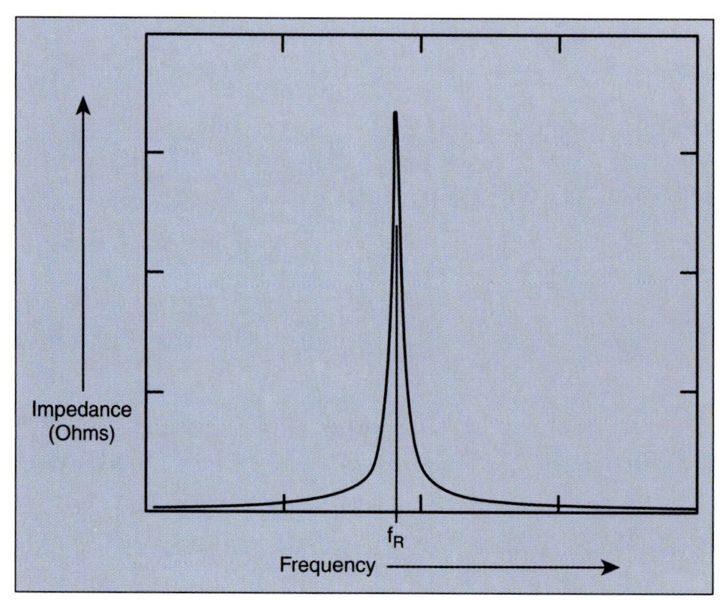

FIGURE 7–17 Impedance versus frequency characteristics for parallel LC circuit

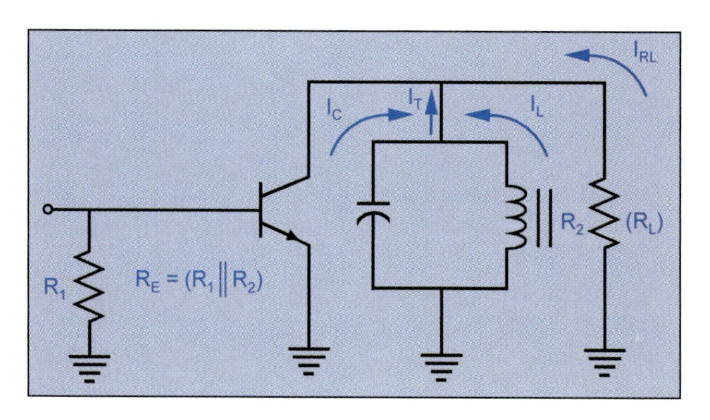

FIGURE 7–18 AC equivalent tuned amplifier

f_R. The impedance starts to decrease above f_R in a mirror image of its behavior below f_R.

This behavior can be explained for the amplifier of Figure 7–16 by referring to Figure 7–18. There are three AC frequency conditions that can be described for the amplifier. Table 7–3 explains the behavior for each.

Table 7–3 Tuned Circuit Amplifier Response

Frequency In	Amplifier Response
$F_{IN} = f_R$	The tank circuit acts as an open circuit (infinite impedance). R_L becomes the only path to ground in the collector circuit. This results in all the current flowing through R_L, which in turn causes the amplifier efficiency to be at maximum.
$F_{IN} < f_R$	I_L increases and I_C decreases as F_{IN} decreases. This causes the tank current (I_T) to increase. As I_T increases, the load voltage decreases, causing a loss in amplifier output efficiency.
$F_{IN} > f_R$	I_C increases and I_L decreases as F_{IN} increases. This causes an increase in the tank current (I_T). As the input frequency increases, the tank current increases. As I_T increases, the load voltage decreases, causing a loss in amplifier output efficiency.

NEGATIVE FEEDBACK

A technique widely used to reduce distortion in amplifiers is called negative feedback. It is also used in transistor circuits to

- Stabilize
- Increase bandwidth
- Improve linearity
- Improve noise performance
- Maintain the amplifier's frequency response

Negative feedback is defined as feeding a portion of the output signal back to the input. The signal is fed back in such a way that it is 180° out of phase with the input signal. This feedback reduces the gain of the amplifier, but it also reduces the amount of distortion in the output signal.

One technique of negative feedback has already been studied—the emitter-to-ground resistor. This resistor is used to stabilize the DC bias of an amplifier. When placed in parallel with a capacitor, the AC signal is bypassed to ground and the effect of the feedback is not felt on the signal, and so the AC gain can be significantly higher.

SUMMARY

In this chapter, we reviewed simple amplifier analysis using the common-emitter class A amplifier. We also learned about the three basic types of amplifier coupling: direct, capacitor, and transformer. Important amplifier characteristics reviewed included: impedance matching, AC and DC equivalent circuits, and transformer tuning. Table 7–4 summarizes the three methods of coupling.

Table 7–4 Summary of Coupling Methods

	Direct Coupling	Capacitor Coupling	Transformer Coupling
Response to Direct Current	Yes	No	No
Provides Impedance Match	No	No	Yes
Pros	Simple design when only a few stages are needed; Darlington pair very popular	Ensures fixed bias by blocking DC; easy to use, inexpensive	High efficiency; ability to tune of selective amplifier design (tank circuit); provide maximum power transfer
Cons	Difficult to design for many stages; temperature sensitive	Low frequency applications may require high values of capacitance	Cost; transformer weight and size
Common Usage	Medical equipment, voltmeters, and DC amplification	Common for coupling of analog signals	Applications where DC needs isolation; where total output impedance to the load needs to be controlled

REVIEW QUESTIONS

1. What is amplifier coupling?
2. Discuss the importance of impedance matching
3. Which of the coupling methods will couple both AC and DC signals between stages?
4. What are the disadvantages of direct-coupled amplifiers?
5. In the ideal RC-coupled amplifier, the capacitor behaves as a short circuit at signal frequencies and an open circuit at nonsignal frequencies. Discuss how the real RC-coupled amplifier behaves.
6. The AC and DC quiescent points (Q-points) are the same for a transformer-coupled transformer. Why?
7. A certain intermediate-frequency amplifier in a radio couples to the next stage using a tuned transformer circuit. Why?
8. The collector curves for a real BJT are not horizontal. Why?
9. Discuss one example of the importance of fidelity in an amplifier. Why is it important in your example?
10. What is negative feedback? What are its uses?

PRACTICE PROBLEMS

1. In Figure 7–18, assume the capacitor is 3 μF and the inductor is 2 mH. What is the resonant frequency (f_R) of the combination?
2. In Figure 7–15, what will the AC instantaneous collector current be for this amplifier with I_B = 0.625 mA?
3. In Figure 7–16, what is the DC quiescent point (I_C and V_{CE}), assuming the following information:
   a. The transistor collector curves are given by Figure 7–15.
   b. $R_E = 100\ \Omega$

c. The collector inductor has a DC resistance (R_I) of 4 Ω.

d. $V_{CC} = 30$ V

e. $V_{CEQ} = 15$ V

f. $\beta = 75$

4. In Figure 7–16, what is the AC quiescent point, assuming all of the same information as question 3 plus the following:

   a. The tank circuit inductance (L_T) = 6 μH.

   b. The tank circuit capacitance (C_T) = 0.4 μF.

5. In Figure 7–16, assume the tank circuit inductance is the primary of a transformer with a 10-to-1 turns ratio, and assume that the amplifier is operating at the resonant frequency of the tank circuit. The load resistance on the transformer secondary is 5 Ω. Draw the AC and DC load lines. Use the same circuit values as those given in problems 3 and 4.

INTRODUCTION AND REVIEW

8.1 The Ideal Amplifier

Amplifiers work by converting some of the DC power supply energy to signal energy. In other words, the amplifier takes power from the DC power supply and "gives" it to the signal that it is amplifying. The ideal power amplifier will deliver all (100 percent) of the power it uses from the power supply to the load. In reality, this never happens. The reason is that the circuit components dissipate some of the power as heat energy.

The following formula is an expression of this loss in terms that can be measured. The Greek letter eta (η) is used as the symbol for efficiency.

$$\text{Amplifier efficiency} = \eta(\%) = \frac{\text{AC output power}}{\text{DC input power}} \times 100$$

8.2 Power Relationships

The power efficiency equation shows that by using less DC input power for the same amount of AC output power, the amplifier can become more efficient. Recall from earlier chapters that DC input power is a function of DC biasing for the amplifier. The lower the Q-point on the DC load line, the lower the DC input power. Figure 8–2a shows a typical DC load line and the corresponding Q-points for the four basic classes of amplifiers. Note that the class A amplifier has a Q-point that is centered on the load line. Figure 8–2b shows that classes AB, B, and C do not conduct for the full 360° of signal. This is why the class A amplifier has the least signal distortion.

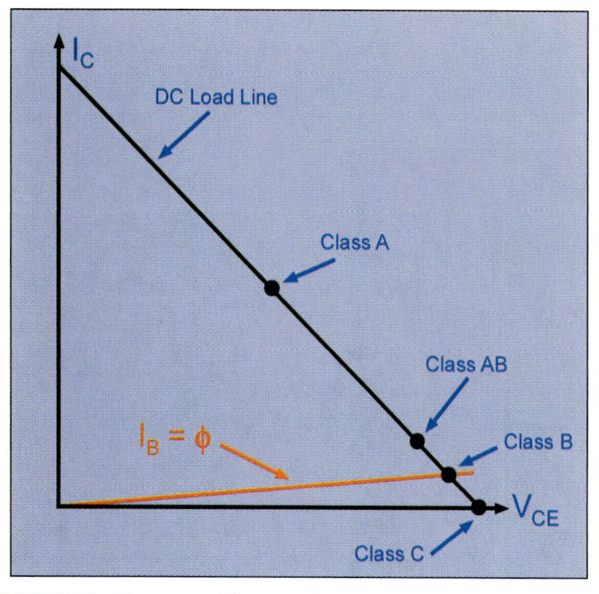

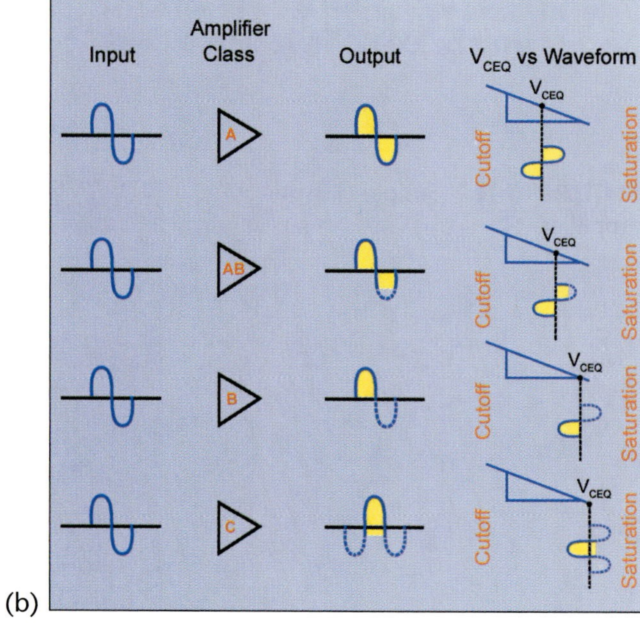

FIGURE 8–2 Amplifier DC load line; a) V_{CE} versus I_C, b) operation characteristics

The DC power that an amplifier uses from its power supply is calculated by

$$P_{DC} = V_{CC} \times I_{CC}$$

Figure 8–3 is an example circuit.

EXAMPLE 1

Analyze this circuit and determine the DC input power.

Solution:
You can start by developing an equivalent circuit for Figure 8–3.

Step 1

Determine the value of I_1.

$$I_1 = \frac{V_{CC}}{R_1 + R_2} = \frac{7\text{ V}}{6{,}100\ \Omega} = 1.15\text{ mA}$$

Step 2

Develop a Thevenin equivalent circuit for the base biasing circuit R_1 and R_2.

$$V_T = V_{CC} \times \frac{R_2}{R_1 + R_2} = 7\text{ V} \times \frac{1{,}100\ \Omega}{1{,}100\ \Omega + 5{,}000\ \Omega} = 1.26\text{ V}$$

$$R_T = \frac{R_1 \times R_2}{R_1 + R_2} = \frac{1{,}100\ \Omega \times 5{,}000\ \Omega}{1{,}100\ \Omega \times 5{,}000\ \Omega} = 901.7\ \Omega$$

Step 3

Redraw Figure 8–3 using V_T and R_T. This is shown in Figure 8–4.

Notice that the collector circuit is shown as a current source of magnitude βi_B, where i_B is the base current. Notice also that the

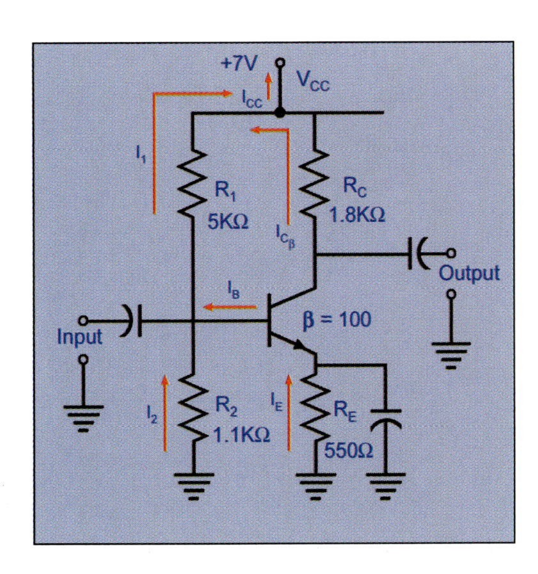

FIGURE 8–3 Calculating I_{CC} for DC power input

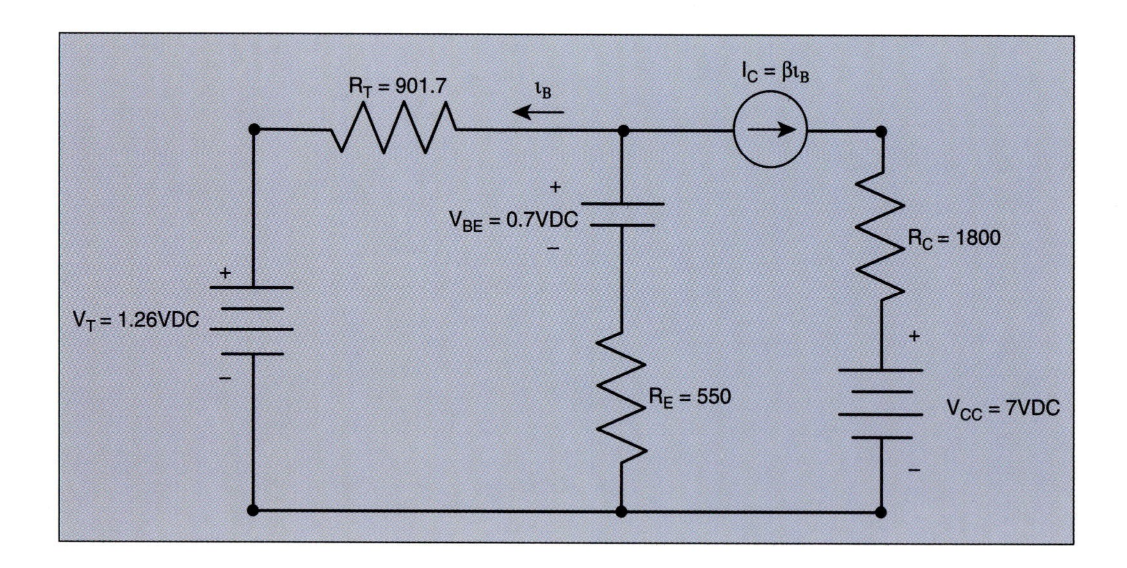

FIGURE 8–4
Equivalent circuit for Figure 8–3

emitter current is equal to $(1 + \beta)i_B$. Writing a voltage equation around the base-emitter circuit of Figure 8–4 yields

$$-(1 + \beta)i_B R_E - V_{BE} - i_B R_E + V_T = 0$$

which simplifies to

$$-(101)i_B\,550\ \Omega - i_B\,903.7\ \Omega = -0.56\text{ V} \Rightarrow i_B = 9.92\ \mu\text{A}$$

Step 4

Calculate I_C.

$$I_C = \beta i_B = 100 \times 9.92\ \mu\text{A} = 0.99\text{ mA}$$

Step 5

Calculate the total DC current supplied by V_{CC}.

$$I_{CC} = I_C + I_1 = 0.99\text{ mA} + 1.15\text{ mA} = 2.14\text{ mA}$$

Step 6

And finally, calculate P_S.

$$P_S = I_{CC} \times V_{CC} = 2.14\text{ mA} \times 7\text{ V} = 15\text{ mW}$$

8.3 AC Load Power

Load power can be calculated using the familiar equation

$$P_L = \frac{V_L^2}{R_L}$$

Where:

V_L = load voltage
R_L = load resistance

Figure 8–5 shows the circuit you worked with earlier, except now a load resistance and a coupling capacitor have been added. For the

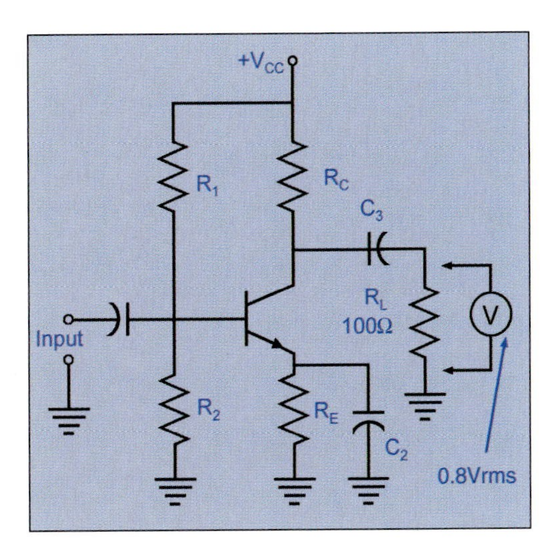

FIGURE 8–5 AC load power calculations

sake of this example, assume that the output voltage read by the meter is equal to 0.8 Vrms. Use the previous formula

$$P_L = \frac{0.8^2}{100} = 6.4 \text{ mW}$$

8.4 Amplifier Efficiency

Amplifier efficiency can now be calculated by the equation developed earlier.

$$\eta(\%) = \frac{P_L}{P_{DC}} \times 100 = \frac{6.4 \text{ mW}}{15 \text{ mW}} \times 100 = 42.67\ \%$$

CLASS B AMPLIFIERS

8.5 Class B Complementary-Symmetry Operation

Basic Operation

The Class B amplifier is biased exactly at cutoff. This means that a single-transistor class B amplifier will not faithfully reproduce its input. Full fidelity can be achieved in several ways. One of the best ways is the use of two transistors connected in what is called a complementary-symmetry connection. Figure 8–6 is an example of such a circuit.

Note that the AC input signal is applied to the base of each transistor. Because each transistor is biased to cutoff, and because Q_1 is PNP and Q_2 is NPN, each transistor will be forward biased on opposite half-cycles of the input waveform. The output waveform is shown in Figure 8–7a.

Because Q_1 and Q_2 are at cutoff with no AC input, the circuit will use no DC power at quiescence. This is the major efficiency advantage that class B amplifiers have. They do not use any DC power unless they are actually amplifying.

The circuit of Figure 8–6 is an emitter-follower type of operation because the load resistor is connected to the emitters of the two transistors.

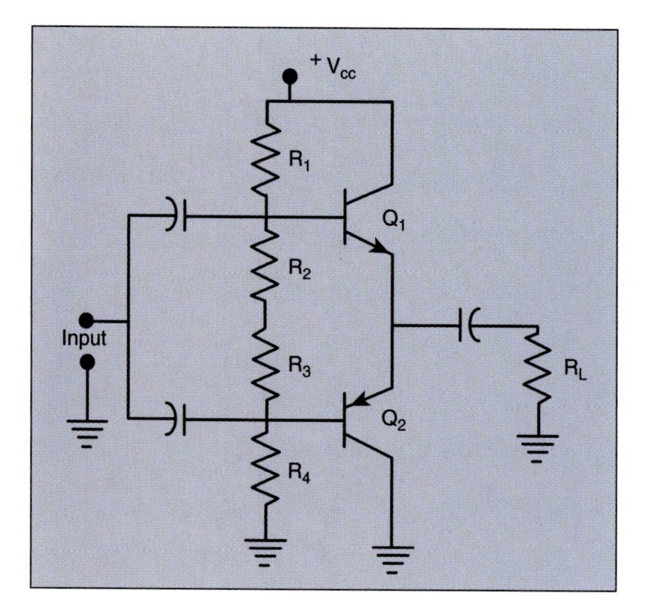

FIGURE 8–6 A class B transistor amplifier using complementary symmetry

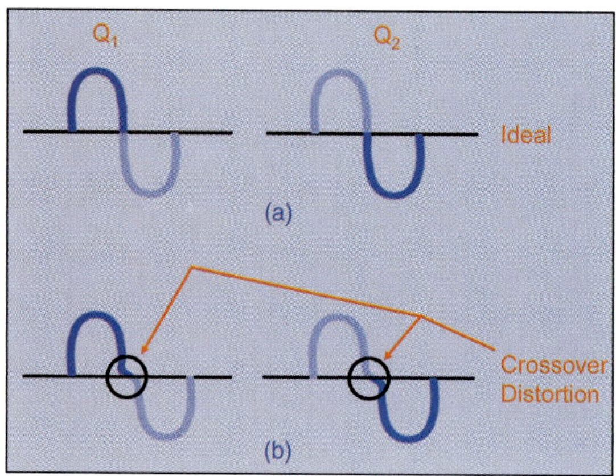

FIGURE 8–7 Output of Figure 8–6; a) ideal, b) with crossover distortion

Crossover Distortion

Crossover distortion is a major problem with class B amplifiers of the type shown in Figure 8–6. Recall that the forward bias, V_{BE}, must exceed the base-emitter forward voltage drop (approximately 0.7 V for a silicon BJT). This means that the input voltage must exceed 0.7 V for each transistor to conduct. The result of this is that the output of the class B will be zero for voltages of ±0.7 V. Such distortion is shown in Figure 8–7b.

To eliminate this type of distortion, the class B amplifier may be biased as a class AB. This biasing operates the transistors slightly above cutoff. This means that Q_1 will not stop conducting until after Q_2 has started, and vice versa. Figure 8–8 shows a circuit that is biased using diodes so that it operates as a class AB.

Diode Biasing

Figure 8–8 shows a circuit that is held above cutoff by the two diodes. These diodes provide a constant 0.7 V drop across each of their PN junctions, for a total 1.4 V. This consistent voltage allows the BE junction of both transistors to maintain a forward bias. This forward bias

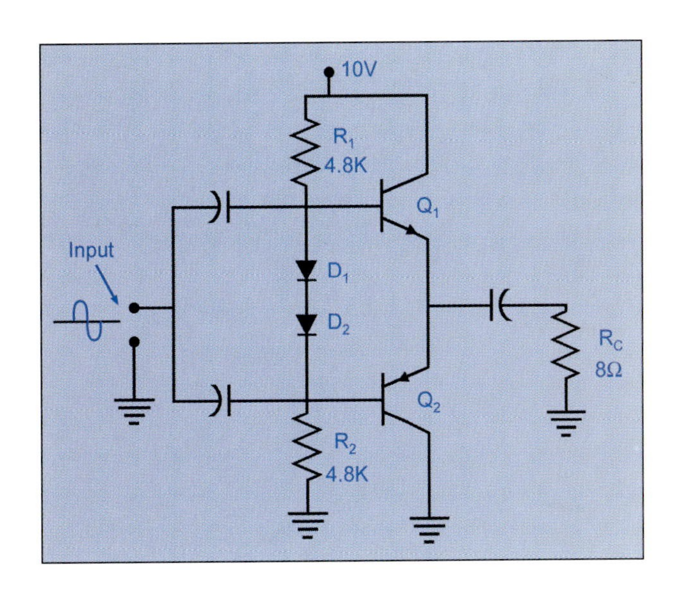

FIGURE 8–8 Diode biasing of a complementary-symmetry amplifier results in class AB operation

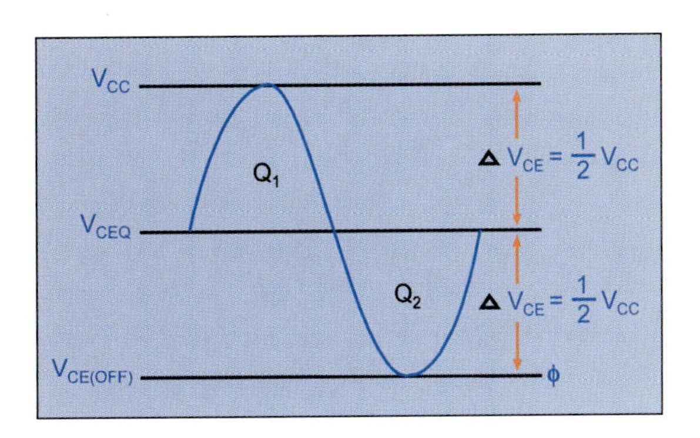

FIGURE 8–9 Output of amplifier shown in Figure 8–6

compensates for the crossover distortion (time) required for the transistor to turn on, as the signal swings above or below the 0 V Q-point bias of both transistors.

AC Operation

Assume that the peak output voltage in Figure 8–6 is at saturation, then $I_{C\,Sat} = \frac{V_{CC}}{2 \times R_L}$. This is because, with both transistors biased to cutoff, the drop across each one is $\frac{V_{CC}}{2}$ at the quiescent point; consequently, the peak voltage across the load resistor is one-half V_{CC} (see Figure 8–9).

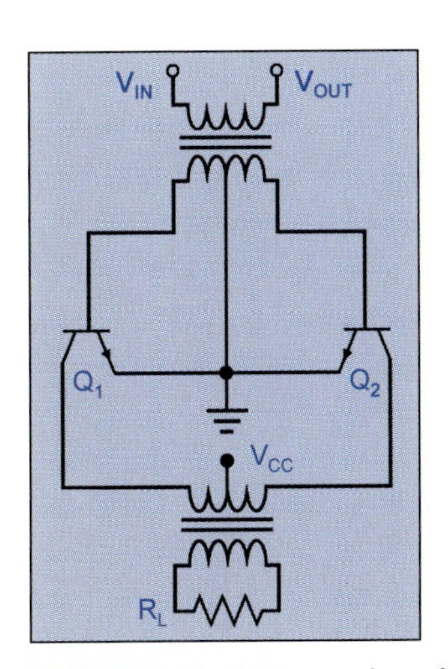

FIGURE 8–10 Push-pull amplifier

8.6 Class B Push-Pull

The class B push-pull amplifier uses two identical transistors and a center-tapped transformer. Figure 8–10 shows a typical class B push-pull amplifier. The disadvantages of this circuit over the complementary pair are that this amplifier needs a center-tapped transformer to

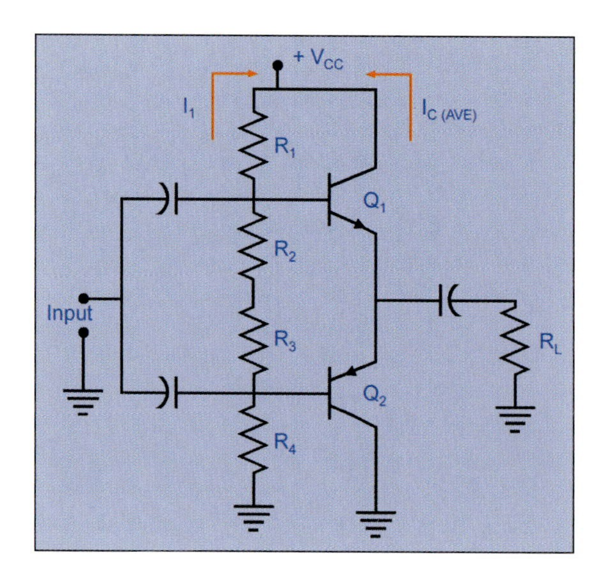

FIGURE 8–11 Calculating I_{CC} for class B amplifier of Figure 8–6

operate, it is more expensive, and it requires much more physical space due to input and output transformers in the circuit.

8.7 Class B Power Calculations

The total input power formula for a class B amplifier is the same as for a class A amplifier, $P_{IN} = V_{CC} \times I_{CC}$. The main difference is in how I_{CC} is calculated. The analysis that follows uses Figure 8–11 and assumes that the amplifier is delivering the maximum sinusoidal output.

$$I_{CC} = \mathrm{I}_1 + I_{C(AVE)}$$

Where:

$$I_{C\,(AVE)} = \frac{I_{pk}}{\pi}$$

This is true because each transistor is producing a one-half sine wave output.

Because you are assuming that the amplifier is producing full output,

$$I_{pk} = I_{C\ Sat}$$

and

$$I_{pk} = I_{C\ Sat} = \frac{V_{CC}}{2R_L}$$

Rearranging the terms, we find that

$$I_{C\,(AVE)} = \frac{V_{CC}}{2\pi R_L}$$

The following example shows how to calculate the maximum power load and efficiency for a class B amplifier using Figure 8–12.

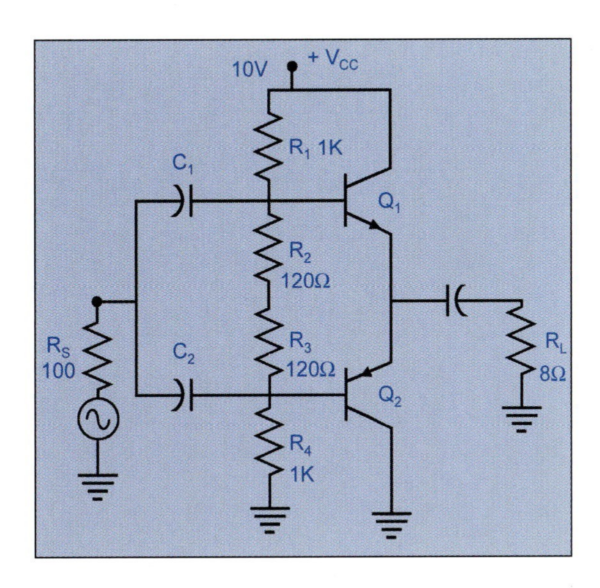

FIGURE 8–12 Class B amplifier total load power

EXAMPLE 1

Calculate the maximum power load.

Solution:

Step 1

Determine the maximum V_{PP} voltage for the amplifier.

This is called the amplifier's compliance. For a class B amplifier, $V_{PP} = 2V_{CEQ}$. And because $V_{CEQ} = \frac{V_{CC}}{2}$, then $V_{PP} \cong V_{CC}$.

Step 2

Find the compliance (V_{PP}) of the class B amplifier.

$$V_{PP} \cong V_{CC} = 10 \text{ V}$$

Step 3

Calculate $P_{L(max)}$.

The peak AC current supplied was calculated earlier as $I_{pk} = I_{C\,Sat} = \frac{V_{CC}}{2R_L}$. This means that the maximum AC power[1] is

$$P_{max} = \frac{1}{2}I_{pk}^2 \times R_L = \frac{1}{2} \times \left(\frac{V_{PP}}{2R_L}\right)^2 \times R_L = \frac{V_{PP}^2}{8R_L}$$

Substituting values gives

$$P_{L\,Max} = \frac{10^2}{8(8)} = 1.56 \text{ W}$$

[1] A more detailed development is given in the summary section of this chapter.

EXAMPLE 2

Calculate the efficiency (η) of the amplifier in Figure 8–12.

Solution:

Step 1

Determine the total power that the amplifier draws from its DC power supply.

$$P_S = V_{CC} \times I_{CC}$$

Where:

$$I_{CC} = I_{C(AVE)} + I_1$$

and

$$I_1 = \frac{V_{CC}}{R_1 + R_2 + R_3 + R_4} = \frac{10\text{ V}}{1\text{ k}\Omega + 120\ \Omega + 120\ \Omega + 1\text{ k}\Omega} = 4.46\text{ mA}$$

$$I_{C(AVE)} = \frac{V_{CC}}{2\pi R_L} = \frac{10\text{ V}}{2 \times \pi \times 8\ \Omega} = 199\text{ mA}$$

$$I_{CC} = 199\text{ mA} + 4.46\text{ mA} = 203.46\text{ mA}$$

Substituting the values, we can now calculate P_S.

$$P_S = V_{CC} \times I_{CC} = 10\text{ V} \times 203.46\text{ mA} = 2.03\text{ W}$$

Step 2

Calculate amplifier efficiency.

$$\eta(\%) = \frac{P_L}{P_S} \times 100 = \frac{1.56\text{ W}}{2.03\text{ W}} \times 100 = 76.8\ \%$$

CLASS C AMPLIFIERS

The class C amplifier in Figure 8–13 has one transistor that conducts for less than 180° of the AC input cycle. To accomplish this, the transistor is biased well into cutoff. The AC input signal causes Q_1 to turn on only during the time T_1 and T_2 are above 0 volts. This is only for a short period of the entire input cycle (much less than 180°).

This short conducting period gives the class C amplifier two distinguishing characteristics:

1. It is extremely efficient. Theoretically, it can have an efficiency rating of up to 99 percent. This is because it uses DC input power only during the short conducting time of the AC input signal.
2. The short conduction time caused by the negative base-emitter bias (refer to Figure 8–13) causes a lot of input signal distortion. For this reason, the class C amplifier is not suited as an audio amplifier. Because of the LC "tank" circuit's resonance frequency, it is used extensively in the radio frequency (rf) range.

FIGURE 8–13 Class C amplifier

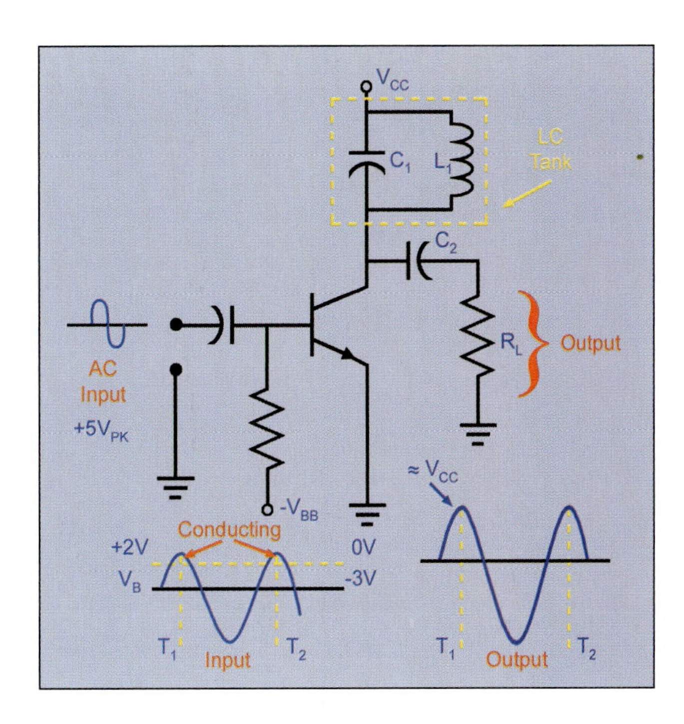

The output signal is generated by the cycling action of the LC "tank" collector circuit. Recall from AC theory that, in a parallel LC circuit, the inductor and capacitor alternate storing (charging) and discharging their energy as the AC signal rises and falls. This cycling action produces a sine wave (see the output signal of Figure 8–13).

At the frequency where the capacitive reactance, X_C, equals the inductive reactance, X_L, a noninductive circuit condition called resonance exists. The total reactance of a "tank" circuit at resonance can be calculated as

$$X_T = \frac{X_L \times X_C}{X_L - X_C}$$

When $X_L = X_C$, the results become

$$X_T = \frac{X_L \times X_C}{0} = \infty$$

Because anything divided by 0 is undefined and usually considered ∞ (in reality, extremely high impedance), at resonant frequency, the "tank" circuit is at maximum impedance. The formula for calculating the resonant frequency of a class C, LC "tank" circuit is

$$f_R = \frac{1}{2\pi\sqrt{LC}}$$

By definition, the class C amplifier shown in Figure 8–13 is a tuned amplifier. The LC circuit is tuned to a specific input signal frequency to provide maximum output impedance. This extremely large output impedance will provide maximum amplifier gain and minimum effect to the load.

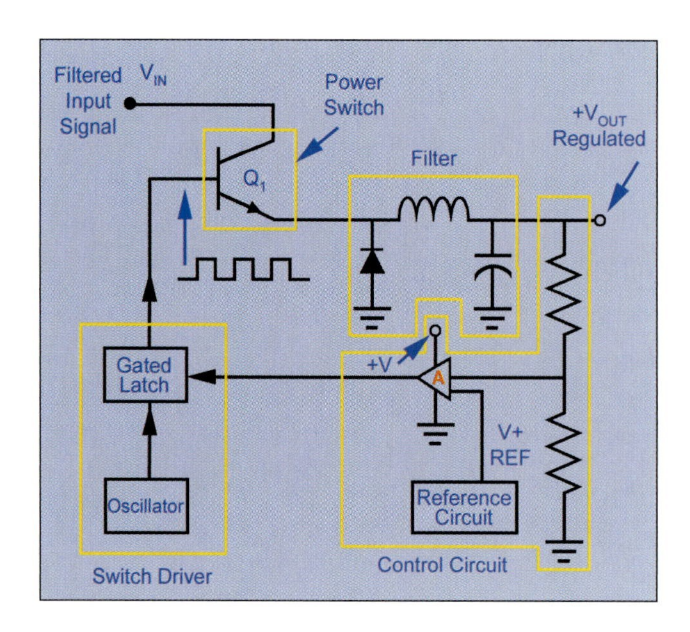

FIGURE 8–14 Typical power-switching circuit

THE POWER SWITCH (CLASS D)

A transistor can be used in a digital switch mode. This on-off configuration is used extensively with inductive motor loads and with voltage regulators. Figure 8–14 shows a typical switching regulator operation. Note that there are four basic circuit areas: the power switch, the switch driver, the filter circuit, and the control circuit.

In Figure 8–14, Q_1 is rapidly switched between cutoff and saturation. When in saturation, the power switch provides a current path between the regulator's input and output. When in cutoff, the path is opened. This results in higher regulator efficiency and higher regulator power-handling capability.

This ability to be on or off for short periods of time increases the overall efficiency and reduces the heat buildup on the transistor at high power loads. This reduction in heat buildup is important, because excess heat due to high power can damage the transistor. Many transistors that routinely handle high power or currents are mounted on heat sinks. A heat sink allows the transfer of the excess heat (dissipates the heat) from the transistor to the sink and the environment.

The circuit in Figure 8–14 operates as follows. When the control circuit senses a change in the regulator's output voltage, it sends a signal to the switch driver. If the change is negative (a lower output voltage), then the oscillator changes the gated latch to increase the signal to the base and increase the conduction of the power switch. Because the oscillator signal is a fixed magnitude and frequency, the gated latch combines the output of the control circuit with the oscillator to change the magnitude (voltage) of the digital pulse to the base. The magnitude of the pulse determines how much the power switch will conduct (determines the forward bias). This will cause the output voltage to rise. The opposite occurs when the control circuit senses an increase in output voltage.

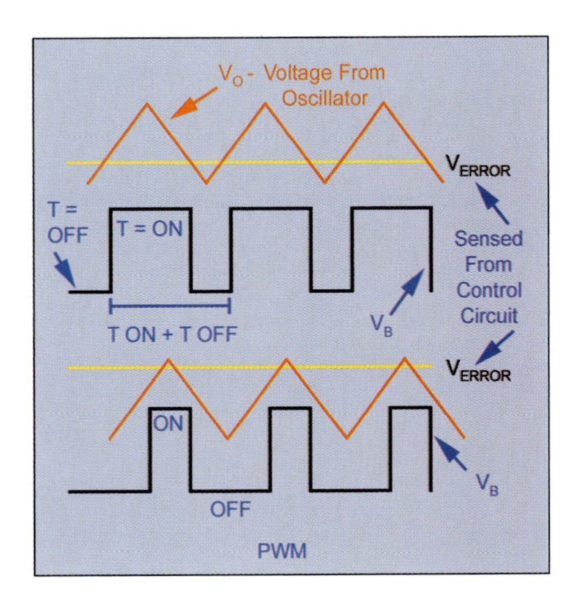

FIGURE 8–15 Pulse-width modulation

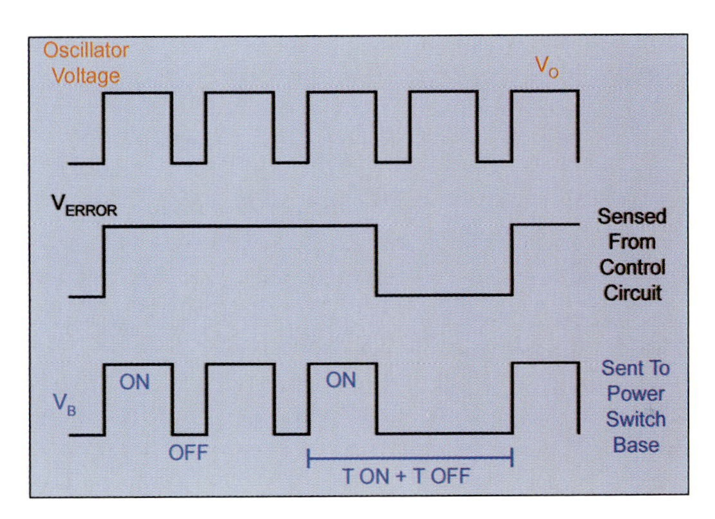

FIGURE 8–16 Variable off-time modulation

Two other methods are also used to control the power switch. The first is pulse-width modulation (PWM). Here the signal from the gated latch varies in pulse width without affecting the cycle time (see Figure 8–15).

The second method uses a stable pulse width but changes the cycle time. This is called variable off-time modulation (see Figure 8–16).

SUMMARY

Amplifiers are often needed in large signal applications. Such applications include audio power amplifiers, rf amplifiers, controllers, variable-frequency drives, regulating power supplies, and a host of other such equipment.

In this chapter, you practiced techniques that you have already learned and applied them to new circuitry, in particular the class B complementary-symmetry amplifier and the class B push-pull amplifier.

Table 8–1 summarizes the various amplifier classes.

Earlier in the chapter, a simple explanation of the formula $P_L = \frac{V_{PP}^2}{8R_L}$ was given. The following information provides a more detailed, and perhaps more interesting, development.

$$P_L = \frac{V_L^2}{R_L}$$

Where:

V_L is the RMS load voltage.

With this case, the V_{PP} must be converted to RMS.

$$V_{RMS} = \frac{V_{PP}}{2} \times \frac{1}{\sqrt{2}}$$

Substituting this into the $P_L = \frac{V_L^2}{R_L}$ formula,

$$P_L = \frac{\left(\frac{V_{PP}}{2} \times \frac{1}{\sqrt{2}}\right)^2}{R_L} = \frac{\left(\frac{V_{PP}}{2\sqrt{2}}\right)^2}{R_L} = \frac{\left(\frac{V_{PP}^2}{4 \times 2}\right)}{R_L} = \frac{V_{PP}^2}{8R_L}$$

REVIEW QUESTIONS

1. Describe the concept of crossover distortion in a class B push-pull amplifier.
   a. What causes it?
   b. If it is severe in an audio amplifier, what would the effect be on the listener?
   c. How is it corrected?
2. How does the class C amplifier shown in Figure 8–13 create a pure sine wave when the transistor conducts for less than half of the input cycle.
3. Discuss efficiency in the various classes of amplifiers.
   a. Which has the highest efficiency?
   b. What are the tradeoffs for high efficiency?
4. Consider Figure 8–3.
   a. Why does the total DC power include I_1 and I_C?
   b. Why does the output capacitor have no effect on the DC input power?
5. The text mentions the use of the class D amplifier in inductive loads and regulators. What other loads might be fed by such amplifiers?

Table 8–1 Summary of Amplifier Classes

Class	Description
A	It has the lowest possible efficiency of all the classes, a maximum of 50%. Uses a single transistor that conducts 360° of the input signal and provides the least signal distortion. The class A amplifier is biased near the center of the load line. The main use of class A amplifiers is in small signal applications.
AB	It has a greater efficiency rating than the class A amplifier and slightly less than the class B. It has two transistors and an improved signal distortion over the class B and is biased near cutoff. The class AB conducts between 181° and 359° of the AC signal input cycle. Diodes are commonly used in this class to provide consistent BE function voltage, therefore maintaining a forward bias for both transistors throughout their respective portion of the input signal. The main use of class AB amplifiers is in high-power stages of coupled amplifiers and in audio and rf applications.
B	The class B has two transistors and has a potential maximum efficiency rating of 78.5%. This amplifier is biased at cutoff and conducts during 180° of the AC signal input cycle. It has a high signal distortion problem due to crossover. The main use of class B amplifiers is in high-power stages of coupled amplifiers and in rf applications. The push-pull amplifier is also a form of class B amplifier. It uses input and output transformers (center tapped) that increase the cost of circuit construction.
C	It has the highest efficiency potential of all the amplifiers, approximately 99%. This amplifier is biased well below cutoff (reverse biased) and conducts during less than 180° of the AC input signal; usually this is around 90°. The signal distortion is very high due to the limited conduction time. The ability of the amplifier to be tuned removes much of the signal distortion. The main use of class C amplifiers is in stages of coupled amplifiers for rf applications.
D	The ability to rapidly switch between saturation and cutoff (on/off) makes the class D an ideal digital switch that provides high regulator efficiency. Its excellent heat dissipation characteristics allow for higher regulator power-handling capability than the other amplifier types. The main uses of the class D amplifier are as power switches for inductive motor loads and as voltage regulators.

chapter 9

Electronic Control Devices and Circuits

OUTLINE

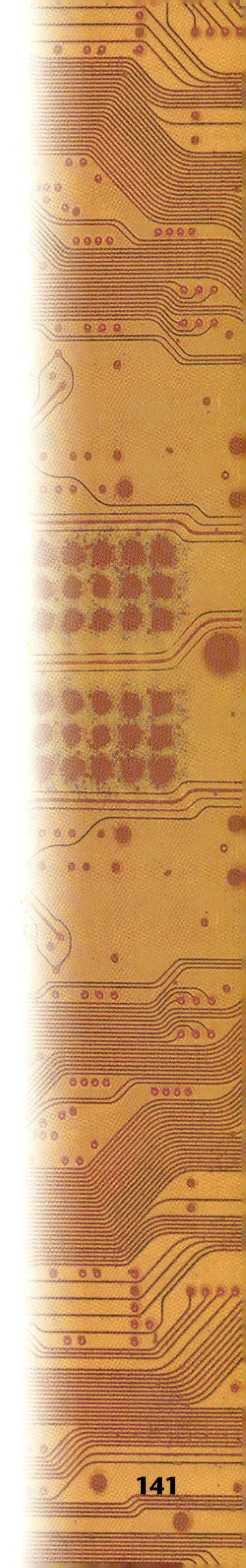

OVERVIEW

Control of current to the load is an area of circuit operation with a lot of significance. For instance, a motor, a lamp, or a heating element may serve as the load in a circuit. By regulating the current in each of these, the speed of the motor, brightness of the lamp, or temperature of the heating element can be controlled.

Control circuits can accurately set the currents and thereby control the load. Although an adjustable resistor (rheostat) can vary or set current, it wastes a lot of energy. Solid-state devices can perform the same job much more efficiently.

In this chapter, you will learn about many of the more important electronic control devices and how they are used in day-to-day applications.

OBJECTIVES

After completing this chapter, the student should be able to:

1. Identify the schematic symbols used for control devices.
2. Explain the AC and DC switching capabilities of the silicon-controlled rectifier (SCR).
3. Describe steps used to test an SCR.
4. Describe the operation of thyristors.

GLOSSARY

Commutation The act of turning an "ON" thyristor to its "OFF" state.

Static switch A switch that uses semiconductor devices such as thyristors to perform the ON and OFF operation.

Thyristor A semiconductor device with three or more junctions that has the ability to turn ON and/or OFF by application of an external signal.

Transient A temporary change in a circuit, usually used when referring to a momentary, high-energy pulse.

CONTROL CIRCUITS

9.1 Rheostat Control

A rheostat circuit used to control the brightness of a lamp is shown in Figure 9–1.

The control device in the circuit is the rheostat. Unfortunately, although very simple, a rheostat control is extremely inefficient, as shown in Table 9–1.

Table 9–1 shows that as the resistance of the rheostat increases, the power dissipated in the rheostat increases and the power dissipated in the load decreases. Full power is delivered to the load only when the rheostat is set at 0 Ω. With increasing values of rheostat resistance, the power delivered to the load decreases because a big portion of the total power is dissipated in the control device itself. Efficiency (η) of the device is calculated by:

$$\eta(\%) = \frac{P_{RL}}{P_T} \times 100$$

Using the values found in Table 9–1 and the formula for efficiency, you can determine that 100 percent efficiency exists only when the rheostat is set at 0 Ω. As the value of the rheostat increases, the efficiency drops. Poor efficiency contributes to high cost of operation due to power loss. Also, the rheostat must be physically large enough to dissipate the large quantities of heat generated. In conclusion, it can be said that rheostat control is inefficient and generally a poor choice for a control application of this type.

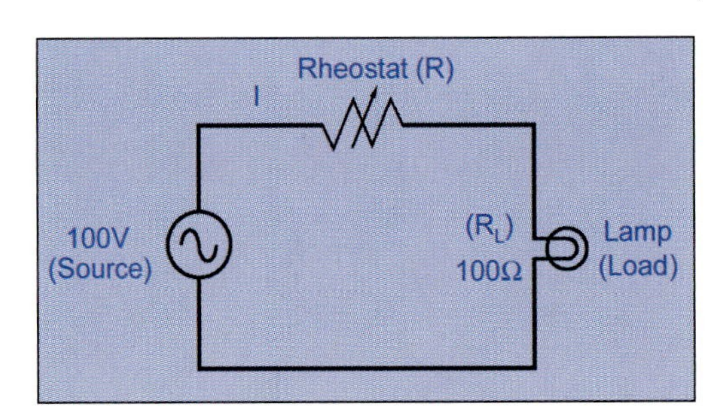

FIGURE 9–1 A rheostat control circuit

Table 9–1 Rheostat Control Circuit Analysis

Rheostat Value	Total Resistance $R_T = R + R_L$	Current $I = V/R_T$	Power Dissipated In Rheostat $P_R = I^2R$	Power Dissipated in Load $P_{RL} = I^2R_L$	Total Power $P_T = P_R + P_{RL}$
100Ω	200Ω	0.5A	25W	25W	50W
50Ω	150Ω	0.66A	22W	44W	66W
0Ω	100Ω	1A	0W	100W	100W

9.2 Voltage Control

An alternative to the rheostat control circuit is voltage control, as shown in Figure 9–2. Here, the voltage applied to the load can be varied from 0–120 V, and the power dissipated in the circuit will be 0–144 W. Note that the only significant resistance in the circuit is the load and the only place for power to be dissipated is in the load. The efficiency of this circuit will always be close to 100 percent. However, controlling the line voltage is a relatively expensive alternative because of the cost of the variable autotransformer.

9.3 Switch Control

The switch control is a form of voltage control; however, it is less expensive and provides for very rapid, automatic control. In order to be efficient, a control device must have very low resistance. A PN junction has very low resistance (ideally zero) when it is conducting (ON) and has high resistance (ideally infinite) when it is nonconducting (OFF). When ON, the low resistance of the junction causes the power dissipation to be almost zero. When OFF, its very high resistance causes almost no current flow in the circuit; hence, the power dissipated in it is almost zero. Thus, there is never any significant power dissipation in the control device. Because of this, a semiconductor device makes an ideal electronic switch.

Figure 9–3 shows a control circuit that employs an electronic switch. At this point in the discussion, it might seem as though the electronic switch just provides ON-OFF control. But imagine a fast switch that can open and close sixty times per second, with the switch being ON for a portion of the entire AC cycle. Because the lamp is connected to the source only for a portion of the AC cycle, it will operate

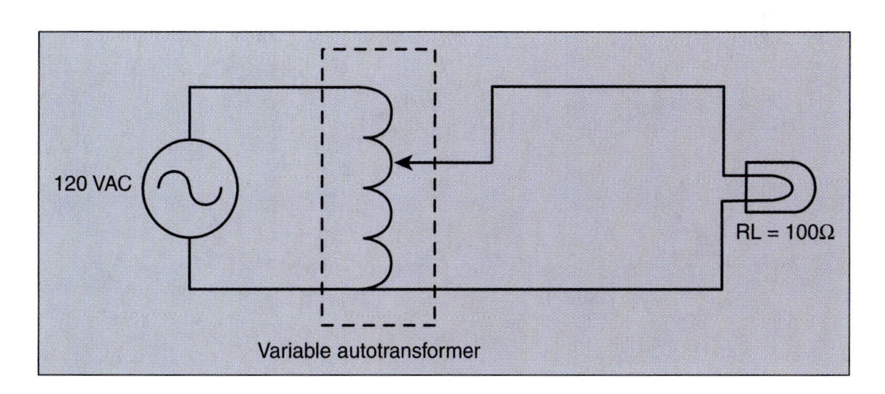

FIGURE 9–2 A voltage control circuit

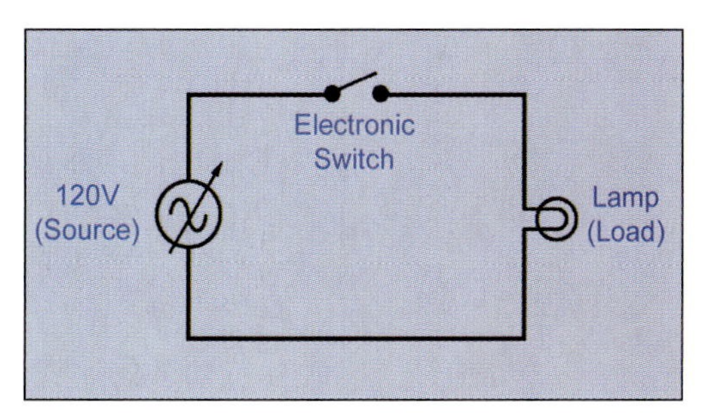

FIGURE 9–3 A switch control circuit

at a lower intensity, and because of the high speed of switching (sixty times per second or more), the lamp will dim without a noticeable flicker.

A mechanical switch used in an application like Figure 9–3 would not be nearly as efficient or rugged. The mechanical switch would generally be much slower, and the contacts would tend to wear out very quickly.

Now that you have a basic understanding of electronic switching and control circuits, let us look at some electronic control devices. The control devices discussed in this chapter belong to a family of solid-state devices called **thyristors**.

THE SILICON-CONTROLLED RECTIFIER (SCR)

9.4 Basic Construction and Operation

The silicon-controlled rectifier (SCR) is the most popular of the control switches. The SCR is a four-layered, three-terminal device also referred to as a four-layer diode. The three terminals are referred to as the anode (A), cathode (K), and gate (G). It is constructed with two P-type regions and two N-type regions in a p-n-p-n sequence, as is shown in Figure 9–4a. It is called a diode because it conducts in one direction and blocks in the other. Figure 9–4b shows the schematic symbol for an SCR. The symbol is that of a solid-state diode with the addition of a gate lead.

To understand the operation of an SCR, consider Figure 9–5a. Here you see an SCR with the middle PN junction split in half. This shows that the SCR can be modeled as two transistors, as shown in Figure 9–5b. The equivalent circuit shows two directly connected transistors, one PNP and the other NPN. Note that the collector of the PNP transistor feeds into the base of the NPN transistor, and vice versa.

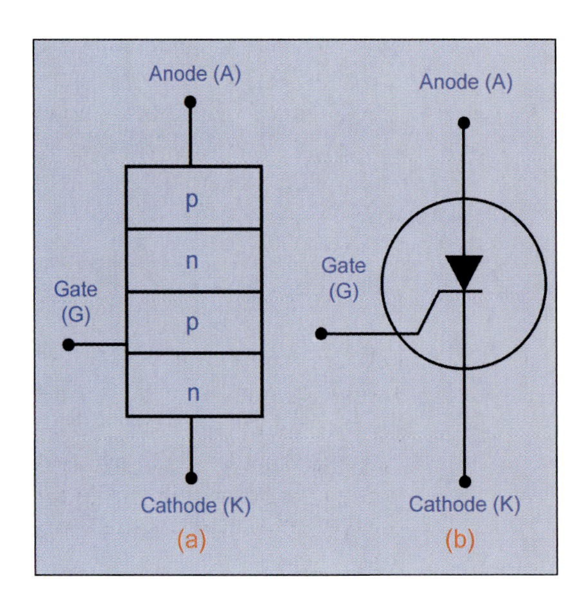

FIGURE 9–4 The SCR; a) construction, b) schematic symbol

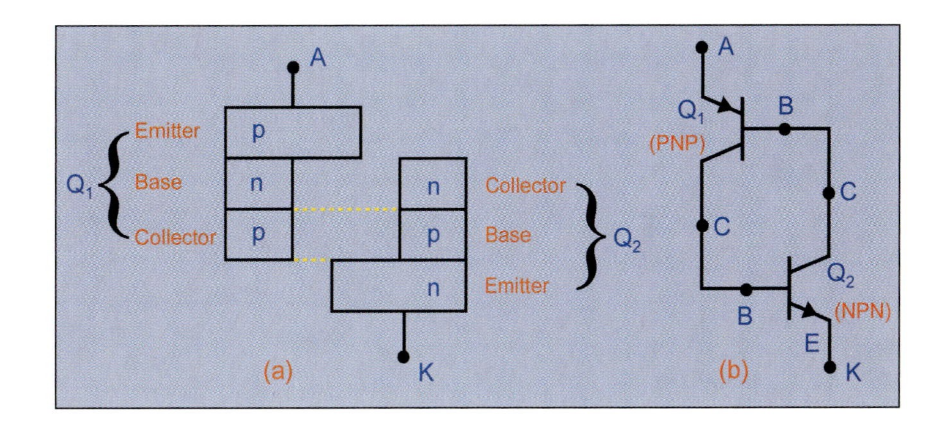

FIGURE 9–5 SCR split construction and equivalent circuit

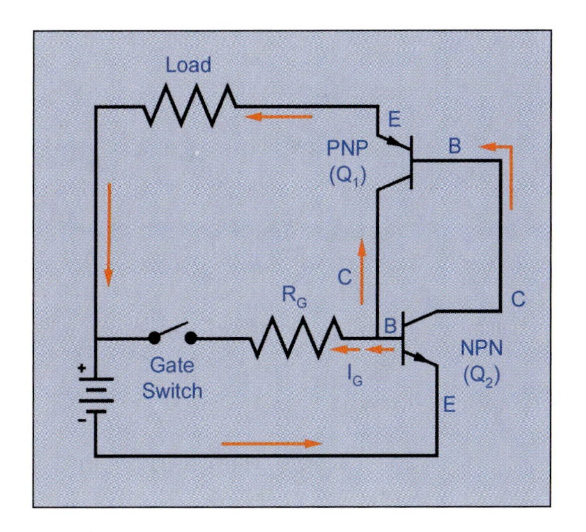

FIGURE 9–6 A two-transistor switching circuit showing the operation of an SCR

Now consider a switching circuit that uses the same type of two-transistor circuit, Figure 9–6. When the gate switch is closed, the positive side of the supply is applied to the base of the NPN transistor. This forward biases the base-emitter junction, and the NPN transistor turns ON. This, in turn, causes current to be supplied to the base of the PNP transistor and, hence, it comes ON. With both transistors ON, the circuit is now complete, and current flows through the load.

When both the transistors are in conduction (ON) and the gate switch is opened, the transistors will not go OFF because they are now supplying each other with base current. Therefore, you can see that once triggered ON by the gating current, the transistors will not turn OFF, even if the gate switch is open and the gate current I_G is cut off. This type of circuit that stays ON once triggered is called a "latch" circuit. Once latched ON, the only way the transistors go OFF is if the source voltage is removed or the load circuit is opened.

The two-transistor equivalent circuit operation gives a basis for understanding the SCR operation. The SCR is a device that conducts when gate current is applied in addition to a positive anode-to-cathode voltage. The SCR then stays latched ON until the current through it is interrupted.

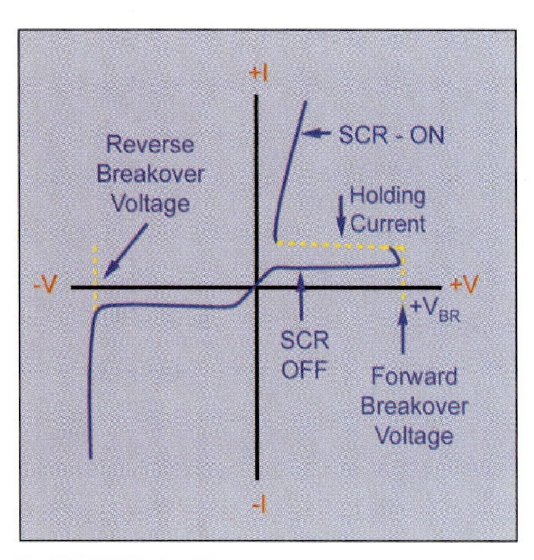

FIGURE 9–7 Volt-ampere characteristics of an SCR

9.5 Characteristics

Figure 9–7 shows the volt-ampere characteristics of an SCR for both forward (+V) and reverse bias (−V). When reverse biased, the SCR behaves like an ordinary diode, in that it is OFF and very little current flows until the reverse breakover voltage is reached. Reverse breakover is avoided by using an SCR with ratings greater than the operating voltage of the circuit.

The SCR differs from an ordinary diode in its forward-biased operation. The SCR stays in its OFF state until forward breakover voltage is reached, at which point the SCR switches to the ON state. The voltage drop across the SCR decreases rapidly, and the current increases. The holding current is the minimum current required to keep the SCR latched ON.

Figure 9–7 does not show the effect of the gate current on the characteristics. The presence of the gate terminal distinguishes an SCR from an ordinary diode. The voltage at which forward breakover occurs is controlled by the gate current. Figure 9–8 shows how the gate current affects forward breakover. As the gate current increases, the forward voltage required for turning ON the SCR decreases. The means that a voltage below maximum breakover can be applied and the SCR will not conduct until gate current is applied.

Normally, the SCR is not operated at high forward breakover voltages. A gate pulse large enough to turn the SCR ON at relatively low forward voltages is chosen. Once triggered ON, the SCR remains in conduction until the current flow through the SCR is reduced below the holding current. The SCR works like a switch—at least for turning it on.

9.6 Commutation

The process of turning an SCR OFF is called **commutation**. Turnoff (commutation) is achieved by reducing the current through the SCR to a value below the holding current level. This is done in practical circuits by either:

1. Opening a series switch that interrupts the current flow in the circuit (Figure 9–9a).

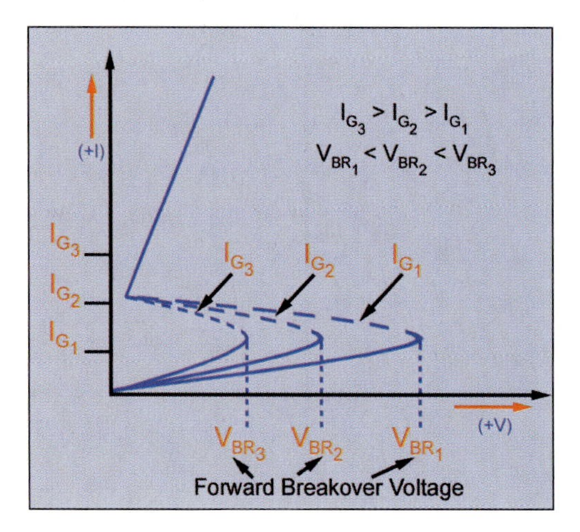

FIGURE 9–8 The effect of gate current on breakover voltage

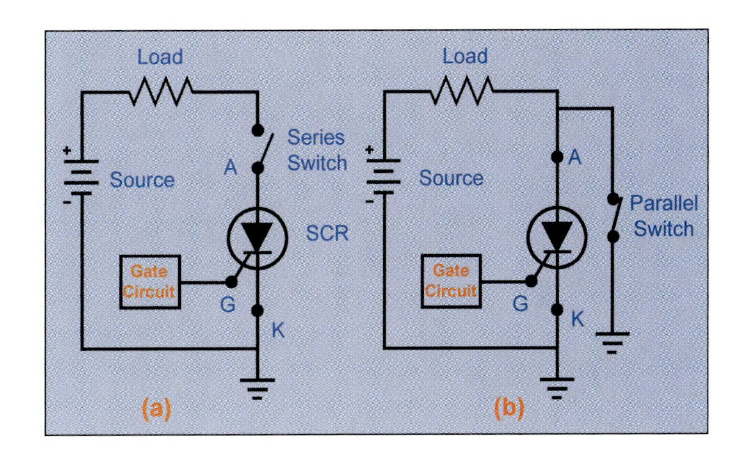

FIGURE 9–9 Commutation of an SCR; a) series switch, b) parallel switch

2. Closing a parallel switch that reduces the forward bias to zero (Figure 9–9b).

Turning the SCR OFF is not complete until all the carriers in the center junctions of the SCR recombine. Recombination is a process by which free electrons occupy holes (valence shells that have a deficiency of electrons) and eliminate the carrier. If forward bias is applied to the SCR before recombination is complete, it may turn ON. Recombination takes some time. The "turnoff" time is defined as the time that elapses after current flow stops and before forward bias can be applied without turning the device ON.

A zero bias between the anode and cathode can turn OFF the SCR. However, the fastest possible turnoff is achieved by applying a reverse bias between the anode and cathode.

9.7 SCR Control Circuits

Figure 9–10 shows how an SCR can be used to control AC power applied to a load. Assume that the load is a lamp and analyze how its brightness can be controlled.

Because the SCR is a unidirectional device, it will conduct only during one-half of the AC cycle. The circuit in Figure 9–10 is therefore a half-wave circuit. The value of the variable resistance, *R*, controls the current in the gate circuit and determines at what instant during the conducting half-cycle the SCR is turned ON. The SCR is reverse biased when the source changes polarity and is turned OFF. The diode, D,

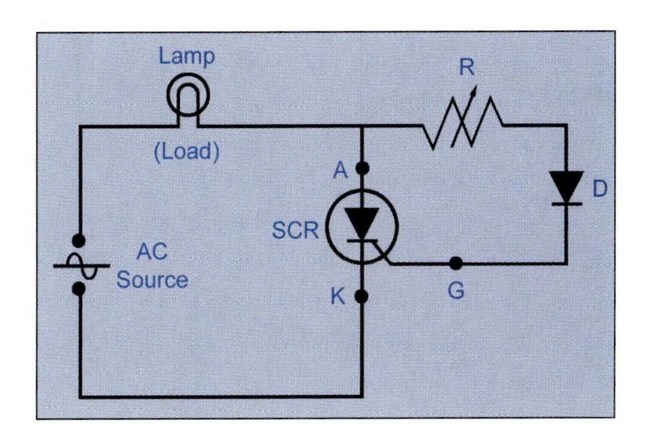

FIGURE 9–10 Half-wave control circuit using an SCR

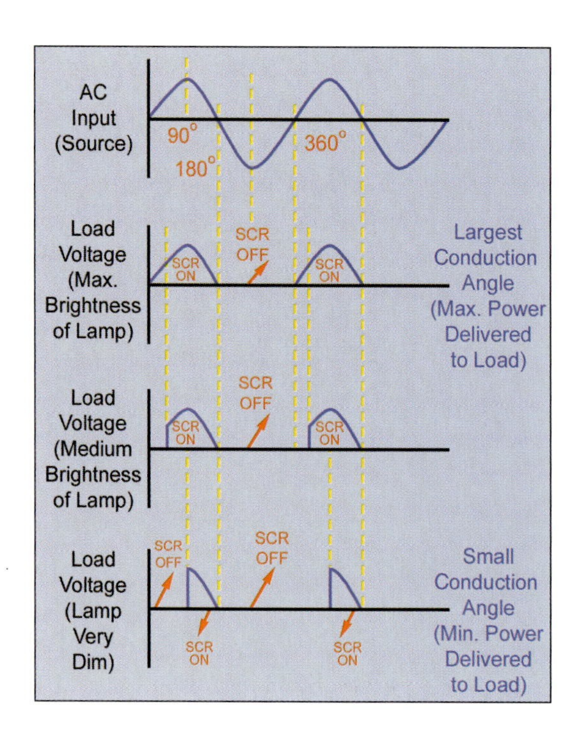

FIGURE 9–11 Control of power in an AC circuit by varying conduction angles

keeps the gate current from flowing when the polarity of the source is reversed.

Figure 9–11 demonstrates how controlling the conduction angle can vary brightness of the lamp. A small conduction angle means that the circuit is ON for a small portion of the AC cycle; conversely, a large conduction angle keeps the circuit ON for a large portion of the AC cycle. If the SCR is gated ON late during the positive half-cycle (small conduction angle), the load power dissipation will be low. On the other hand, if the SCR is triggered earlier during the positive half-cycle (larger conduction angle), the load power dissipation increases.

The SCR control circuit is very efficient because most of the power is dissipated in the load. The forward resistance of the SCR is very small, and the power dissipated in the control device (SCR) is only a fraction of the total power.

An SCR can be used to obtain full-wave control by combining it with a rectifier circuit. Figure 9–12 illustrates the output of a full-wave rectifier. This output can then be fed to the SCR, as shown in Figure 9–13. When the pulsating DC waveform reduces to zero, there is zero forward bias applied to the SCR, thus dropping the holding current to zero and turning the SCR OFF.

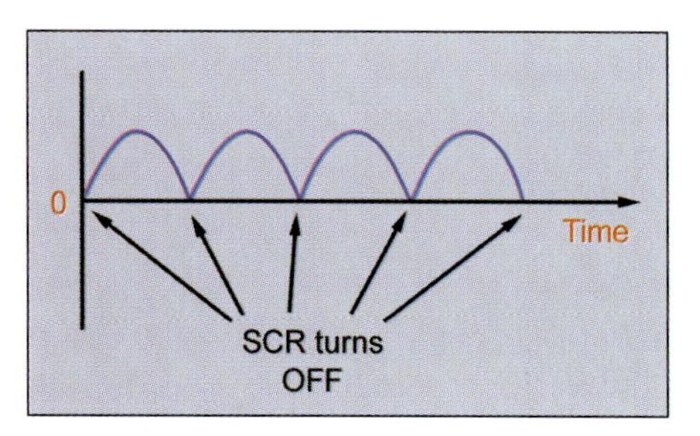

FIGURE 9–12 Full-wave pulsating direct current

Figure 9–13 shows a full-wave control circuit using SCR. A center-tapped full-wave rectifier with diodes D_1 and D_2 is used to obtain full-wave, pulsating DC voltage across the SCR. Resistor R sets the gate current, which controls where in the cycle the SCR comes ON. Thus, full-wave power control to the load can be achieved by controlling the conduction angle of the SCR.

A similar result could be obtained by substituting two SCRs for diodes D_1 and D_2 and triggering them on the alternate half-cycles.

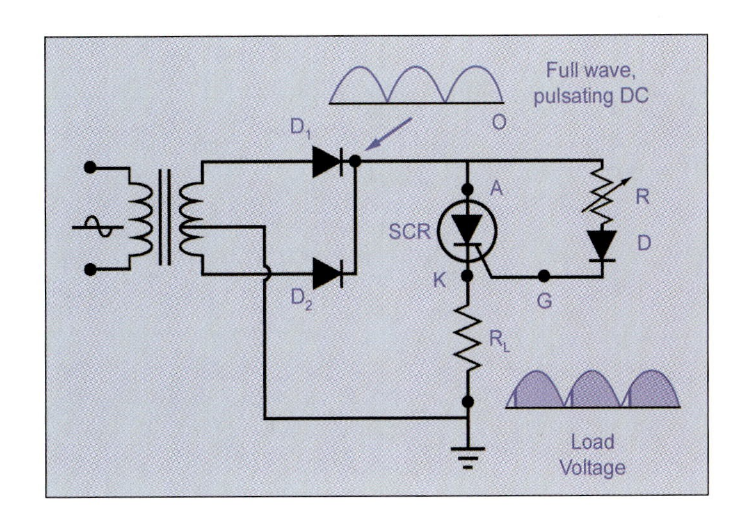

FIGURE 9–13 Full-wave control circuit using an SCR

THE TRIAC

9.8 Basic Construction and Operation

The SCR, as discussed in the previous section, is a unidirectional device. A triac (triode AC semiconductor switch) is a four-layered, bidirectional device and accomplishes the function of two SCRs. It can be considered as two SCRs connected in antiparallel. When one of the SCRs conducts, the other one does not, and vice versa. A triac is therefore a full-wave device and can be viewed as a bidirectional SCR. Both the SCR and triac are thyristors.

Figure 9–14a shows the construction of a triac. The triac connections are called main terminal 1 (MT1), main terminal 2 (MT2), and the gate. The MT2 can be positive or negative with respect to MT1 when triggering occurs. Also, the triggering pulses applied by the gate can be positive or negative with respect to the MT1. This leads to four possible combinations by which the triac can be turned ON. They are:

1. MT2 (+), gate (+)
2. MT2 (+), gate (−)

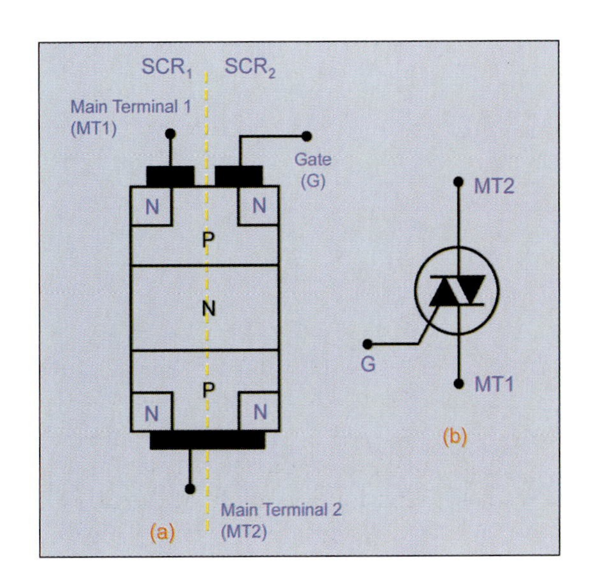

FIGURE 9–14 The triac; a) construction, b) schematic symbol

3. MT2 (−), gate (+)
4. MT2 (−), gate (−)

Of the combinations listed above, 1 and 4 are the ones that are generally employed to trigger the triac. Note that when a positive triggering pulse is applied to the gate, MT2 is positive; and when a negative triggering pulse is applied to the gate, MT2 is negative. By changing the amount of gate current, the instant at which the triac comes ON can be controlled.

Figure 9–14b shows the schematic symbol for a triac. Keep in mind that triacs have smaller current and voltage ratings than SCRs. Triacs are suitable for small- and medium-power AC applications, whereas SCRs are capable of handling high-power applications.

9.9 Triac Control Circuits

Figure 9–15 shows a circuit that uses a triac to control AC power. The gate pulses allow the triac to be triggered during both the positive and negative alteration of the AC source. Just as discussed in SCRs, the timing of these trigger pulses can be used to control the conduction angle and the power delivered to the load.

Figure 9–16 illustrates the conduction angle control in a triac control circuit. By comparing Figure 9–16 with Figure 9–11, you can see that the triac is a full-wave device, whereas the SCR is a half-wave device.

9.10 Static Switching

SCRs and triacs can be used to perform the task of static switching of AC loads. A **static switch** is a switch with no moving parts. Switches that contain moving parts tend to wear out faster and are subjected to contact bounce and arcing. Such problems are eliminated by electronic circuits that use static switches such as triacs and SCRs. The triac functions well as a static switch at lower frequencies (50 to 400 Hz), whereas the SCR is designed to operate up to frequencies of 30 kHz.

A three-position static switching circuit is shown in Figure 9–17.

- In position 1, no gate signal is applied and the triac is OFF.
- In position 2, the presence of the diode causes the gate signal to be applied only during the positive alteration of the AC source. Hence,

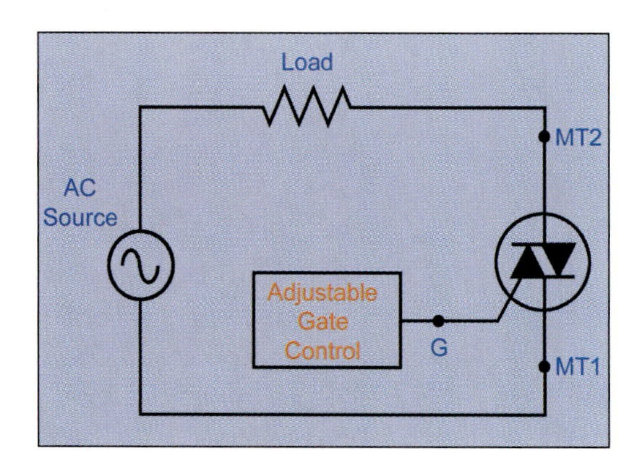

FIGURE 9–15 Control circuit using a triac

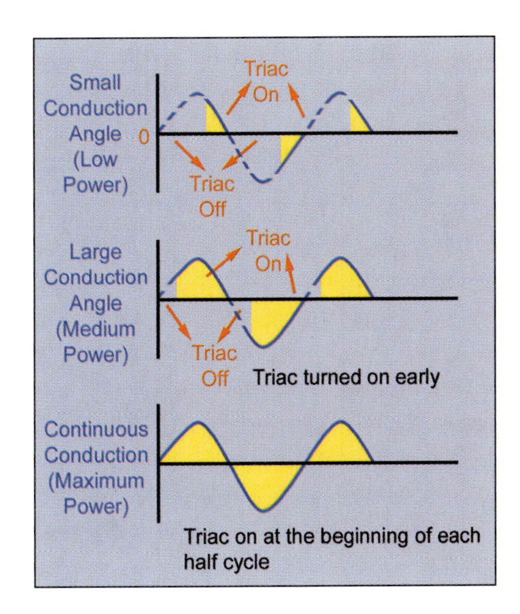

FIGURE 9–16 Conduction angle control in a triac control circuit

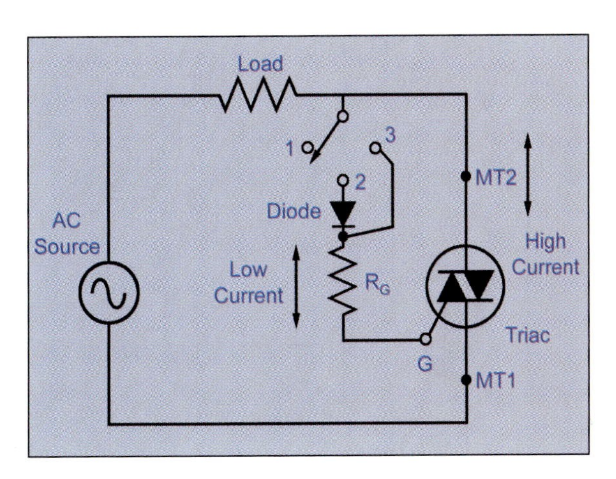

FIGURE 9–17 A three-position static switching circuit

the triac is ON only for half the AC cycle and half power is delivered to the load.

- In position 3, the gate signal is applied to the triac during both the alterations of the AC cycle. Hence, the triac conducts during both the alterations of the AC cycle and full power is delivered to the load.

Although the three-position switch shown in Figure 9–17 has a mechanical switch, note that it operates in the gate circuit, which carries low currents. As a result, the mechanical switch is not subject to arcing and related contact problems.

9.11 Commutation

The triac is a bidirectional device, which means that commutation is different from an SCR. When the AC power becomes zero, the triac has to go OFF. We know that power hits the zero point twice in one AC cycle. If the triac does not turn OFF at these points, power control cannot be achieved.

When the load is resistive, voltage and current in the circuit are in phase. The zero-power point is reached when the current is zero, the voltage is zero, and the triac is turned OFF with no problems. However, if the load is reactive (inductive or capacitive), the voltage and current in the circuit are out of phase and turning the triac OFF may be a problem.

Consider an inductive load such as a motor, where the motor winding offers an inductive reactance. The current in the circuit lags the voltage. When the current becomes zero, the voltage is not zero and is applied across the triac. This voltage may be sufficient to turn the triac ON, resulting in false triggering. Thus, commutation is not achieved and power control may be lost.

In addition to inductive loads, transients can affect triacs (and SCRs). The word **transient** is used to refer to a large voltage change occurring for a very short time. The transient voltages are usually associated with noise signals and can be of magnitudes high enough to turn the device ON.

Thyristors are made of PN junctions and, when not conducting, these junctions have a depletion region. The depletion regions are insulators and can act as the dielectric of a capacitor. Hence, we can associate internal capacitances with the nonconducting thyristor. These capacitances respond to sudden changes in voltages caused by transients and draw charging currents. The charging currents may serve as gate currents and turn the device ON, thereby resulting in false triggering. Hence, power control will be lost if the thyristor responds to transients.

9.12 Snubber Network

As was discussed in the previous section, inductive loads and transients can trigger the triac ON and can cause the power control to be lost. A snubber circuit reduces false triggering in both triac and SCR control circuits. Figure 9–18 shows an RC snubber network connected to a triac control circuit.

The property of the capacitor to oppose changes in voltage is used in the RC snubber circuit. The RC circuit is connected between MT2 and MT1 of the triac (anode and cathode of the SCR). The snubber network bypasses the charging currents from the nonconducting thyristor. The resistor, R, is included to limit the discharge current of the capacitor, C, when the thyristor turns ON.

FIGURE 9–18 Snubber network

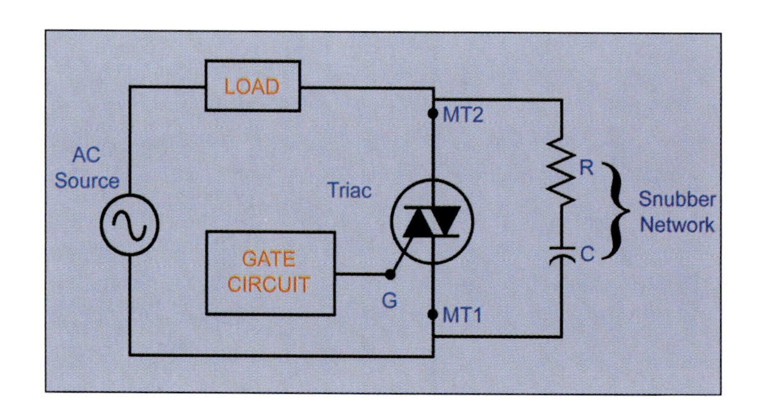

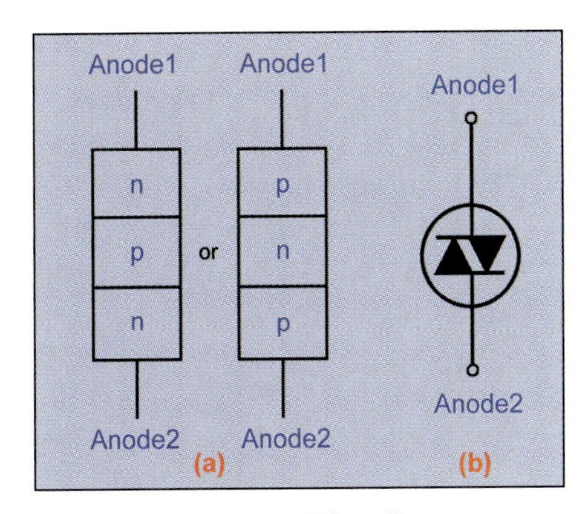

FIGURE 9–19 The diac; a) construction, b) schematic symbol

THE DIAC

9.13 Basic Construction and Operation

The diac is a three-layer, bidirectional device that has only two external terminals. The construction of the diac is shown in Figure 9–19a, and schematic symbol is shown in Figure 9–19b. The terminals of the diac are called anode 1 and anode 2 and can be connected interchangeably.

The diac's construction is very similar to that of an NPN or PNP transistor. There are, however, crucial differences.

1. A diac has no base connection.
2. The diac's three regions are identical in size and level of doping. In the BJT, the base is extremely narrow in dimension and very lightly doped.

The operating characteristics of a diac are shown in Figure 9–20. As you can see, the forward operating characteristic is identical to the reverse characteristic. In either direction, the diac remains open (OFF) until the breakover voltage is reached. At this point, the diac is triggered and conducts (ON) in the appropriate direction. The diac continues to conduct until the current drops below the holding current value. The operating principles are same for an NPN and PNP diac.

9.14 Diac-Triac Control Circuit

The bidirectional diac is very well suited for triggering the triac. The following sections show how the two devices work together.

Example One

Figure 9–21 shows an example of a diac-triac control circuit. The diac is used in the gate circuit of the triac.

Operation in the positive half-cycle In the positive half-cycle (Figure 9–21a), diode D_1 is forward biased and D_2 is reverse biased. A positive voltage appears at point P. When this voltage becomes greater

FIGURE 9–20 The diac operating characteristic curve

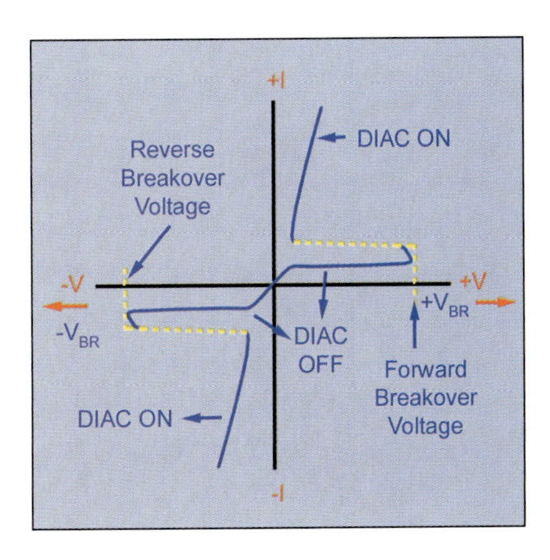

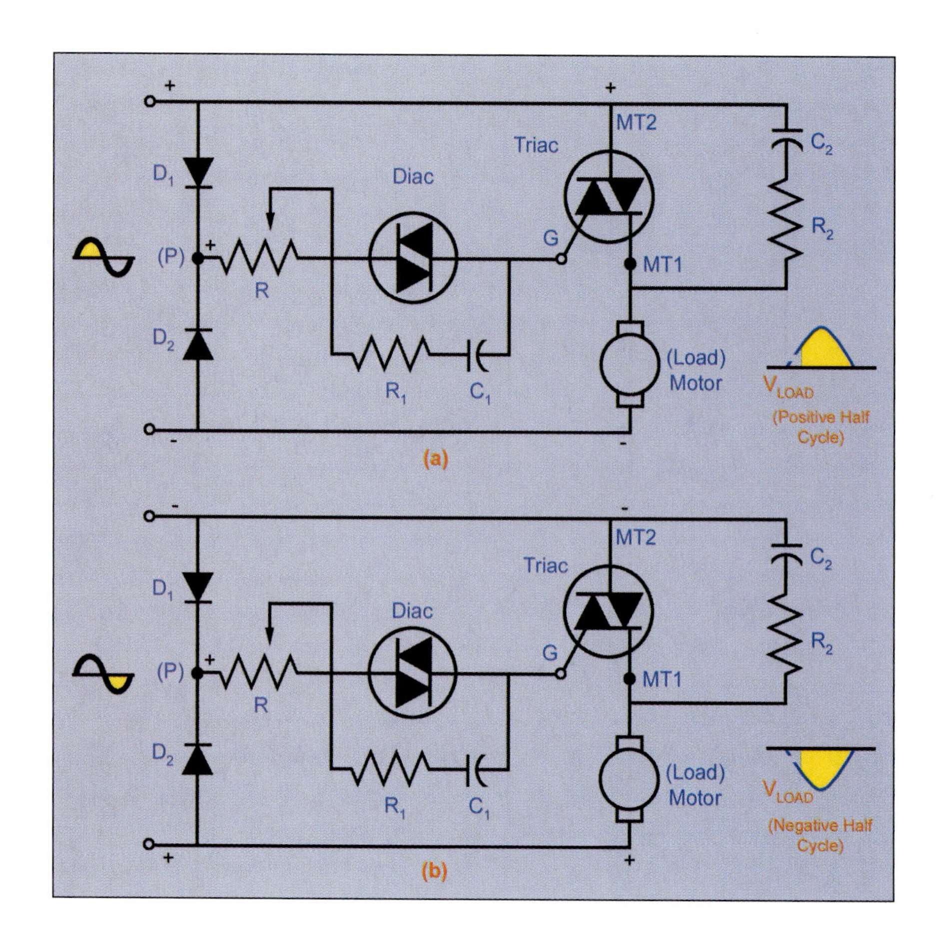

FIGURE 9–21 Diac-triac control circuit

than the breakover value of the diac, the diac conducts. When the voltage is high enough, resistor R passes enough current to trigger the triac. Thus, triggering occurs in the positive alternation of the AC cycle.

Operation in the negative half-cycle During the negative half-cycle (Figure 9–21b), the diode D_2 is forward biased and D_1 is reverse biased. Point P in the circuit is still positive, but MT2 is negative. This causes the triac to be triggered in the negative alternation of the AC cycle.

The rheostat R is used to change the gate current through the triac, which in turn controls the conduction angle. The load in the circuit is a motor, whose speed of rotation is influenced by the conduction angle. The greater the conduction angle, the higher the power delivered to the motor, and hence the faster the motor will rotate. Note the presence of RC snubber network R_1C_1 across the diac and R_2C_2 across the triac.

Example Two

Figure 9–22 shows another example of a diac-triac control circuit. The resistors R_1 and R_2 determine the time it takes for the capacitor C_3 to charge. When the C_3 has charged to the breakover point of the diac, the diac conducts, thus providing a discharge path for the capacitor C_3 through the gate circuit of the triac. The discharge current from C_3 causes the triac to conduct.

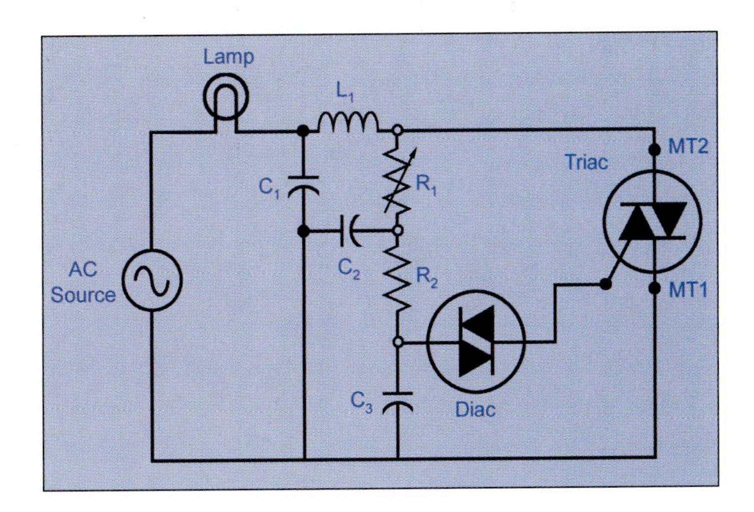

FIGURE 9–22 Another diac-triac control circuit

The lamp that serves as the load in the circuit can be made to glow with varying brightness by changing the amount of power applied to it. Resistor R_1 determines the timing of the gate current applied to the triac, which in turn controls the conduction angle of the triac. If the value of R_1 is set to a low value, C_3 charges to the breakover point quickly, and the triac comes into conduction earlier in the AC cycle. This will contribute a larger amount of power delivered to the lamp, which will glow with increased brightness. Conversely, decreasing the value of R_1 leads to a chain of events that would reduce the brightness of the lamp.

L_1C_1 and R_1C_2 are used for filtering. When the LC low-pass filter allows AC line frequency to pass, it may also allow a wider band of frequencies to trigger the triac. The R_1C_2 filter provides a narrow frequency trap at the low band end and adds frequency stability to the triggering circuit.

The components L_1 and C_1 also suppress radio frequency interference that may be generated in the circuit. When triacs switch from OFF to ON, the sudden increase in current produces harmonic noise. Harmonic energy emitted by the triac circuits can range up to several MHz and may be radiated into the surrounding area by the load wiring. AM radio receivers operating nearby can pick up the signals, and this contributes to noise in the reception.

L_1 and C_1 form a low-pass filter, which, as the name suggests, allows the relatively low AC line frequency signals to pass through to the load. However, signals in the radio frequency range (MHz) will be blocked by the filter and will not reach the load. The presence of the low-pass filter helps to eliminate interference to any AM radio receiver located nearby.

COMPONENT TESTING

9.15 SCR Testing

If an SCR is suspected to be the cause of a malfunction, it should be removed and tested using the ohmmeter test circuit shown in Figure 9–23. The internal battery of the ohmmeter applies different polarities

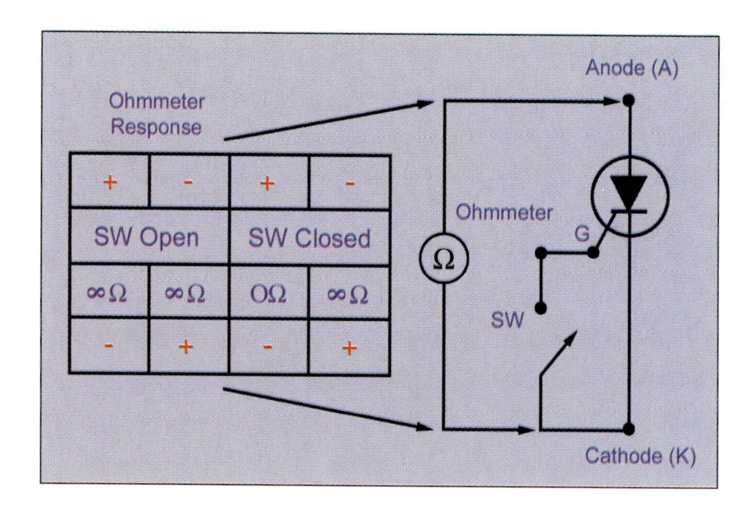

FIGURE 9–23 Testing SCRs

to the terminals of the SCR. The switch SW in the circuit is used to apply or disconnect gate current.

The ohmmeter response table shown in Figure 9–23 can be explained as follows:

- When switch SW is open, the ohmmeter's battery is not connected to the gate circuit, thus making the gate current zero and the SCR OFF. At this setting of SW, the resistance between the anode and cathode should read extremely high—theoretically infinite Ω. This is true no matter what polarity is applied to the cathode.
- When switch SW is closed and the anode is made positive with respect to the cathode, the gate will also be positive through SW. This will turn the SCR ON, and 0 Ω (very low resistance) will be measured between the anode and cathode. If the anode is made negative with respect to the cathode, the SCR is reverse biased and the ohmmeter should read infinite Ω (or a high resistance).

If any one of the ohmmeter readings does not correspond to that listed in the ohmmeter response table of Figure 9–23, the SCR is probably defective and should be replaced.

9.16 Triac Testing

If a triac is suspected to be malfunctioning, it should be removed and tested using the ohmmeter test circuit shown in Figure 9–24. As in Figure 9–23, the ohmmeter's internal battery is used to apply different polarities to the terminals.

The ohmmeter response table shown in Figure 9–24 can be explained as follows:

- With switch SW open, there is no gate current through the triac, so the triac is OFF; consequently, the resistance between MT2 and MT1 will read very high no matter what polarity is applied between the two.
- When the switch SW is closed, the gate of the triac will receive a trigger current from the battery. Because the triac is bidirectional

FIGURE 9–24 Testing triacs

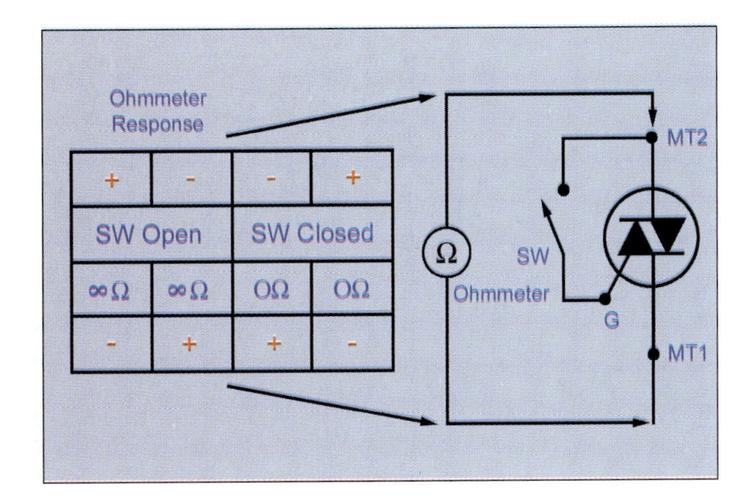

and operates with either a positive or a negative trigger, the triac should turn ON no matter what polarity is applied between MT1 and MT2. Therefore, 0 Ω (low resistance) should be read between MT2 and MT1 either way.

If any one of the ohmmeter readings does not correspond to that listed in the ohmmeter response table of Figure 9–24, the triac is probably defective and should be replaced.

9.17 Diac Testing

If a diac is suspected to be malfunctioning, the diac should be removed and checked with an ohmmeter. The diac is basically two diodes connected back to back and hence should measure 0 Ω (or low resistance) between its terminals, no matter what polarity is applied between them. If the ohmmeter reads a high resistance in either of the connections, the diac is probably defective (see Figure 9–25). Of course, the voltmeter must apply a high enough voltage to trigger the diac.

FIGURE 9–25 Testing diacs

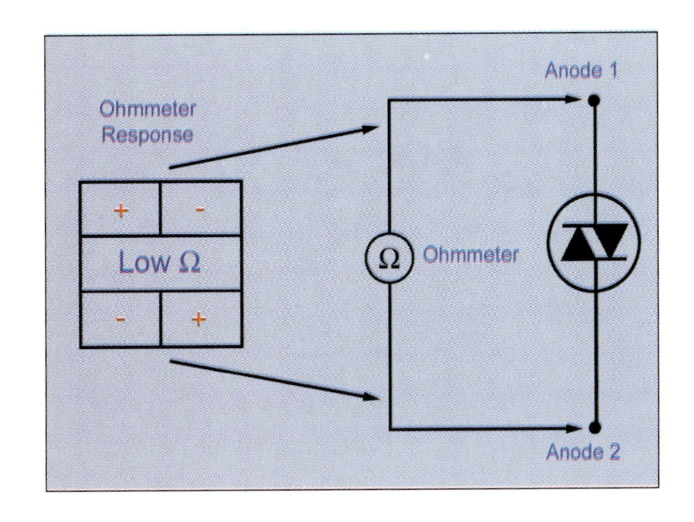

SUMMARY

The discussion in this chapter familiarizes you with electronic control devices (SCR, diacs, and triacs) and how they can be used in circuits to control the amount of power delivered to the load. This lesson should enable you to identify the schematic symbols of these devices and understand their basic operation. The lesson also helps you analyze control circuits and appreciate the switching capabilities of thyristors. Finally, the methods of testing an SCR, triac, and diac addressed in the discussion should provide you with techniques to identify malfunctioning components in circuits. Table 9–2 summarizes the semiconductors discussed in this chapter.

Table 9–2 Summary of Thyristor Devices

	SCR	TRIAC	DIAC
Symbol	Anode (A) Gate (G) Cathode (K)	Main Terminal 2 (MT2) Gate (G) Main Terminal 1 (MT1)	Anode 1 (A1) Anode 2 (A2)
Current Flow	Unidirectional	Bidirectional	Bidirectional
Common Applications	Large current	Modest current. Recent improvements allow for increased current capacity	Triggering device for triacs
Turn On	Forward biased & forward breakover current (which is controlled by the gate current)	Gate pulse in both alterations Note: 4 combinations of lead termination are possible	Forward and reverse breakover current
Turn Off (commutation)	1. Open circuit or 2. Reduce forward bias to 0 Note: recombination has to be complete	AC power becomes 0 in both positive & negative alteration	Current drops below holding current value
Drawing	K G A SCR	MT2 G MT1 TRIAC	A2 DIAC A1

REVIEW QUESTIONS

1. What are the advantages of using semiconductor switches for control circuits as opposed to
   a. mechanical switches?
   b. resistor controls?
2. Describe the ohmmeter method for testing SCRs, triacs, and diacs.
3. Look at Figure 9–13. How should resistor R be adjusted to set the load current to zero? To maximum?
4. Which of the circuits discussed in this chapter might be used to make a simple light dimmer?
5. The amount of energy supplied to a load is proportional to the amount of the AC waveform that is passed to it. How does this statement apply to the material covered in this chapter?

current. To find the wattage rating for R_S, simply apply the power formula $P = I^2R$.

EXAMPLE 1

What is the resistance and wattage rating of a limiting resistor in a 12 VDC circuit for an LED that requires a 2 V forward bias with a 20 mA current rating?

$$R_s = \frac{V_s - V_D}{I_D} = \frac{12\text{ V} - 2\text{V}}{20\text{ mA}} = 500\ \Omega$$

and

$$P = I^2R = (20\text{ mA})^2 \times 500\ \Omega = 200\text{ mW}$$

Therefore, a standard size ¼ watt resistor is required. (Note: ¼ watt = 250 mW)

Multicolor LEDs

Multicolor LEDs are available that will

1. Emit one color light when the supply voltage is one polarity
2. Emit a second color when the polarity is reversed
3. Emit a third color when the bias polarity is rapidly switched

Figure 10–10 shows the schematic symbol for multicolor LEDs. Multicolor LEDs are usually two LEDs connected in antiparallel; that is, the anode of each diode is connected to the cathode of the other. Each LED can emit light only when forward biased. So, when voltage of either polarity is applied, one LED is forward biased and emits its native color.

The two LEDs most commonly used are red and green in color. The green LED is normally used to indicate whether something is functioning properly, and the red LED is used to indicate there is a problem. If the multicolor LED is rapidly switched between the two polarities, the red/green LED appears to produce a third color, which is yellow.

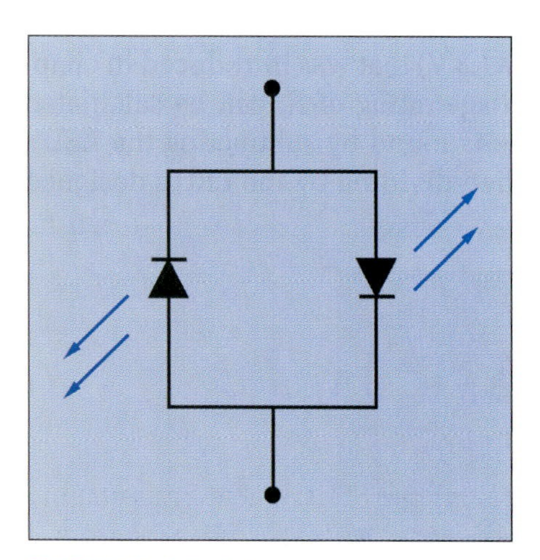

FIGURE 10–10 Multicolor LEDs

Infrared-Emitting Diode (IRED)

As discussed earlier in this chapter, the infrared band (30 THz to 400 THz) falls below the frequencies that the human eye can detect. Diodes made of gallium arsenide release energy by way of heat and infrared light. Such a diode is called an infrared-emitting diode (IRED). IREDs are used for home electronics such as remote controls, fiber optic communications, discriminating organic solvents in the field of medicine, and other such applications.

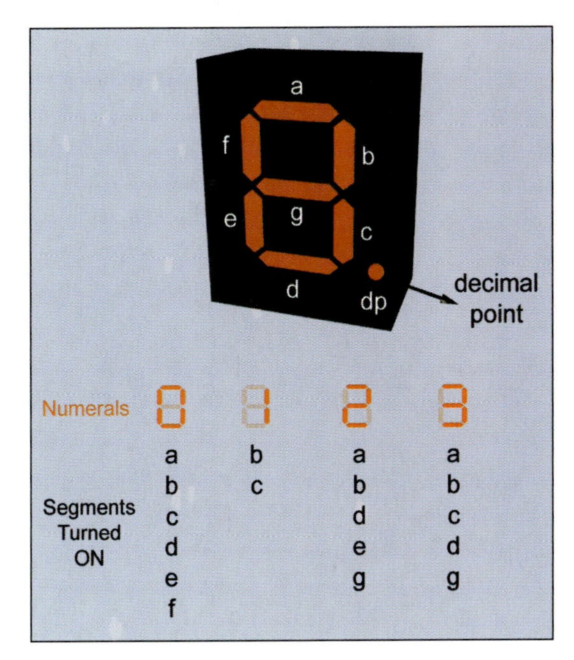

FIGURE 10–11 Seven-segment display

FIGURE 10–12 Sixteen-segment display

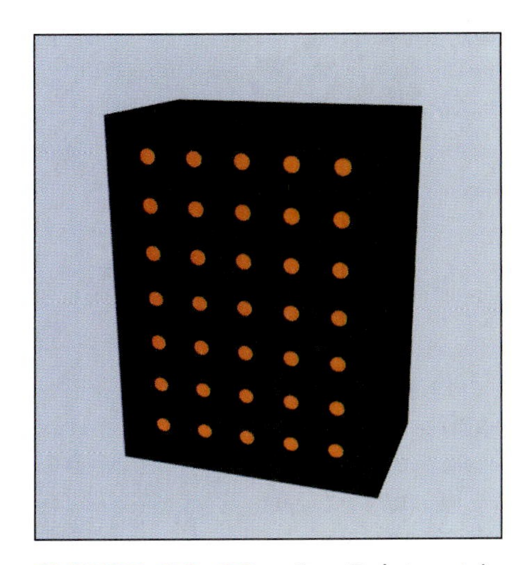

FIGURE 10–13 5 × 7 dot matrix display

Multisegment LED Displays

LEDs are very widely used in multisegment displays. Figure 10–11 shows the most commonly used multisegment display. Its seven segments are labeled as a, b, c, d, e, f, and g. The LED labeled *dp* is used to display the decimal point. By lighting a combination of different LEDs, any number from 0 to 9 can be displayed.

The seven-segment display cannot be used very efficiently to display all alphabets, so other multisegment displays have been developed to satisfy these requirements. The sixteen-segment display is illustrated in Figure 10–12, and the 5 × 7 dot matrix display is shown in Figure 10–13.

10.5 LED Testing

LEDs are usually damaged by excessive current flowing through them. Such damaged LEDs can be identified by a discoloration on the casing of the LED, due to the burnt junction.

The LED can be tested with a DMM (digital multimeter). In the diode check position, the DMM supplies about 2.5 V at very low current. If, at this setting, the cathode of the LED is connected to the negative lead of the DMM and the anode of the LED is connected to the positive of the DMM, the supplied 2.5 V should be sufficient to forward bias the LED and cause it to glow dimly.

LASER DIODE

10.6 Principles of LASER Generation

LASER is an acronym for **L**ight **A**mplification by **S**timulated **E**mission of **R**adiation. Most light sources are multichromatic; that is, they comprise many different colors. Each color corresponds to a different frequency of

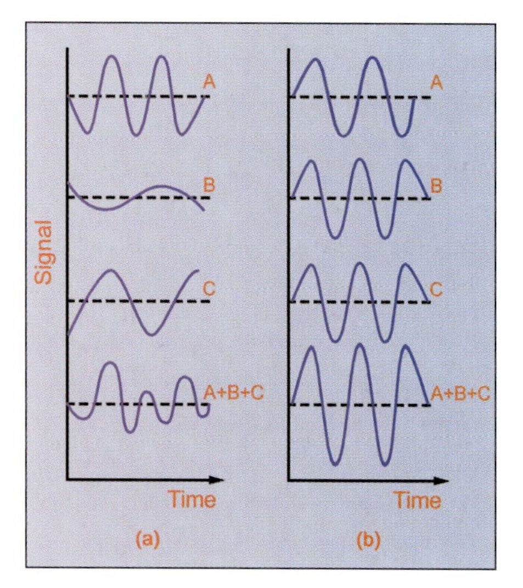

FIGURE 10–14 Light sources; a) incoherent, b) coherent

light. In an *incoherent* light source, most of these waves are out of phase with one another. Figure 10–14a shows how these waves, A, B, and C, add and strengthen each other at some points, while opposing and weakening each other at the other points. Most ordinary light sources diffuse, or spread out in all directions, so the light beam cannot travel very far.

Laser light sources, on the other hand, are monochromatic and have only a single color (frequency). Because all the waves are in phase, they constructively aid each other and form a coherent light source, as shown in Figure 10–14b. A laser beam can stay highly focused and travel very large distances.

In order for laser generation to take place, light waves must be of the same frequency, phase, and direction. This kind of emission of energy by the electron is termed stimulated emission. When a large number of electrons are subjected to stimulated emission, the amplification of light, as demonstrated in coherent light, takes place.

10.7 Basic Operation and Construction

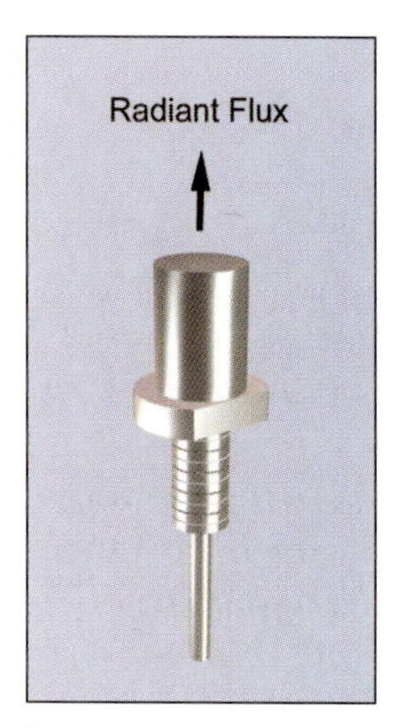

FIGURE 10–15 A typical ILD package

Laser diodes are light-emitting devices, capable of emitting laser beams. One of the commonly used laser diodes is the injection laser diode (ILD). A typical ILD package is shown in Figure 10–15.

The laser diode works exactly like an ordinary diode. Figure 10–16 shows the structure of an injection laser. The P- and N-type materials are made of AlGaAs (aluminium gallium arsenide). When forward biased, light is emitted at the PN junction. But, unlike the LED, the emitted light is coherent and monochromatic. Also, in the LED, the light emitted is scattered in all directions. But, in the laser diode, the light emitted can escape only from the end faces because the edges are roughened. In most laser diodes, one of the end faces is coated with a reflective material so that the radiation of the laser is emitted only in one direction.

The laser diode emits light when forward biased and can withstand only relatively small reverse-biased voltages. A high reverse voltage can damage or destroy the laser diode.

The schematic symbol for the laser diode is shown in Figure 10–17. Note that the light emission arrows are zigzagged rather than straight, as in the symbol for an ordinary LED.

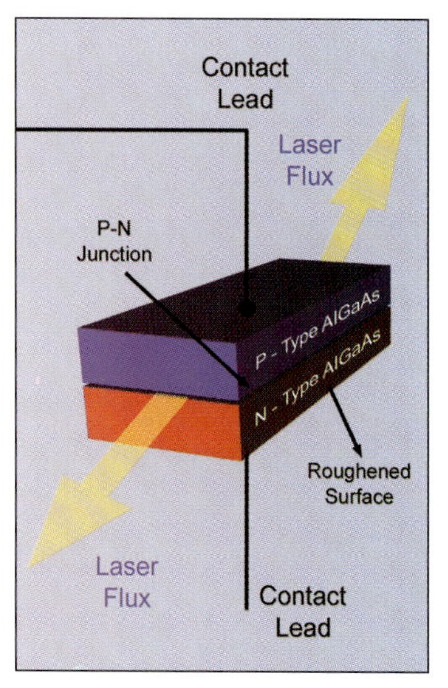

FIGURE 10–16 Structure of the injection laser diode (ILD)

10.8 Applications of the LASER Diode

Laser diodes are used primarily in fiber optic communications. It is the only device that is capable of producing optical energy high enough in concentration to pass through lengthy fiber optic cables. However, it also has some major disadvantages as compared to LEDs.

- They cost ten times more than LEDs.
- Their life expectancy is ten times smaller than that of LEDs.
- They require elaborate power supplies and consume much more power than LEDs.

Laser diodes do not compete with LEDs on the cost or reliability basis. Therefore, LEDs are used as much as possible in fiber optic systems, and laser diodes are used only when absolutely necessary.

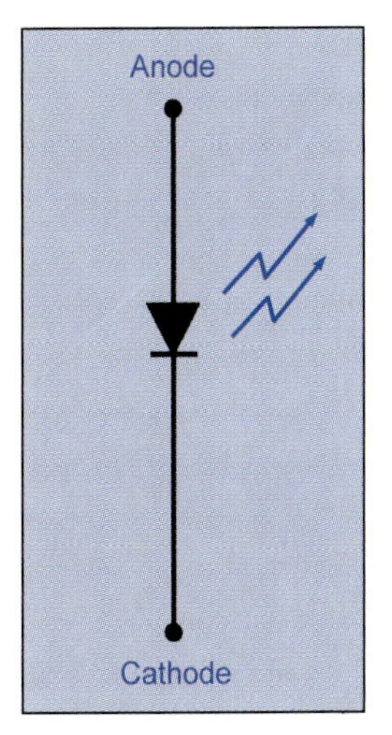

FIGURE 10–17 Schematic symbol for a laser diode

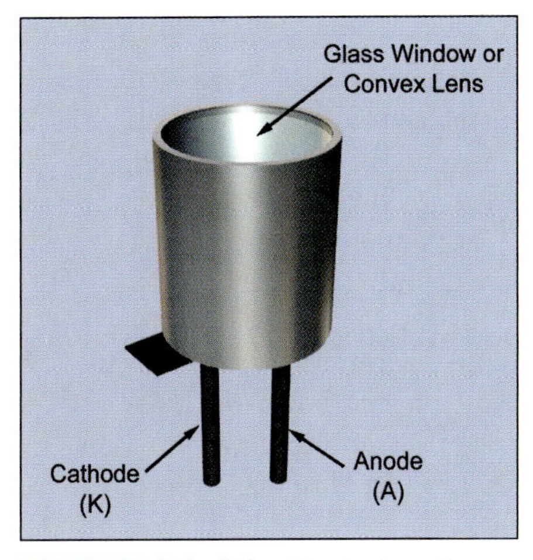

FIGURE 10–18 Typical package of a photodiode

PHOTODIODE

10.9 Basic Operation and Construction

A photodiode is a light-receiving device that contains a semiconductor PN junction. Figure 10–18 shows a typical photodiode package. A glass window or convex lens allows light to enter the case and strike the photodiode that is mounted within the glass case.

Figure 10–19 shows the construction of a photodiode. It is constructed basically with a P-type region that is diffused into the N-type region. The metal base makes connection between the cathode terminal and the N-type region, and the metal ring makes contact between the anode and the P-type region. Light enters the photodiode through a hole in the metal ring, which accommodates the glass or convex lens.

Figure 10–20 shows the most commonly used schematic symbol of a photodiode. Note that the two arrows point toward the photodiode, indicating that it responds to light.

Photodiodes can operate in two modes:

- Photovoltaic mode
- Photoconductive mode

Photovoltaic Mode

When operating in the photovoltaic mode, the photodiode generates a voltage in response to light. The incidence of light on the photodiode creates electron-hole pairs. The electrons generated in the depletion region are attracted to the positively charged ions in the N-type material, and the holes are attracted to the negatively charged ions in the P-type material. This creates a separation of charges, and a small voltage drop of about 0.45 volts is developed across the diode.

Figure 10–21 shows the photovoltaic mode of operation. In this mode of operation, the photodiode acts as a solar cell. If a load resistor is connected across the voltage source, a small current will flow from the cathode to the anode.

FIGURE 10–19 Construction of the PN photodiode

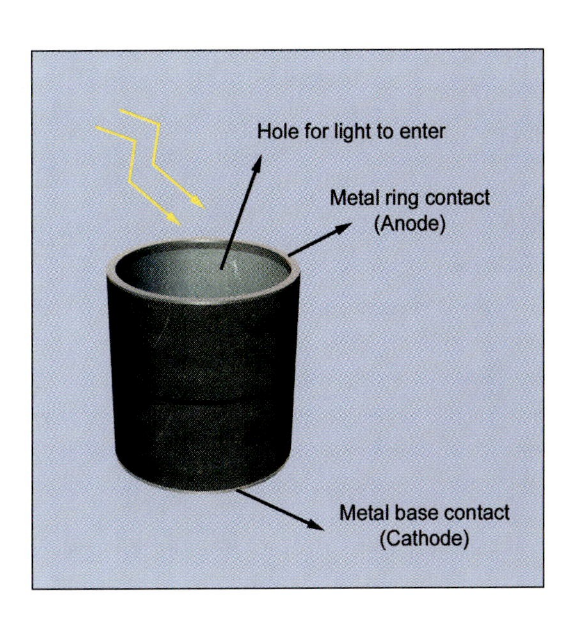

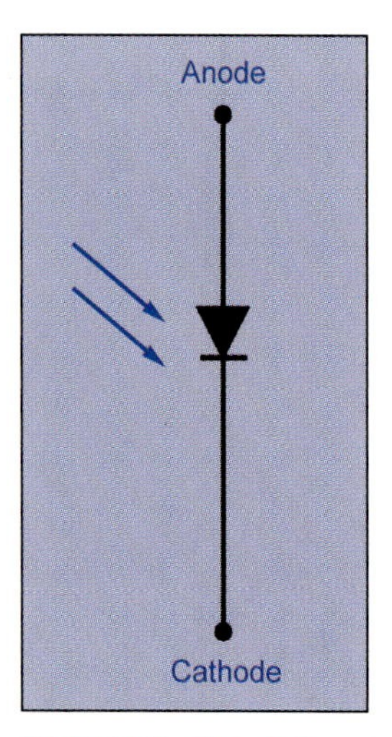

FIGURE 10–20 Schematic symbol of a photodiode

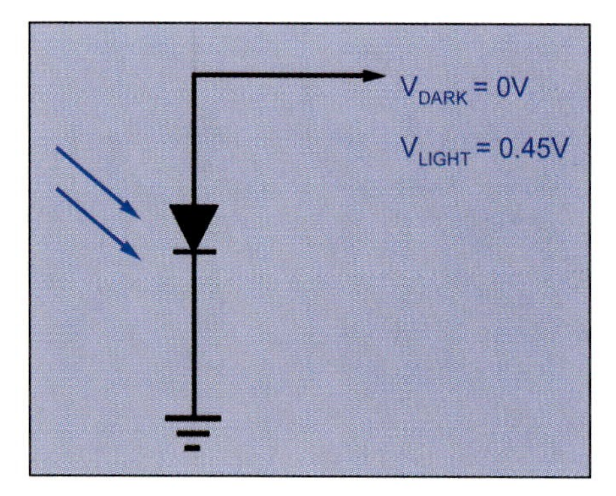

FIGURE 10–21 Photovoltaic mode of operation

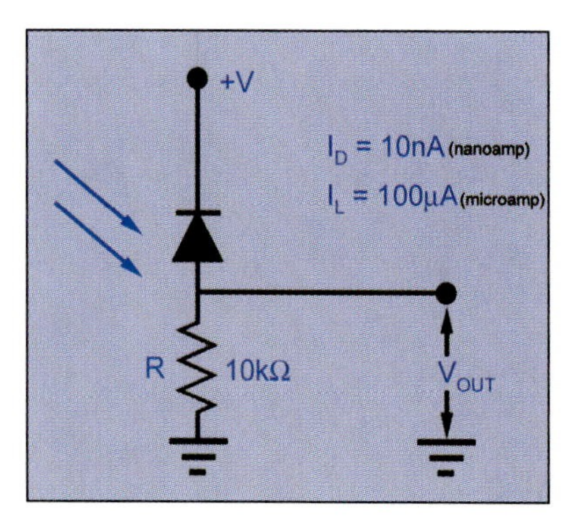

FIGURE 10–22 Photoconductive mode of operation

Photoconductive Mode

In the photoconductive mode, the conductance of the diode changes when light is applied. In this mode, the photodiode is reverse biased. Figure 10–22 shows a reverse-biased photodiode in the photoconductive mode. The depletion region of the reverse-biased photodiode is very wide, the resistance of the diode is high, and hence there will be only a small reverse current through it. This reverse current that flows through the diode when there is no light being applied is called dark current (I_D).

When light is applied, electron-hole pairs are generated. The electrons are attracted to the positive bias voltage, and the holes are attracted to the negative bias voltage. This movement of electrons and holes causes a considerable reverse current to flow through the photodiode. The resistance of the photodiode is very low when light is applied. If the intensity of light is increased, the resistance decreases and, therefore, the reverse current increases. The current that passes through the photodiode when light is being applied is called the light current (I_L).

The conductivity of the photodiode is low when there is no light applied, and the conductivity increases as the intensity of light increases; consequently, the magnitude of the dark current is very much smaller than that of the light current. Consider the example in Figure 10–22. Assume that the dark current (I_D) flowing through the diode is 10 nA and the light current (I_L) is 100 mA. The output voltage (V_{out}) is dependent on the amount of current flowing in the circuit.

With no light present,

$$V_{out} = I_D \times R = 10 \text{ nA} \times 10 \text{ k}\Omega = 100 \ \mu\text{V}$$

With light present,

$$V_{out} = I_L \times R = 100 \ \mu\text{A} \times 10 \text{ k}\Omega = 1 \text{ V}$$

10.10 Applications

Solar Cells

A solar cell is a photodiode operated in its photovoltaic mode. In its most common application, it is used to convert light energy to electrical energy and thereby serves as a DC voltage source. In solar power applications, solar cells are connected in series/parallel arrays. Parallel connections increase the total current supplied by the array, whereas series connections increase the voltage. For example, in Figure 10–23, if each solar cell is capable of delivering about 0.8 A of current at 0.5 V of voltage, the entire 5 × 4 array would be capable of supplying a total of 4 A of current at a voltage of 2.0 V.

The current generated in the solar panels can also be used to charge batteries during the day, so that the stored power can be consumed when it is dark. Solar cells thus can be used to store energy during the day to light streetlamps at night.

Optocouplers

Optocouplers are devices that use light to optically couple a signal between two electrically isolated points. Optocouplers are also called op-

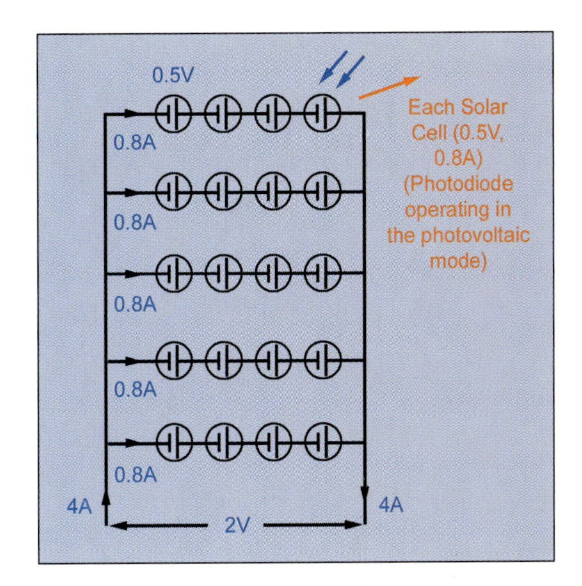

FIGURE 10–23 Solar panel

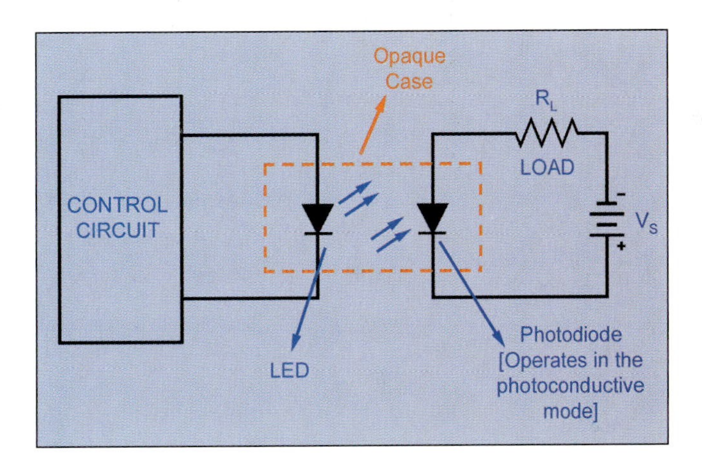

FIGURE 10–24 An optocoupler used to control load current

toisolators. These devices include a light-emitting device and a light-sensing device, both available in one package.

Figure 10–24 shows an optocoupler that uses an LED and a photodiode. The two devices are placed together in an opaque case. The photodiode operates in the reverse-biased photoconductive mode. A control circuit forward biases the LED. When the LED lights up, the light strikes the photodiode, increasing its conductivity. This results in current flowing through the load resistor R_L. When the LED is turned off, the conductivity of the photodiode is low and there is no current through the load. The load current is controlled by the control circuit even though there is no direct electrical connection. Such a system is particularly useful for isolating control circuit noise from a sensitive electronic load circuit.

PHOTORESISTOR

10.11 Basic Operation and Construction

A photoresistor is a light-detecting device and is also called a photoconductive cell or light-dependent resistor. It is a passive device composed of a semiconductor material that changes resistance when its

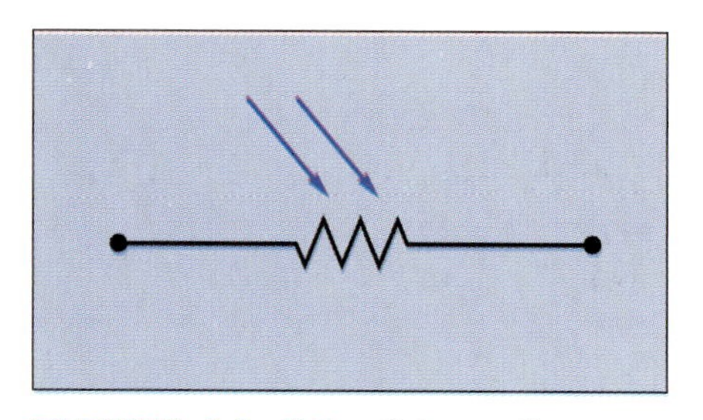

FIGURE 10–25 Schematic symbol of a photoresistor

surfaced is exposed to light. The schematic symbol for a photoresistor is shown in Figure 10–25.

In conventional semiconductor devices, electron-hole pairs are created by heat energy. In photoresistors, the semiconductor material is light sensitive and free electrons are created by light energy. With the creation of free electrons, the resistance of the material drops. The greater the intensity of light, the greater will be the number of free electrons and the smaller will be its resistance.

Photoresistors are the simplest and the least expensive in the class of optoelectronic devices. A typical construction is shown in Figure 10–26. The light-sensitive semiconductor material is arranged in a zigzag strip whose ends are attached to external terminals. The semiconductor material is either cadmium sulfide (CdS) or cadmium selenide (CdSe). A glass or transparent cover is attached for light to pass through to the device. The resistance of a photoresistor is largest when it is not conducting and is called the dark resistance.

10.12 Applications

Photoresistors are used in lighting controls, automatic door openers, alarm-activating circuits, and other such applications. An example of a photoresistor streetlight control circuit is shown in Figure 10–27. The opening and closing of the relay contact of Figure 10–27 is controlled by the photoresistor. A minimum amount of energizing current flowing through the relay coil is required to activate the switching.

The relay contact is normally closed. At night, when there is no light incident on the photoresistor, its resistance is large, making the current through the photoresistor and the coil very small. The relay is therefore not activated, and the contact remains in its normally closed position. This causes the streetlight to be connected to the AC source.

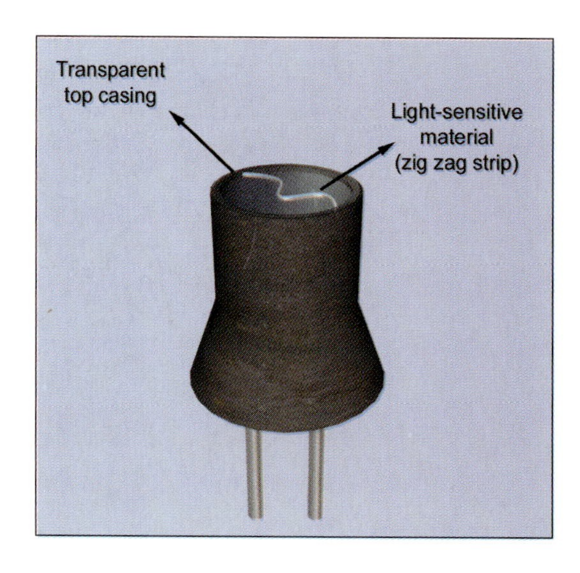

FIGURE 10–26 Structure of a photoconductive cell

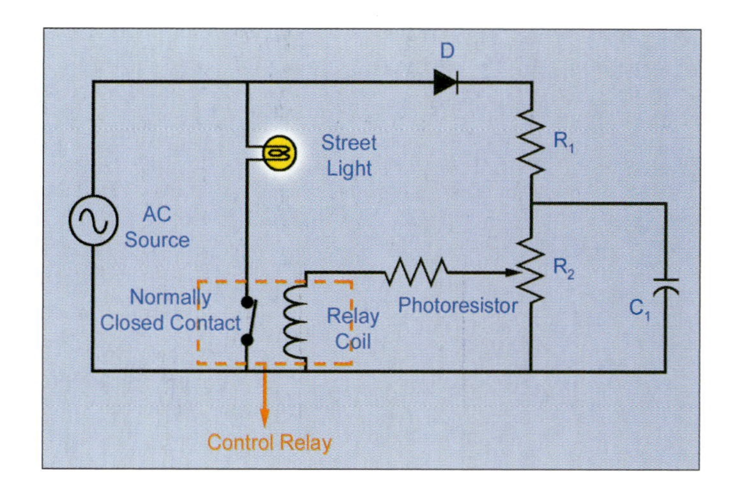

FIGURE 10–27 Photoresistor used to control a streetlight

However, during daylight, when there is sufficient light striking the photoresistor, its resistance drops, thus increasing the current through the coil. This current is sufficient to energize the relay coil, and therefore the relay contact changes its position from the normally closed position to the open position. This results in the streetlight being cut off from the source. Note that the diode D, resistors R_1 and R_2, and capacitor C_1 form a half-wave rectifier circuit with filter that supplies a DC voltage to the photodiode and relay circuit.

OTHER OPTOELECTRONIC DEVICES

10.13 Phototransistor

A phototransistor is a light-detecting device that is also called a photo sensor. It is a transistor whose base current is supplied by the carriers generated due to the incident or striking light. The collector current of the transistor is thus controlled by the intensity of light. The schematic symbol of a phototransistor is shown in Figure 10–28.

Although the phototransistor is a two-terminal device like the photodiode, it produces a higher output current than the photodiode. The photodiode, however, has a faster response time. Thus, the phototransistor is preferred in high-current applications, whereas the photodiode is used in high-speed operations.

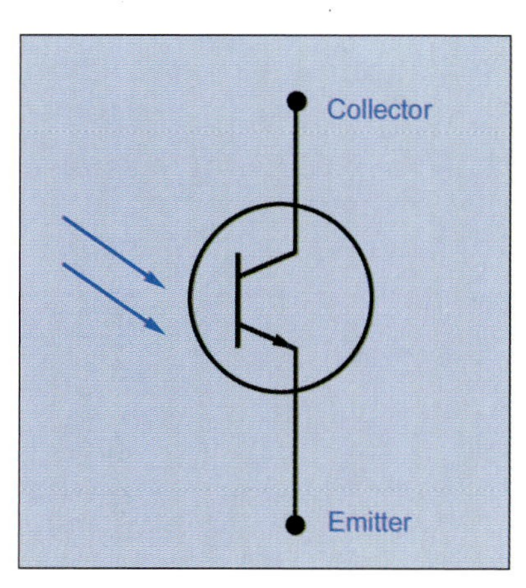

FIGURE 10–28 Schematic symbol of a photoresistor

10.14 Photodarlington

A photodarlington is a phototransistor packaged with another transistor connected in the Darlington configuration. The schematic symbol of the photodarlington is shown in Figure 10–29. Because of its large current gain, the photodarlington produces greater output current than either the photodiode or phototransistor. However, the response of the photodarlington is slower than both the photodiode and the phototransistor.

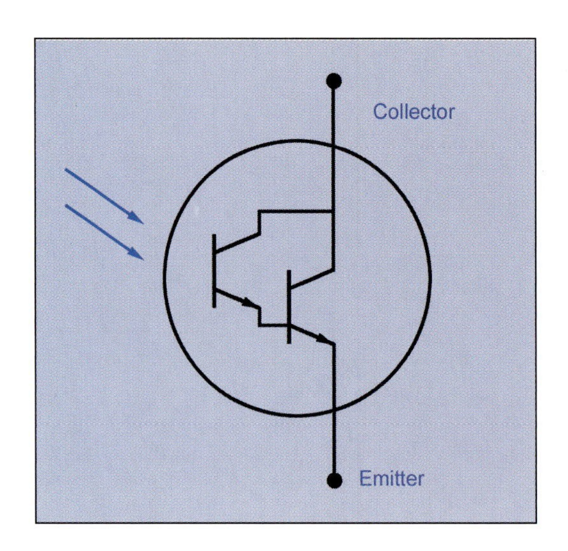

FIGURE 10–29 Schematic symbol of a photodarlington

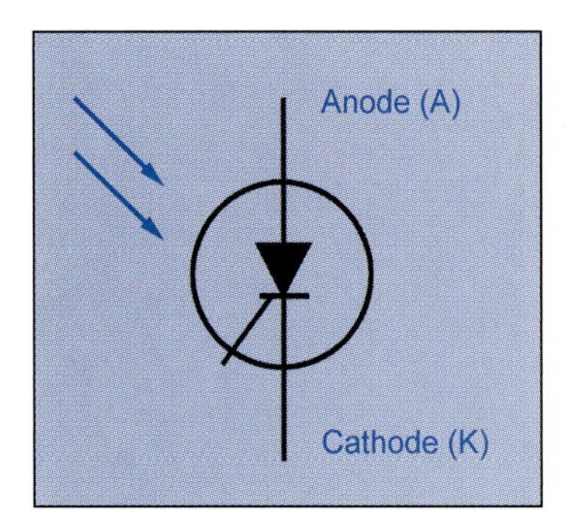

FIGURE 10–30 Schematic symbol of the LASCR

10.15 LASCR

LASCR is an acronym for **L**ight-**A**ctivated **S**ilicon-**C**ontrolled **R**ectifier. The incoming light strikes the photosensitive surface of the device, which serves as the gate signal that triggers the LASCR. The LASCR can be used in an optically coupled phase control circuit. Figure 10–30 shows the schematic symbol of the LASCR.

SUMMARY

This chapter introduces you to some of the most commonly used optoelectronic devices. The lesson should enable you to identify optoelectronic devices and describe their basic operations. The discussion in this chapter also acquaints you with applications of light-emitting and light-detecting devices.

Some of the key topics in this chapter include:

- Light-generation devices—those that create and/or modify light
- Light-detecting devices—those whose characteristics are changed or controlled by light
- Light is a form of electromagnetic radiation or electromagnetic waves
- Wavelength is defined by the formula $\lambda = \frac{c}{f}$
- Light-emitting diode basic operation, construction, and identification
- Multicolor LED's operation based on the polarity of the biasing
- Seven-segment, sixteen-segment, and 5 × 7 dot matrix displays
- LASER (**L**ight **A**mplification by **S**timulated **E**mission of **R**adiation) operation and construction
- Incoherent and coherent light sources
- Photodiode in both photovoltaic and photoconductive modes
- Voltage and current summation of solar cell arrays
- Input isolation of a control circuit by means of optocoupler devices
- Photoresistors and how their resistance varies with the intensity of light
- Phototransistor and photodarlington transistors versus photodiodes
- LASCR (**L**ight-**A**ctivated **S**ilicon-**C**ontrolled **R**ectifier)

REVIEW QUESTIONS

1. What is the optical light spectrum?
   a. Frequency and wavelength
   b. Visible versus invisible
   c. Infrared and ultraviolet
2. The infrared band is given as 30 THz to 400 THz. What is this in wavelength? Answer the same question for visible light and ultraviolet light.
3. Define, in your own words, the differences between coherent and incoherent light.
4. Discuss the operation of the LED.
   a. How does it emit light?
   b. When does it emit light?
   c. How does the circuit of Figure 10–9 work?
5. A certain piece of equipment shows a green light when it is working properly and a red light when it is not working properly. The light appears to come from the same device. How might this work?
6. A particular fiber optic application will use a very long length of fiber optic cable. Would this application use an LED or a laser diode? Why?
7. Discuss the two modes of operation of a photodiode.
8. A certain control circuit is being used to control the operation of a high-power motor starter. The motor starter generates enormous amounts of noise that might interfere with the control system. Discuss how the circuit might be designed to isolate the noise from the controls.
9. Redraw the circuit of Figure 10–27 so that it works with a normally open contact instead of normally closed.
10. You wish to operate a door opener with a light source. Draw a block diagram of how this might be done.

Fiber Optics and Fiber-Optic Cable

OUTLINE

OVERVIEW

Light travels in a straight line when not bothered by an outside influence such as water, oil, or reflective surfaces. Microscopes, telescopes, and cameras are examples of devices that use the "straight line" properties of light. Other applications such as periscopes require that light be bent around corners or refracted at an angle.

William Wheeler patented the idea of "piping" light in 1880. His idea was to use a bright arc light and pipes with highly reflective inside surfaces. Figure 11–1 shows how he "piped" light to different rooms in a house using a single light source.

Ideas and uses for fiber optics developed from these first inventions; however, until recently, the use of light to carry information or data was limited to flashers and other such direct visual devices. Then, in 1977, GTE and AT&T started using fiber cables to carry telephone signals on a specially modulated light beam. Today, fiber optics are used in virtually all areas of health, business, industry, communications, and government. Uses vary from internal patient examinations **(endoscopy)**, machine process control, and monitoring, to satellite communications.

A fiber-optic cable is made of spun glass or a special transparent plastic. It does not conduct electricity, but rather it conducts "electrically encoded" light signals—such as those generated by a laser diode. Fiber-optic signals are virtually immune to electrical interference and noise caused by high-power applications or high-magnetic fields. Because they do not use an electrical signal, the shock hazard that might be present with copper wire is virtually eliminated.

OBJECTIVES

After completing this chapter, the student should be able to:

1. Describe the components of a fiber-optic system.
2. Explain the mechanism used by fiber-optics to 'transmit' electrical signals.
3. Identify types of fiber optic cables and connectors.
4. Explain the uses and applications of fiber optics in an electrical system.

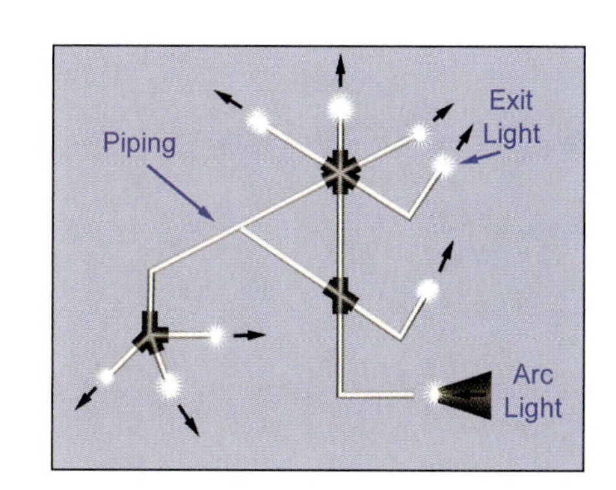

FIGURE 11–1 Patented light-piping system for lighting a house

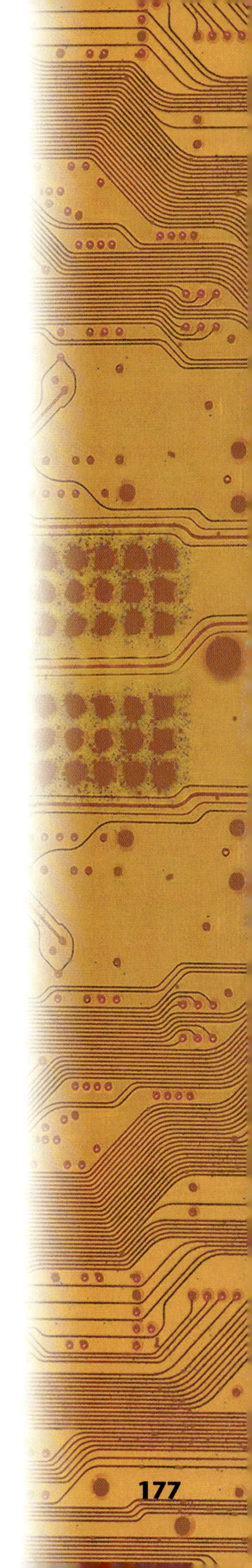

GLOSSARY

Attenuation Reduction of signal strength, usually through losses.

Endoscopy Endoscopy is a minimally invasive diagnostic *medical* procedure used to evaluate the interior surfaces of an organ by inserting a small scope in the body, usually through a natural body opening.[1]

Modulation To attach information to a signal by varying its frequency, amplitude, or other characteristic. The information is attached at a transmitter and recovered at the receiver.

Reflect To throw or bend back (light, for example) from a surface.[2]

Reflection The act of reflecting or the state of being reflected.[3]

Refraction The turning or bending of any wave, such as a light or sound wave, when it passes from one medium into another of different density.[4]

[1]Excerpted from *Wickipedia.org*
[2]Excerpted from *American Heritage Talking Dictionary.* Copyright © 1997 The Learning Company, Inc. All Rights Reserved.
[3]Excerpted from *American Heritage Talking Dictionary.* Copyright © 1997 The Learning Company, Inc. All Rights Reserved.
[4]Excerpted from *American Heritage Talking Dictionary.* Copyright © 1997 The Learning Company, Inc. All Rights Reserved.

FIBER-OPTIC FUNDAMENTALS

11.1 Principles of Operation and Construction

Reflection and Refraction

In free space, light waves travel in a straight line at approximately 300,000,000 m/s (or about 186,000 mi/s). When light traveling in one medium (e.g., air) strikes a different medium (e.g., water or glass), some of it is rejected by the new medium and **reflects** back. The remainder of the light passes into the new medium; however, because light travels at a different speed in the new medium, it is refracted or bent.

Consider Figure 11–2. An incident light beam is directed at the surface of a swimming pool. When the light beam strikes the water, some of it is reflected back, as shown by the reflected light beam. Notice that the angle of incidence (45°) is equal to the angle of **reflection**.

Some of the light passes into the water; however, because the light travels more slowly in water than in air, it is bent. The angle of **refraction** (26.6°) is different than the angle of incidence. The actual angle of refraction depends on the materials, the angle of incidence, and the refractive index of the material.[5] For an example of refraction, think about looking at a fish from the docks of a clear lake. The actual location of the fish is deceiving to your eyes because light is refracting in the water.

Total Internal Reflection

Total internal reflection is the fundamental principle upon which fiber optics is based. When light crosses an interface into a medium with a higher index of refraction (n), it bends toward the normal. This is shown in Figure 11–2, where water has a higher index of refraction

FIGURE 11–2 Properties of reflected and refracted light

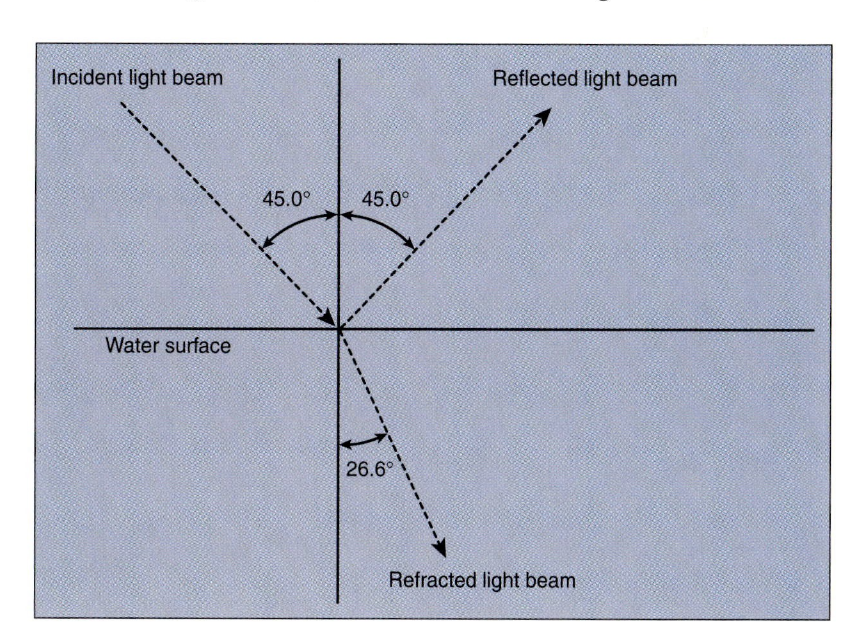

[5]The refractive index (n) is given by the formula $n = \frac{c_v}{c_m}$, where c_v is the speed of light in a vacuum and c_m is the speed of light in the material that it is entering.

than air. Conversely, light traveling across an interface from higher *n* to lower *n* will bend away from the normal. This means that at some angle, known as the critical angle, light traveling from a medium with higher *n* to a medium with lower *n* will be refracted at 90°; in other words, refracted along the interface. If the light hits the interface at any angle larger than this critical angle, it will not pass through to the second medium at all. Instead, all of it will be reflected back into the first medium, a process known as total internal reflection. This means that light can be confined inside glass or any other transparent substance, provided that the material has a higher index of refraction than the material that surrounds it.

In 1841, Swiss physicist Daniel Colladon developed an example of how this refractive index (material density) could contain a light signal. Figure 11–3 shows a drawing of the experiment that he designed.

The key to making this "total internal reflection" principle work was cladding. Cladding is a method of covering the glass fiber to ensure that all the light is internally reflected. If the cladding has a lower refractive index than the carrier, total internal reflection will occur. The first successful glass cladding was accomplished in 1956 by Larry Curtiss from the University of Michigan. Figure 11–4 shows the difference between clad and unclad fibers carrying a light signal.

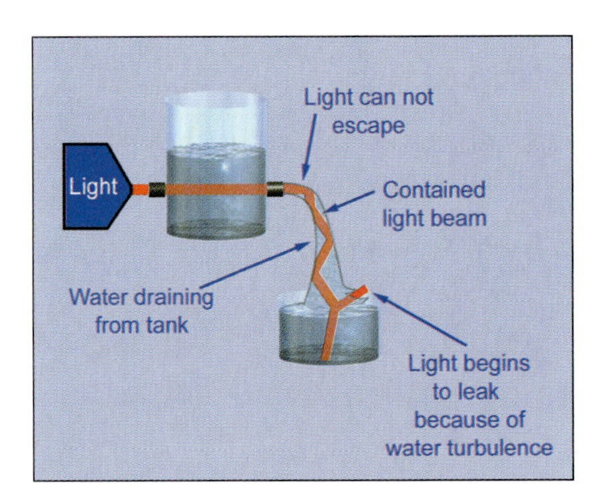

FIGURE 11–3 Total internal reflection

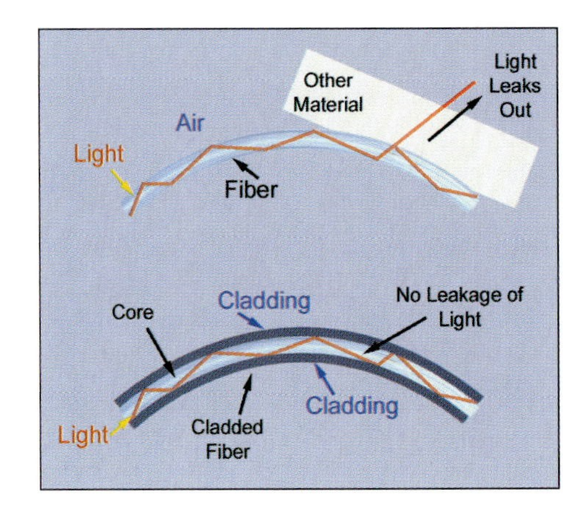

FIGURE 11–4 Clad and unclad fibers

Optical Fiber Construction

The typical fiber-optic cable is made up of several components (see Figure 11–5). The core and the cladding are the components that actually channel the light. Ambient environment protection, moisture protection, and strength are provided by the other components.

As discussed earlier, the glass fiber itself operates on the principle of total internal reflection. Light entering the end of the glass (or plastic) fiber is carried through the glass and is received on the other end. The light is directed down the core. The cladding, which surrounds the core, is also glass but has a higher index of refraction and conducts no useful light. The light-carrying portion (center) of the cable is very small and is typically measured in micrometers (μm). One common cable size, for example, is 62.5/125, which means the core diameter is 62.5 μm and the cladding diameter (outside) is 125 μm. By comparison, a human hair is approximately 100 μm. Figure 11–6 is a diagram of such a cable.

The difference in refraction between the core and the cladding is what allows fiber-optic cable to work. As long as the angle at which the light strikes the boundary (the angle of incidence) is greater than the critical angle, light striking the core-to-cladding boundary is reflected back into the fiber. The angle of incidence is determined by measuring the angle to the axis perpendicular with the surface of the fiber and angle of the incoming light (see Figure 11–7). The critical angle varies among different types of glass (indexes of refraction) and different wavelengths (color) of light. Note, the angle of reflection (∡ R) is always equal to the angle of incidence (∡ I).

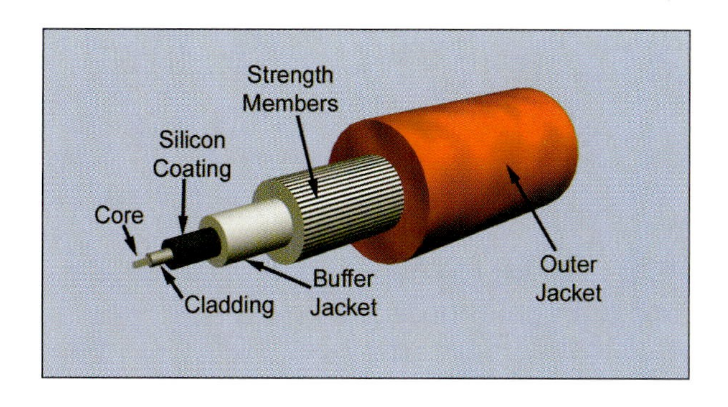

FIGURE 11–5 Construction of fiber-optic cable

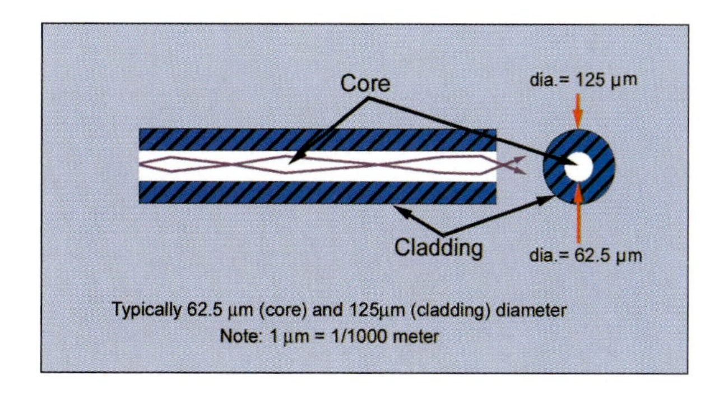

FIGURE 11–6 Internal reflection

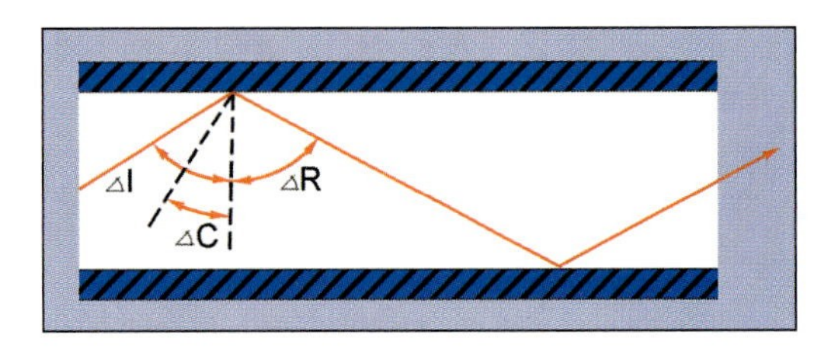

FIGURE 11–7 Critical angle, angle of incidence, angle of reflection

All light striking the core-cladding boundary will be reflected back into the core, as long as the angle of incidence (∡ I) is greater than the critical angle (∡ C).

Optical fiber cables are designed in two broad categories: single mode and multi mode. The physical difference between these two is primarily in the diameter of the core. A typical single-mode fiber, for example, is 8/125, where the core diameter is only 8 μm. Single-mode fiber is more expensive to install and requires more sophisticated equipment to communicate; however, single-mode fiber has considerably lower loss at its design frequency. Normally, single-mode fiber is used only when signals are traveling long distances or when extremely low loss is required. The single-mode fiber core is so small that only a very few modes (rays) of light can travel down the fiber. This mostly straight-shot effect results in minimum modal dispersion and a clean nondistorted signal at the other end. Single-mode fiber is the most expensive configuration in regard to electronic components and terminations, but it provides the maximum amount of information.

Multimode fiber, in addition to having a larger core, is also subdivided into two categories, depending on how the core-to-cladding boundary is implemented. One type of fiber, called step-index multimode fiber, has a single transition between the core and the cladding glass. The light in the core reflects off the cladding and propagates (travels) along the fiber. The drawback is that several light rays enter into the core at different angles, therefore traveling various paths and various distances. As illustrated in Figure 11–8, these different light rays arrive at different times due to their different distances traveled. The different arrival times at the end of the fiber cable lower the ability of the electronic equipment to decipher the signal, therefore resulting in a slower communication rate.

The second type of multimode fiber is the graded-index multimode fiber. Graded-mode fiber has a refractive index that changes gradually from the core to the cladding. As you will study in the next section,

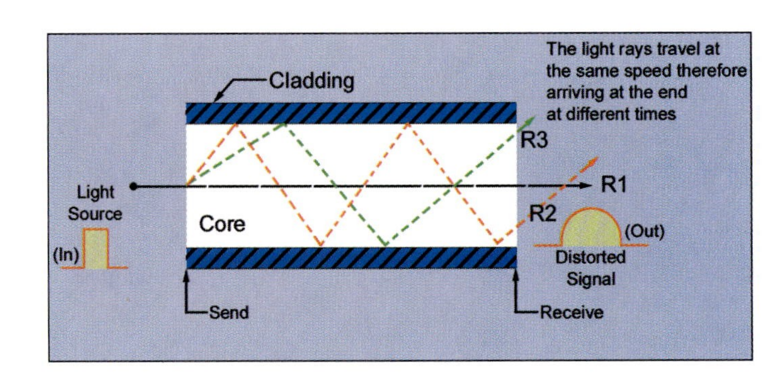

FIGURE 11–8 Step-index multimode fiber

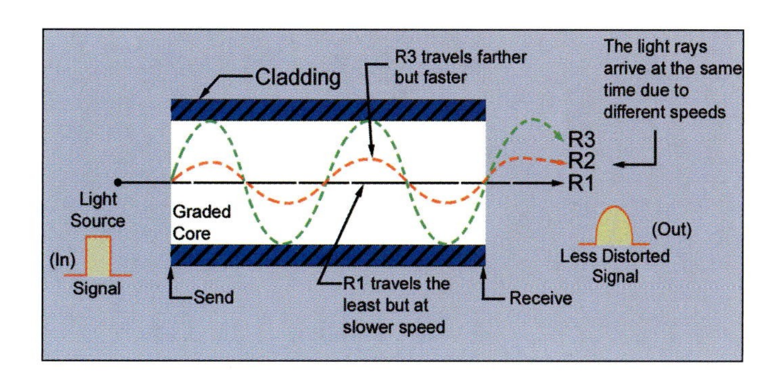

FIGURE 11–9 Graded-index multimode fiber

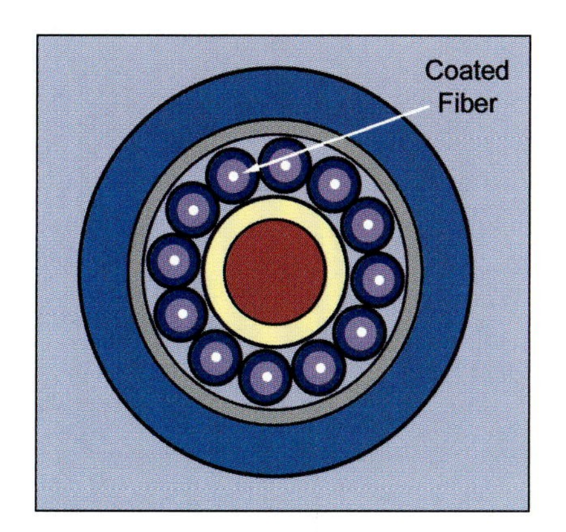

FIGURE 11–10 Fiber-optic bundles

light rays will travel at different speeds through different levels of refractive indexes. As indicated in Figure 11–9, different light rays entering into the core travel at different speeds throughout the length of fiber as it passes through gradual levels of refractive indexes. This has the effect of giving the slower modes a shorter distance to travel. The result is that all the rays reach the other end of the fiber at nearly the same time, therefore improving signal quality.

The fibers are made of pure glass. The glass may have small levels of doping called impurities to increase or decrease the refractive index. The glass is usually made of silicon dioxide (silica, SiO_2). Some special fibers are made of plastic. These fibers are less clear than glass but are more flexible and easier to handle.

Fibers usually come in bundles. Bundles are of two types: flexible or rigid. The flexible bundle is usually surrounded by a protective plastic coating, and at the ends of the cable the individual fibers are tied or joined together. In the rigid bundle, the individual fibers are melted together into a single rod and are shaped during the manufacturing process. Figure 11–10 shows a flexible fiber-optic bundle.

11.2 Optical Fiber Characteristics

To understand how a fiber-optic system operates, you need a fundamental knowledge of three areas: optics, electronics, and communications. In physics, light is treated as either electromagnetic waves or as photons (electromagnetic energy particles). For this discussion, we will concentrate on the electromagnetic wave characteristics of light. The light spectrum (light measured as a wave or electromagnetic frequency) is quite small when compared to the entire spectrum range. Figure 11–11 is a chart of the electromagnetic spectrum.

As you can see, there is only a small area of the spectrum that we will consider when dealing with fiber optics—the so-called optical spectrum from infrared to ultraviolet frequencies.

As shown earlier in this chapter, light bends as it passes from one material to another. The bending is caused by the change in speed. The change in speed is the way that the refractive index is calculated, as given by the formula

$$n = \frac{c_v}{c_m}$$

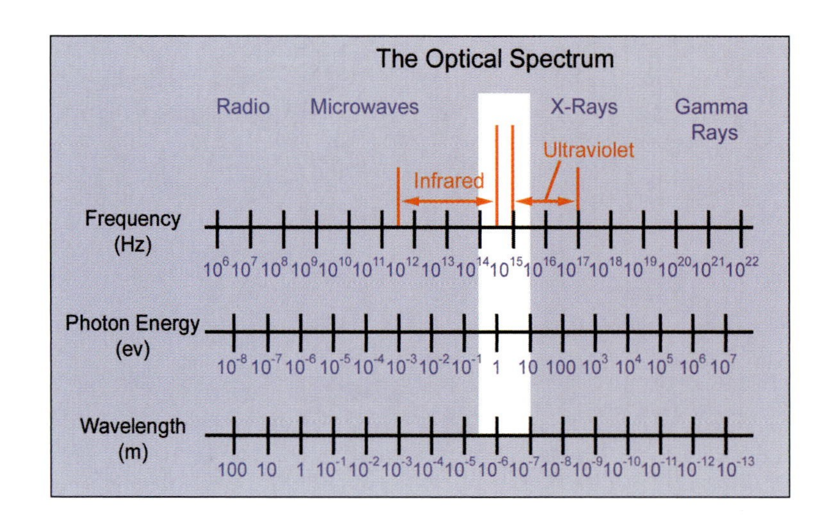

FIGURE 11–11 Electromagnetic spectrum

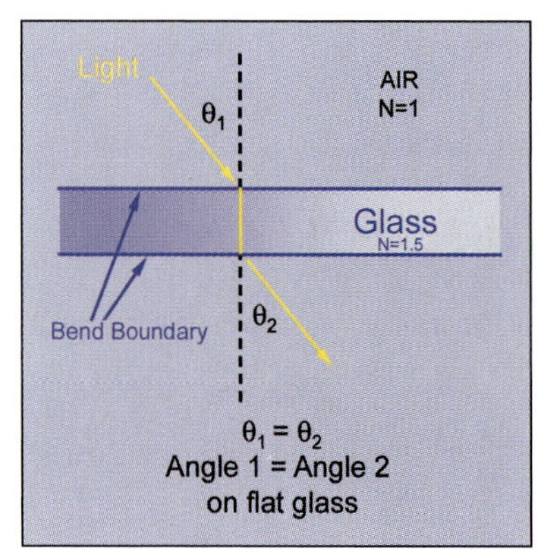

FIGURE 11–12 Light refraction through glass

Where:

c_v is the speed of light in a vacuum

c_m is the speed of light in the other material

This bending is shown in Figure 11–12.

Another critical characteristic of an optical fiber is its acceptance angle, which makes up its cone of acceptance (see Figure 11–13). Only light (or light signals) that enter within the cone of acceptance (the required angles) will be transmitted through the optical fiber. Light or signals trying to enter outside these angles will be rejected. The acceptance angle is called the *NA*, or numerical aperture, and is calculated by using the following formula:

$$NA = \sqrt{n_0^2 - n_1^2}$$

Where:

n_0 is the core refractive index

n_1 is the cladding refractive index

Typically, *NA* values vary from 0.2 to 0.5.

The following example shows how to find NA for an optical fiber cable.

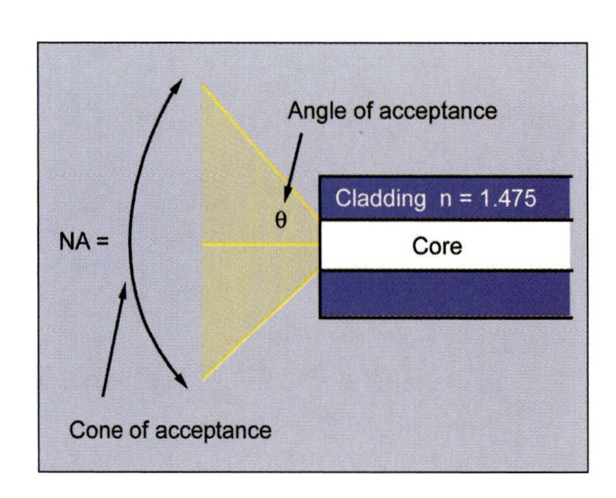

FIGURE 11–13 Optical fiber's acceptance angle

EXAMPLE 1

The glass core of a fiber cable has an n of 1.5, its cladding has an n of 1.475. Find the NA for the fiber.

$$NA = \sqrt{n_0^2 - n_1^2} = \sqrt{1.5^2 - 1.475^2} = \sqrt{2.25 - 2.18} = \sqrt{0.7} = 0.26$$

FIBER-OPTIC SIGNALS

11.3 Signal Transmission

From the discussion on *NA* (numerical aperture), you might guess that light transmission is not 100 percent efficient. The signal is attenuated because of light loss. This is caused by a number of factors, including:

- Light scattering in the core
- Light leakage from the core to the environment
- Light absorption by the fiber material

Recall that **attenuation** is loss of signal strength. To measure the signal strength loss, we use a familiar unit of measurement, the decibel (dB). The two formulas that relate decibels to power levels and power levels to decibels are

$$\text{dB}_{\text{loss}} = 10 \times \log_{10}\frac{P_{\text{out}}}{P_{\text{in}}}$$

and

$$\frac{P_{\text{out}}}{P_{\text{in}}} = 10^{\frac{\text{dB}}{10}}$$

For instance, if a 1-mile length of fiber cable has a 10dB/mile loss, the amount of energy output compared to the input can be calculated as

$$\frac{P_{\text{out}}}{P_{\text{in}}} = 10^{\frac{\text{dB}}{10}} = 10^{\frac{-10}{10}} = 0.1$$

Note that the dB loss is given a minus sign (−). From this formula, you can see that only 10 percent of the light entering the cable will come out the other end. If you add another mile to the cable and assume the same loss rate, only 1 percent of the signal is present at the end of 2 miles. Expressed another way, there was a 20 dB loss over the 2-mile cable length.

Low attenuation or signal loss is possible at frequencies of light near the infrared spectrum (1300 nm to 1550 nm). When transmitting a light signal at the high end of this frequency, 1 percent of the light is still available after 50 miles. Electric wires (copper) and coaxial cable have higher attenuation rates as the frequency of the electrical signal increases. This is not the case for the optical fiber.

Figure 11–14 shows the comparison graph between the coaxial cable and fiber-optic cable. The reason light loss (attenuation) is unaffected

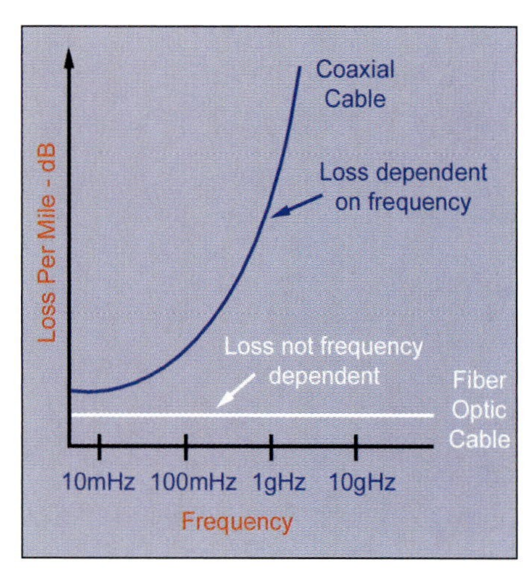

FIGURE 11–14 Different cable signal losses

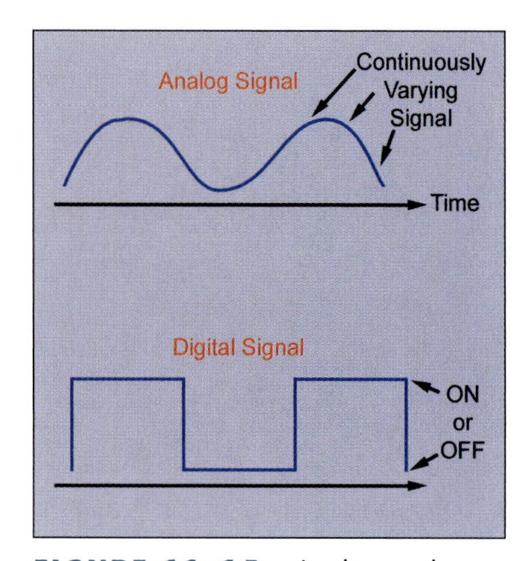

FIGURE 11–15 Analog and digital signals

by signal frequency changes is that the light is modulated to carry the signal, but the light essentially operates within its frequency spectrum and maintains a constant attenuation or loss rate. This is not true of copper or coaxial cable. At higher electrical frequencies, signal loss is caused by counterelectromagnetic fields, heat loss, and increased skin resistance.

11.4 Signal Modulation

Recall that there are two basic types of signal **modulation**. One type varies the intensity or amplitude of the carrier to encode the signal. The other type varies the frequency of the carrier to encode the signal. For radio or TV, the carrier is typically a single frequency transmitted steadily. For optical signals, it is a beam of light. Both the electrical signal and the light signal can be modulated using an analog or digital signal. Today, most fiber-optic systems use digital modulation.

To transmit a fiber-optic signal, the intensity of the light source must be modulated. This is done in one of two ways. The input power to the light source can be changed, and thus the intensity of the light beam is changed, or the intensity of the light beam can be changed after it leaves the light source. Figure 11–15 shows the difference between an analog and a digital signal.

Light Sources

There are three major types of commercially used light sources for fiber-optic signal generation, as shown in Table 11–1.

Direct Modulation

Direct modulation is very simple and works very well for LEDs and semiconductor lasers. The principle of operation is based on changing the current through the semiconductor. Turning the semiconductor on

Table 11–1 Light Sources for Fiber-Optic Systems

Type	Use
Visible light red LEDs	Used with plastic fibers because plastic transmits the visible light wavelengths better than the glass. LEDs are used with systems that transmit short distances at slow to medium speeds.
Near-infrared LEDs Gallium arsenide (GaAs) lasers	These LEDs and lasers use glass fibers and are good for short distances and medium data speeds.
Indium gallium arsenide phosphide (InGaAsP) lasers	These lasers are the most common for telecommunications and transmit at 1,300 nanometers with a dB loss of only .35 dB/km.

FIGURE 11–16 Input current versus output light

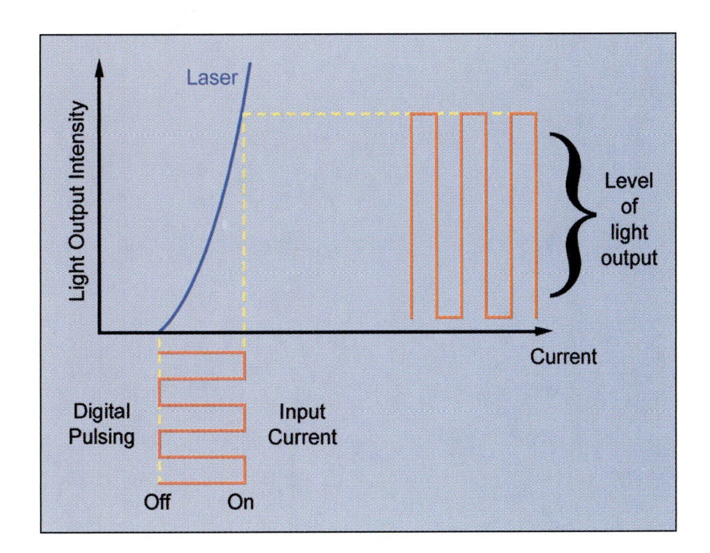

and off with the digital signal creates a series of light pulses that are identical to the electrical pulses. Figure 11–16 shows how this relationship works.

FIBER-OPTIC SYSTEMS

11.5 System Overview

Figure 11–17 shows a block diagram of a simple fiber-optic system. Note that the incoming electrical signal is converted to a light signal, transmitted a distance, received, and converted back to an electrical signal. Fiber-optic cables used for data transmission typically carry light signals at levels of 100 microwatts or less.

Fiber-optic system components that generate light are called light emitters. LEDs and semiconductor lasers are light emitters. Light receivers and devices that convert light back to electrical energy are called light detectors or photodetectors. Photodiodes, phototransistors (photodarlington), and light-activated SCRs are some of the common photodetectors. Figure 11–18 shows the schematic symbols for some common photodetectors.

One application of the coupling the output of an LED to the input of a phototransistor (optocoupling) is shown in Figure 11–19. Here, the opaque case (opaque means that light is not allowed to penetrate the case) allows the light from the LED to activate the phototransistor only

FIGURE 11–17 Simple fiber-optic system

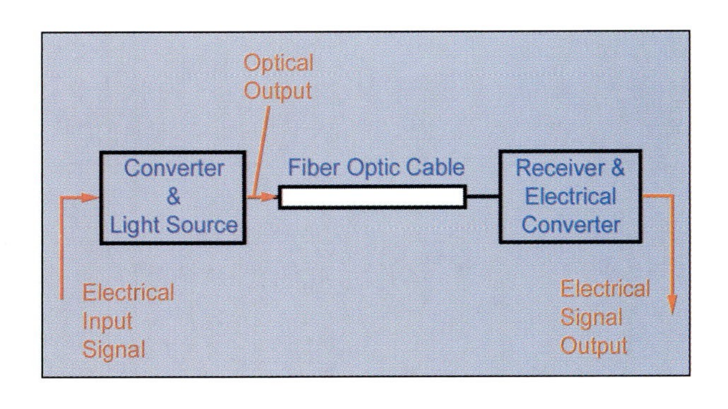

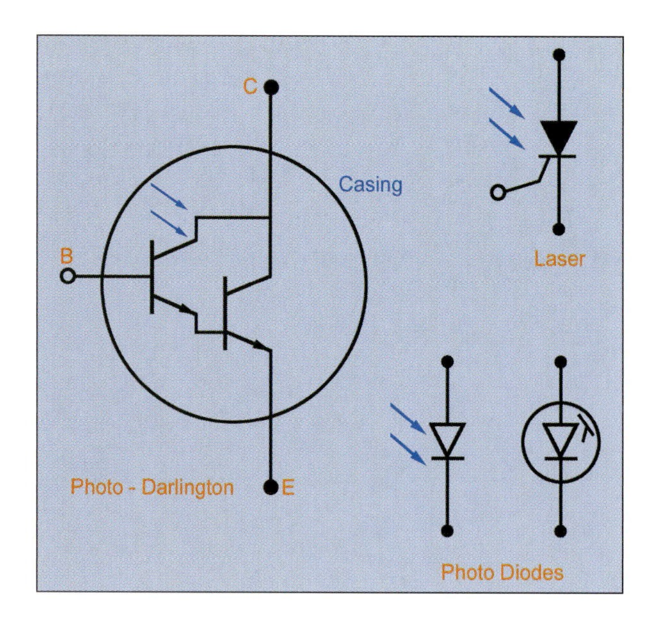

FIGURE 11–18 Schematic symbols for photodetectors

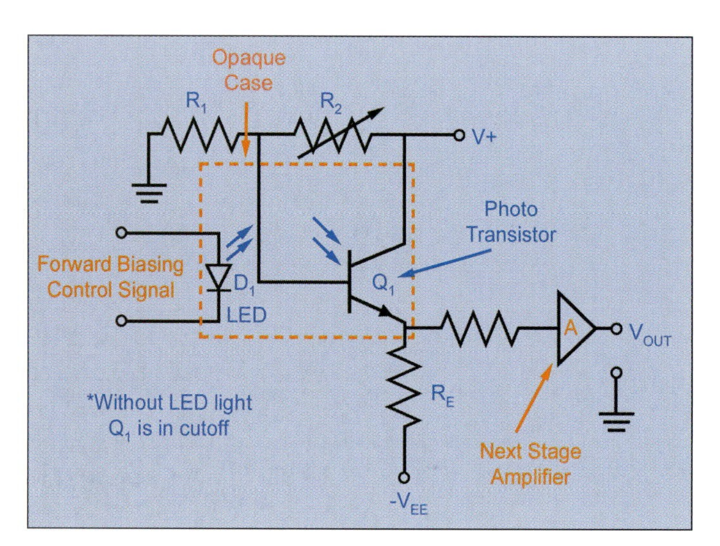

FIGURE 11–19 Optocoupling

when the LED is forward biased by the incoming electrical signal. When the phototransistor receives the light from the forward-biased LED, it becomes forward biased and conducts. The output current is then received by the next amplifier stage and an electric output signal is generated.

11.6 Optoisolators and Optointerrupters

Two special configurations for optocouplers are optoisolators and the optointerrupters. Optoisolators are six-pin DIPs (dual in-line packages), as shown in Figure 11–20. The circuitry inside the optoisolator is shown in Figure 11–19.

A common application of the optoisolator is that of a solid-state relay. This type of relay uses a DC input voltage to pass or block an AC signal. Because the isolation medium is light, the optoisolator can be designed to attain an equivalent isolation rating of several thousand volts.

FIGURE 11–20 Typical six-pin DIP

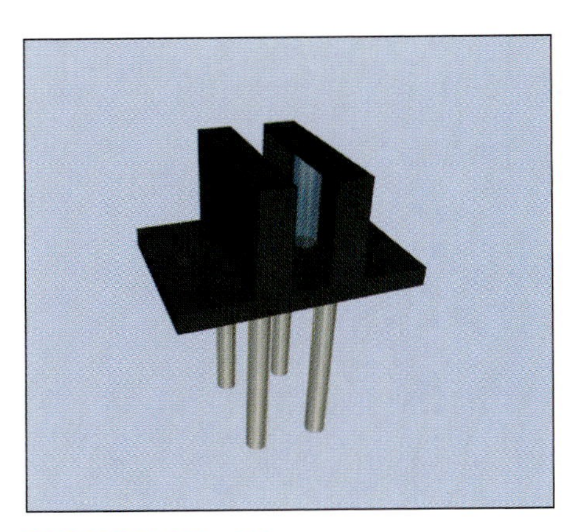

FIGURE 11–21 Optointerrupter casing (slotted optical switch)

The optointerrupter is used as an optical switch. The optointerrupter is designed to have an external object block the light beam path between the photoemitter and the photodetector. A common case or physical construction of the optointerrupter is shown in Figure 11–21. The write-protect tab on a 3½-inch floppy drive serves the same purpose. When it is in place, it blocks the light signal. When not in place, the disk is write protected because the sensing light is allowed to activate the detector.

FIBER-OPTIC CABLE

11.7 Cable Construction

Optical fiber cable cannot be made from just pure silica (glass). Recall that fibers require at least two refractive indexes—one for the core and one for the cladding. Pure silica has just one refractive index. Doping with impurities is required to get a different cladding index. The most common doping element is germanium. Germanium has very low light absorption and also forms a glass.

The actual fiber is constructed by placing a rod of high-index glass into a tube with a lower refractive index. These two are melted into each other to form one rod. This new rod is called a preform. The rod is heated at one end, and a single, very thin fiber is drawn from it.

Some plastic fibers are used in communication. Their primary advantages are lighter weight, lower cost, and greater flexibility. Their major disadvantage is higher signal loss. They are used for short communication links such as in office buildings and cars.

Another limitation of plastic fiber-optic cable is the high degree of degradation over time in environments that have high operating temperatures. This is true of sensors and monitoring processes in industrial manufacturing.

In reality, fiber-optic cables look much like conventional metal cables. Polyethylene is used on both fiber-optic and metal outdoor cables to protect them against the environment. In theory, fiber cables are stronger than their copper cable counterparts. When pulled, the fiber cable will not stretch like copper but will extend a little, and when released, it will spring back to its original size. A fiber cable will stretch only about 5 percent before breaking; however, a copper cable can stretch as much as 30 percent before breaking.

Figure 11–22 shows a cross-sectional view of different grades of fiber-optic cable.

There are many fiber cable types and specifications. Table 11–2 shows NEC cable specifications.

11.8 Cable Connectors

Single-fiber connections are, of course, easiest, but seldom do you find just one fiber to connect to another. The main problem with connections is that of alignment. Figure 11–23 illustrates what happens when connecting fibers are misaligned.

During the past 10 years, many of the industry associations and the telecommunications industry have tried to standardize connector

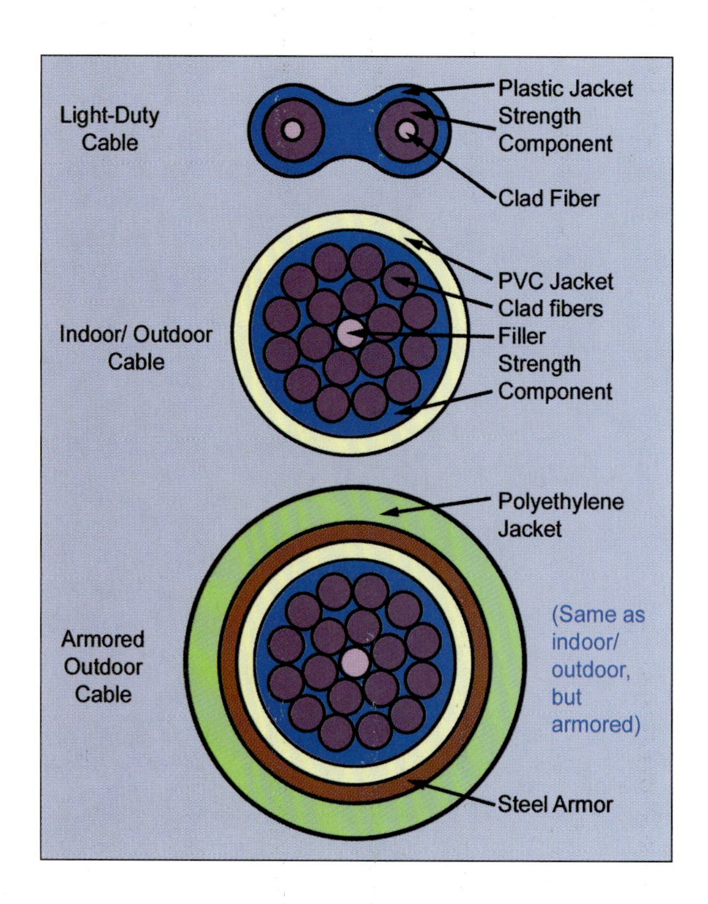

FIGURE 11–22 Types of fiber-optic cable construction

Table 11–2 Cable Specifications Under U.S. National Electrical Code

Cable Type	Description	Designation	UL Test
General purpose (horizontal)—fiber only	Nonconductive optical fiber cable	OFN	Tray/1581
General purpose (horizontal)—hybrid (fiber/wire)	Conductive optical fiber cable	OFC	Tray/1581
Riser/backbone—fiber only	Nonconductive riser	OFNR	Riser/1666
Riser/backbone—hybrid	Conductive riser	OFCR	Riser/1666
Plenum/overhead—fiber only	Nonconductive plenum	OFNP	Plenum/910
Plenum/overhead—hybrid	Conductive plenum	OFCP	Plenum/910

types. The following are some examples of the types of connectors now in use:

- The snap-in single-fiber connector (SC) was developed by Nippon Telegraph and Telephone of Japan. It uses a cylindrical ferrule that holds the fiber and plugs into a coupling receptacle or housing (see Figure 11–24a).

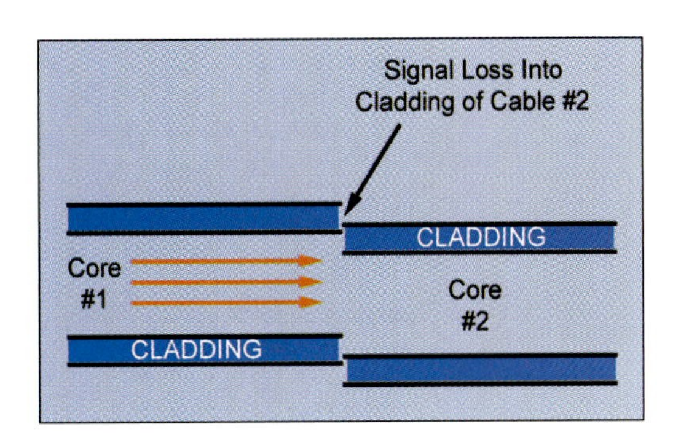

FIGURE 11–23 Misaligned fiber-optic cables

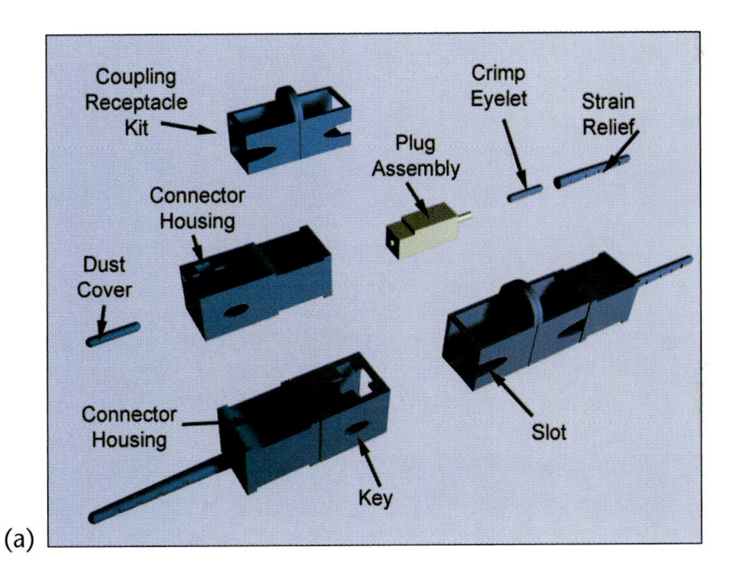

(a)

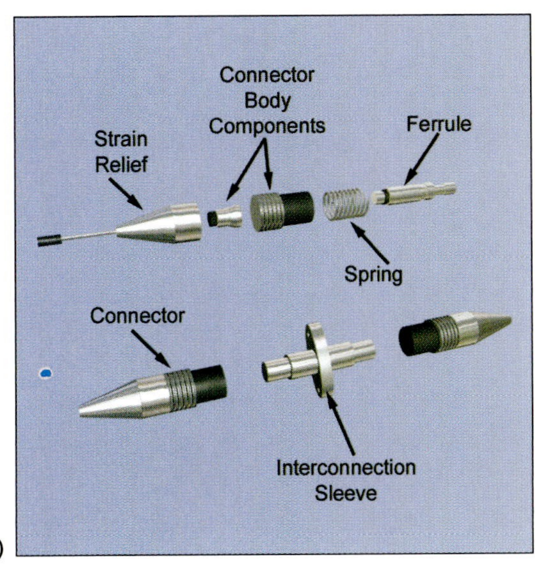

(b)

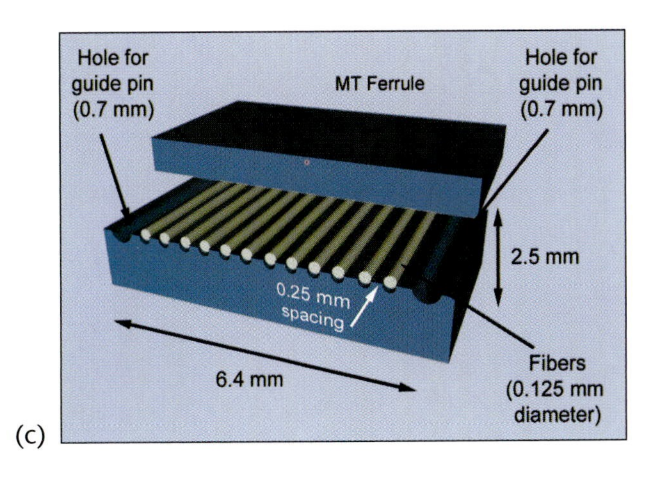

(c)

FIGURE 11–24 Fiber connectors; a) snap-in single-fiber connector, b) twist-on single-fiber connector, c) multifiber connector

- The twist-on single-fiber connector (ST and FC) was originally designed for copper cable connections. It also uses a cylindrical ferrule; however, the ferrule connects with an interconnection sleeve by twisting into a locked position. The FC type uses threads and is screwed into place instead of twisted into place. See Figure 11–24b.
- Multifiber connectors use an MT ferrule. This connector can align up to twelve fibers. This connector is used with optical fiber ribbon cables and is shown in Figure 11–24c.

SUMMARY

The original idea of "piping light" for the purpose of room illumination has given way to the use of light for information transfer. The ability to put information on a light signal and then send it down a fiber-optic channel has started to open so-called wide-bandwidth information channels that provide everything from industrial data and process control to television and communications channels into our homes.

The whole process works because of the fact that light will reflect and/or refract from some surfaces. This allows the light to be "bounced" down a long fiber strand, emerging from the remote end with little or no attenuation as compared to other information-transfer methods.

Available in both multimode and single-mode construction, fiber-optic cable has the ability to transmit information for many miles from its source. The original information is captured in electronic form and then converted to modulated light impulses by light sources such as visible red LEDs, near-infrared LEDs, and lasers—either GaAs or InGaAsP semiconductors.

The source then injects the light into the fiber-optic cable, and it travels to the end, where an optocoupler changes the light back into electrical signals for processing.

The low attenuation, insensitivity to interference, high data rates, and safer (no electrical shock hazard) installations all combine to make fiber-optic cable the basis of the future of data transmission.

REVIEW QUESTIONS

1. Diagram a cross-section of fiber-optic cable showing:
   a. The core
   b. The outer jacket
   c. The cladding
   d. The strength members
   e. The silicon coating
   f. The buffer jacket
2. Define the following terms:
   a. Angle of incidence
   b. Angle of reflection
   c. Critical angle
   d. Attenuation
3. What is the refractive index? How does it affect the transmission of light down an optical fiber?
4. The acceptance angle of a particular optical fiber is 20°. What will happen to an optical signal that strikes the input of the cable at an angle of 25°?
5. If you are standing on the edge of a swimming pool and see a quarter on the bottom, should you dive straight towards it to recover it? Why or why not?
6. What is modulation?
7. If two fibers are not aligned correctly when they are spliced, what will happen to the signal?
8. In one particular fiber-optic cable, the refractive index of the core (n_0) is 1.25. The refractive index

of the cladding (n_1) is 1.225. What is the acceptance angle (NA) of this cable?

9. Discuss the three types of connectors described in chapter 11. What are the advantages and disadvantages of each?
10. In electric power distribution systems, fiber-optic cable is often routed right along with electric power conductors. Sometimes, in utility overhead line applications for example, it may be actually wrapped around the energized electric conductor.
    - a. What are the advantages to such a practice? Disadvantages?
    - b. What safety issues might be involved in such practices?
    - c. Why is the use of electric signal cables limited in such applications?

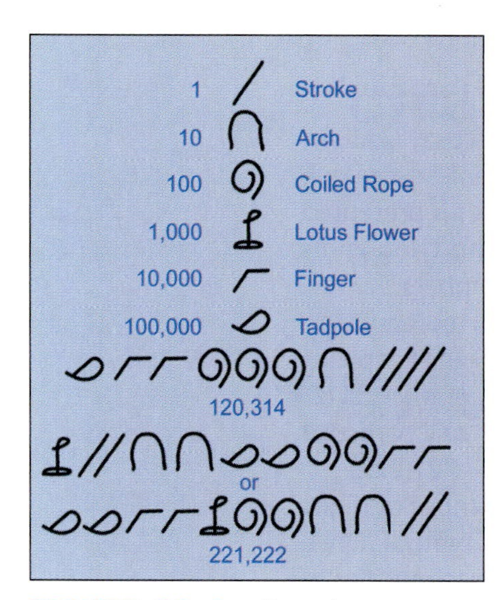

FIGURE 12–1 Egyptian hieroglyphics used for counting

DECIMAL AND BINARY NUMBER SYSTEMS

12.1 The Decimal System

The number of digits or symbols used in a number system is called its **base** or **radix**. The decimal system that we use every day has ten digits. It is the most familiar to most of us. Table 12–1 compares the number of strokes to the symbols used in the decimal system.

For example, the number 9,652 is actually shorthand for a value that is equal to a quantity of 1s, 10s, 100s, and 1,000s. This number can be written according to its positional weight.

$$(9 \times 10^3) + (6 \times 10^2) + (5 \times 10^1) + (2 \times 10^0) =$$
$$9{,}000 + 600 + 50 + 2 = 9{,}652$$

This is because the 9 is in the third position, the 6 is in the second position, the 5 is in the first position, and the 2 is in the zeroth position. Even though we do not think about it, what we do with each of the shorthand numbers is multiply each number by 10 raised to the number's position weight and then add all the numbers together.

The previous example considers only whole numbers—numbers to the left of the decimal point. It is often necessary to work with fractional numbers. Decimal fractions are numbers whose positions have weights that are negative powers of 10. The numbers to the right of the decimal point start in the position called the −1 position. Thus,

$$10^{-1} = \frac{1}{10} = 0.1$$

$$10^{-2} = \frac{1}{10^2} = 0.01$$

$$10^{-3} = \frac{1}{10^3} = 0.001$$

Table 12–1 Comparison of Strokes to Symbols in the Decimal Number System

Strokes	Symbols
None	0
/	1
//	2
///	3
////	4
/////	5
//////	6
///////	7
////////	8
/////////	9

$$10^{-4} = \frac{1}{10^4} = 0.0001$$

$$10^{-5} = \frac{1}{10^5} = 0.00001$$

$$10^{-6} = \frac{1}{10^6} = 0.000001$$

Notice that in each case the number 1 falls in the same position as the value of its power. For example, for 10^{-6}, the number is the decimal followed by five zeroes and then the number.

A decimal point for base 10 separates the integer and the factional parts of the number. The whole part of the number is left of the decimal and the fractional part is right of the decimal. To illustrate this procedure, consider the number 328.75.

$$(3 \times 10^2) + (2 \times 10^1) + (8 \times 10^0) + (7 \times 10^{-1}) + (5 \times 10^{-2}) = 300 + 20 + 8 + 0.7 + 0.05 = 328.75$$

In this example, the number to the far left, 3, carries the greatest weight and is designated the **MSD (most significant digit)**. The number to the far right (5) is the **LSD (least significant digit)** because it has the lowest weight in determining the overall value of the number.

The convention is to write only the digits and deduce the corresponding powers of 10 from their position. A decimal number with a decimal point is represented by a string of coefficients:

$$\ldots A_5\ A_4\ A_3\ A_2\ A_1\ A_0\ A_{-1}\ A_{-2}\ A_{-3} \ldots$$

Each A_n is one of the ten digits used in the decimal system, with the subscript "n" providing the position of the digit. Therefore, the coefficient A is multiplied by 10^n for each position, and each total is added to obtain the number they represent.

12.2 The Binary System

The decimal number system is a wonderful system that has served us well in our everyday lives. Digital and electrical systems, however, are natural binary number systems where only two states exist. Either the system is on, or it is off.

Digital systems need to accept a variety of number systems, depending upon the type of measurement or input instruments. These numbers are processed within the digital system, and the proper number system is selected for the output device. The input number system, the digital number system, and the output number system may all be the same, or they may all be different. Figure 12–2 illustrates this concept.

The binary number system is the simplest of those that use positional notation. The two digits used in the binary system are 0 and 1. Table 12–2 indicates the number of strokes to the symbols for the binary digits.

The binary number system has a radix of 2 and uses the digits 0 and 1. The number is expressed with a string of 1s and 0s and a possible binary point. The value of the digit is set by its position in the string.

FIGURE 12–2 Digital system

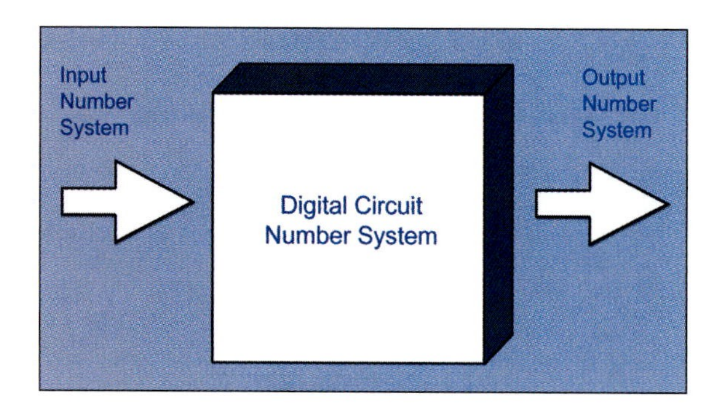

Table 12–2 Comparison of Strokes to Symbols in the Binary Number System

Strokes	Symbols
None	0
/	1

Table 12–3 Comparison of Strokes and Digits between the Binary and the Decimal Number Systems

Strokes	Binary Number	Decimal Number
None	0	0
/	1	1
//	10	2
///	11	3
////	100	4
/////	101	5
/////	110	6
///////	111	7
////////	1000	8
/////////	1001	9

Table 12–3 shows three ways to count to nine by comparing the number of strokes to the binary and decimal system of numbers.

To evaluate the total value of the number, the specific bits and the weights of their positions must be considered. As in the decimal system, the first digit to the left of the binary point has a power of 0, with sequential powers to the left of 1, 2, 3, 4, . . . The difference exists in the base number. The base is now 2 instead of 10 used in the decimal system. Remember that any number raised to the zero power is equal to 1.

A condensed listing of the powers of 2 is given in Table 12–4, with equivalent numbers in the decimal system.

Table 12–4 Decimal Values of the Powers of 2

$2^0 = 1_{10}$	$2^6 = 64_{10}$
$2^1 = 2_{10}$	$2^7 = 128_{10}$
$2^2 = 4_{10}$	$2^8 = 256_{10}$
$2^3 = 8_{10}$	$2^9 = 512_{10}$
$2^4 = 16_{10}$	$2^{10} = 1{,}024_{10}$
$2^5 = 32_{10}$	$2^{11} = 2{,}048_{10}$

Thus, the binary number 101010 can be evaluated using the same approach that we used earlier for decimal numbers:

$$101010_2 = (1 \times 2^5) + (0 \times 2^4) + (1 \times 2^3) + (0 \times 2^2) + (1 \times 2^1) + (0 \times 2^0) = 32 + 0 + 8 + 0 + 2 + 0 = 42_{10}$$

In the binary number system, the radix point is called the binary point, as opposed to the decimal point in the decimal system. Fractional binary numbers are expressed as negative powers of 2, just as fractional decimal numbers are expressed as negative powers of 10. Following is a partial list of the values of the binary number to the right of the binary point, with their decimal equivalents.

$$2^{-1} = \frac{1}{2^1} = 0.5_{10}$$

$$2^{-2} = \frac{1}{2^2} = 0.25_{10}$$

$$2^{-3} = \frac{1}{2^3} = 0.125_{10}$$

$$2^{-4} = \frac{1}{2^4} = 0.0625_{10}$$

$$2^{-5} = \frac{1}{2^5} = 0.03125_{10}$$

$$2^{-6} = \frac{1}{2^6} = 0.015625_{10}$$

$$2^{-7} = \frac{1}{2^7} = 0.0078125_{10}$$

$$2^{-8} = \frac{1}{2^8} = 0.00390625_{10}$$

The decimal equivalent of the binary number 0.1011 is calculated as

$$(1 \times 2^{-1}) + (0 \times 2^{-2}) + (1 \times 2^{-3}) + (1 \times 2^{-4}) = 0.5_{10} + 0_{10} + 0.125_{10} + 0.0625_{10} = 0.6875_{10}$$

There is another way to evaluate this fraction. If the binary point is ignored, the number 1011_2 is equal to

$$11_{10}\ \{1011_2 = (1 \times 2^3)_{10} + (0 \times 2^2)_{10} + (1 \times 2^1)_{10} + (1 \times 2^0)_{10} = 8_{10} + 0_{10} + 2_{10} + 1_{10} = 11_{10}\}$$

Four bits are possible, providing a maximum count of $2^4 = 16$. $\frac{11_{10}}{16_{10}} = 0.6875_{10}$

12.3 Decimal-to-Binary Conversion

Even though most scientific calculators will easily convert from one number system to another, it is helpful to understand the methods that are used. For example, one method of finding the binary value of a decimal number is to determine a set of binary weight values whose sum is equal to the decimal number; consequently, the number 14 can be expressed as the sum of binary weights.

$$14 = 8 + 4 + 2$$

By placing a 1 under the appropriate weight, the binary-equivalent number is calculated, as shown in Table 12–5.

Using this procedure, 14_{10} is equal to 1110_2.

A more systematic method for converting from a decimal number to a binary number is to use the repeated division-by-2 process. This method is illustrated in Figure 12–3.

Table 12–5 Converting (Encoding) a Binary Number from a Decimal Number

Power Position	3	2	1	0
Decimal (14)	8+	4+	2+	0
Power of 2	$1 \times 2^3\vert_{10}$	$1 \times 2^2\vert_{10}$	$1 \times 2^1\vert_{10}$	$0 \times 2^0\vert_{10}$
Binary	1	1	1	0

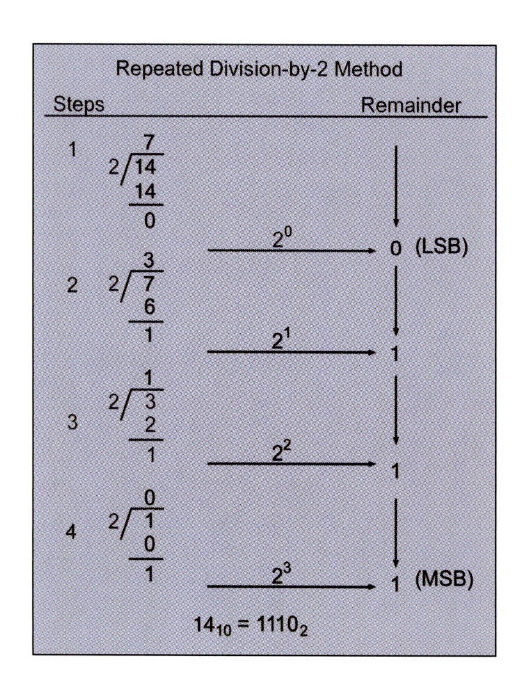

FIGURE 12–3 Repeated division-by-2 method to convert from decimal to binary number

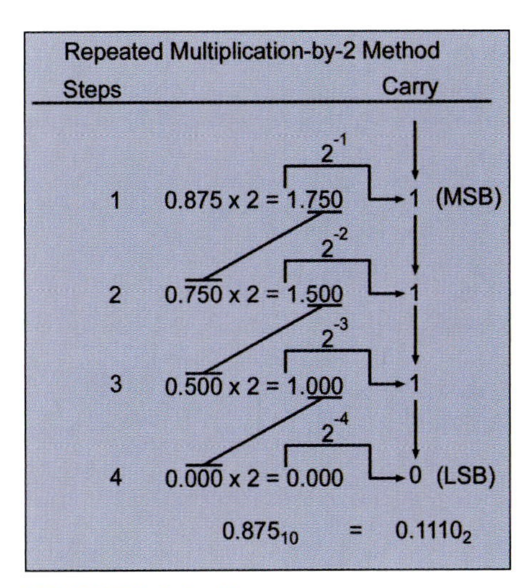

FIGURE 12–4 Repeated multiplication-by-2 method to convert from decimal to binary number

To convert the decimal number 14 to binary, the 14 is divided by 2. The remainder is carried over as the least significant binary number. In this position, 2^0, it represents 0. The quotient, 7, of the first division by 2 is then divided by 2 in step 2. Two divided into 7 is 3 with a remainder of 1. The second binary number in the 2^1 position is therefore 1. In step 3, the quotient 3 is divided by 2, yielding a quotient of 1 and a remainder of 1 in the 2^2 position. In step 4, the quotient 1 from step 3 is divided by 2, yielding a quotient of 0 and a remainder of 1 in the 2^3 position. The results of this process show that 14_{10} is equal to 1110_2.

Decimal fractions can be converted to binary fractions using the sum-of-weights method. For example:

$$0.875_{10} = 0.5 + 0.25 + 0.125 = 2^{-1} + 2^{-2} + 2^{-3} = 0.111_2$$

Decimal fractions can also be converted to binary fractions using a more systematic process called the repeated multiplication-by-2 method. Figure 12–4 shows this process.

In step 1, 0.875 is multiplied by 2. Because the result is greater than 1, the 1 carries as the most significant digit, 2^{-1} position, of the binary fraction. The remaining fraction, after the 1 is removed, is then multiplied by 2 in step 2. Once again, the result is greater than 1, so the 1 carries into the 2^{-2} position of the fraction. The remaining fraction of 0.500 is again multiplied by 2 in step 3 to yield 1.000. The 1 carries to the third position of the fraction, 2^{-3}.

The result of the exercise is that

$$0.875_{10} = 0.111_2$$

12.4 Binary Number Sizes

Binary numbers are also known as binary words; consequently, an eight-bit number is also an eight-bit word. The term **byte** is used to mean an eight-bit word. Digital equipment normally uses a fixed word size. The size of the word determines the magnitude and resolution of numbers processed with the system. The number of bits in a word determines the number of discrete states that can exist and the maximum decimal number value that can be calculated.

In the binary system, the number of states can be calculated with the following formula:

$$N = 2^n$$

where N is equal to the total number of states and n is equal to the number of bits in the word. For example, a three-bit word will have 8 states and a four-bit word will have 16 states.

$$N = 2^3 = 8 \quad \text{and} \quad N = 2^4 = 16$$

The maximum value of a number that can be represented is one less than the number of bits in a word. The largest numbers that can be represented by three and four bits are

$$N = 2^3 - 1 = 7(111_2 = 7_{10})$$
$$N = 2^4 - 1 = 15(1111_2 = 15_{10})$$

Note that 2^4 still represents a total of 16 numbers including $0 \rightarrow 0, 1, 2, 3, 4, 5, 6, 7, 8, 9, 10, 11, 12, 13, 14, 15$.

The number of bits in the binary system needed to represent a given number (N) can be determined using the equation:

$$B = 3.32 \times \log_{10} (N + 1)$$

The common logarithm can be obtained from a set of tables or from almost any scientific calculator. For example, to calculate the maximum number of bits when the largest number that needs to be processed is 700

$$N = 3.32 \times \log_{10} (700 + 1)$$
$$N = 3.32 \times \log_{10} 701$$
$$N = (3.32) \times (2.8457) = 9.448$$

Because fractional bits cannot be implemented, the number of bits required will be the next highest number; thus, to represent the number 700, you need 10 bits. This can be confirmed with the previously given equation for determining the largest number that can be calculated with a fixed number of bits.

$$N = 2^{10} - 1 = 1{,}023$$

If 9 bits had been used, the maximum number would had been

$$N = 2^9 - 1 = 511$$

12.5 Binary-Coded Decimal (BCD)

You have seen that numbers can be represented by binary digits. Letters and other symbols can also be represented by 1s and 0s. Combinations of binary digits that correspond to numbers, symbols, or letters are called digital codes. In many applications, special codes are used to indicate error detection and corrections.

Consider the arrangement shown in Figure 12–5. When switch A is up and all others down, the number represented is 4. In sequence, switch B alone will input a 3, switch C alone a 2, and switch D alone a 1. The numbers 0, 1, 2, 3, or 4 can be sent to the computer using these switches. Zero occurs when all the switches are in their down positions.

With a single switch, the highest number is 4. Suppose the number 5 is required. This could be obtained by putting switch A (4) and switch D (1) in their up positions. It is also possible to obtain a 5 by putting switches B (3) and C (2) up. This could cause some confusion in a computer system where more than one combination can produce a common result.

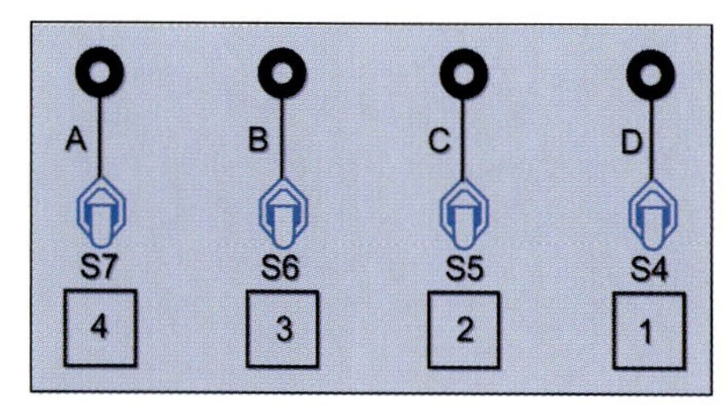

FIGURE 12–5 Numerical values switches (4, 3, 2, 1)

For the number 3, switches C (2) and D (1) in their up positions could be used. The number 3 can therefore be eliminated from the code. This means that switch A has a value of 4 and switches C and D have values of 2 and 1 respectively. Unfortunately, this combination can count only from 0 to 7. To be compatible with the decimal system, it is necessary that the count goes to 9 to include all the integers. To fix this problem, the value of switch A is transferred to the vacant switch B. In

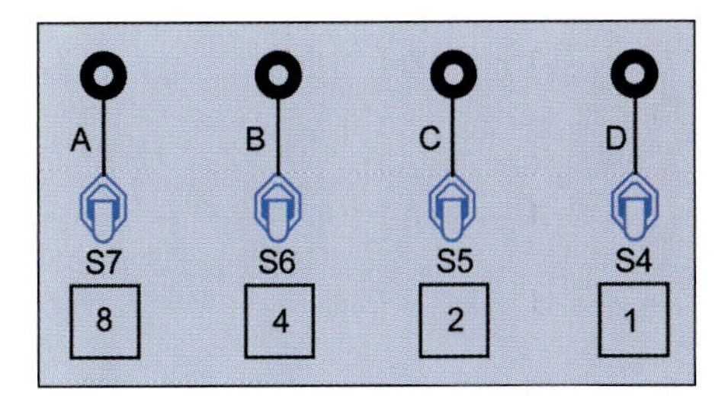

FIGURE 12–6 Binary-coded decimal switches (8, 4, 2, 1)

place of the 4 value for switch A, the new value of 8 is assigned. The new coding for the switches is shown in Figure 12–6.

Binary-coded decimal (BCD) means that the binary code represents all the numbers used in the decimal code, 0–9. It is designated the 8421 code, where the position of the digit is weighted in term of these numbers. The 8421 code is the predominant BCD code.

With four bits, sixteen numbers (2^4) numbers can be represented. In the 8421 code, only ten of these numbers are used, and 1010, 1011, 1100, 1101, 1110, and 1111 are invalid. To express any number in BCD, simply replace each decimal digit with the appropriate four-bit code.

THE OCTAL NUMBER SYSTEM

Octal numbers are often used with microprocessors and have a radix of 8. The primary application of octal numbers is representing binary numbers. The digits 0, 1, 2, 3, 4, 5, 6, and 7 are used in the octal number system. These digits have the same numerical values as the decimal system.

As with the decimal and binary number systems, each digit in the octal number system carries a positional weight. The weight of each position is determined by some power of 8. For example, the octal number 234.01_8 can be written

$$243.01_8 = (2 \times 8^2) + (3 \times 8^1) + (4 \times 8^0) + (0 \times 8^{-1}) + (1 \times 8^{-2}) = 128 + 24 + 4 + 0 + 0.015625 = 156.015625_{10}$$

The decimal value of the octal number is determined by multiplying each digit by its positional weight and adding the results. The octal (radix) point separates the integer from the factional part of the number.

The decimal values of the positional integers from the octal point left are

$$8^0 = 1_{10}$$
$$8^1 = 8_{10}$$
$$8^2 = 64_{10}$$
$$8^3 = 512_{10}$$
$$8^4 = 4{,}096_{10}$$
$$8^5 = 32{,}768_{10}$$
$$8^6 = 262{,}144_{10}$$
$$8^7 = 2{,}097{,}152_{10}$$

Fractional values in the decimal system from the octal point right are

$$8^{-1} = \frac{1}{8} = 0.125_{10}$$

$$8^{-2} = \frac{1}{8^2} = 0.15625_{10}$$

$$8^{-3} = \frac{1}{8^3} = 0.001953125_{10}$$

$$8^{-4} = \frac{1}{8^4} = 0.000244140625_{10}$$

Table 12–6 Decimal, Octal, and Binary Equivalents

Decimal	Octal	Binary
0	0	000
1	1	001
2	2	010
3	3	011
4	4	100
5	5	101
6	6	110
7	7	111

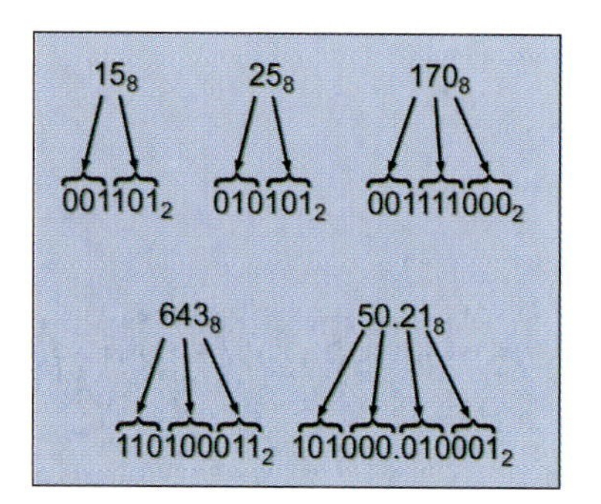

FIGURE 12–7 Octal-to-binary conversion

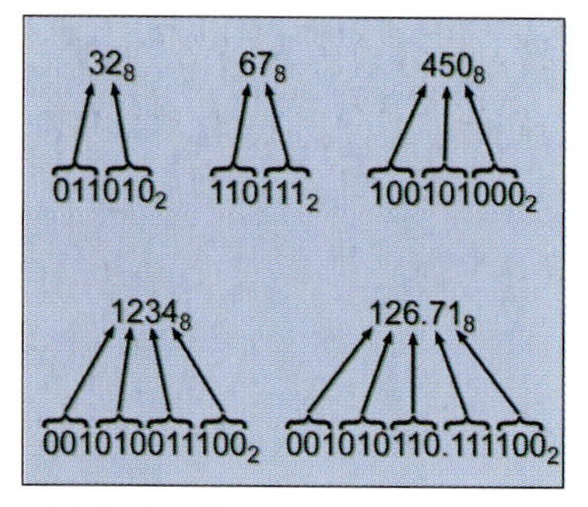

FIGURE 12–8 Binary-to-octal conversion

Each octal number can be represented by a three-bit binary code. The binary, octal, and decimal equivalent values are shown in Table 12–6.

To convert an octal number to a binary number, simply replace each octal digit with the equivalent three-bit binary code. Figure 12–7 shows examples of this approach.

Figure 12–8 shows examples of converting from a binary number to an octal number. Begin at the binary point and break the binary number into groups of three bits, then convert each group to the equivalent octal digit. With binary whole numbers, the binary point is understood to be to the right of the lowest significant binary, LSB.

Zeros can be added with the MSB and LSB where needed to complete the group of three digits. These zeros are sometimes not in the original numbers. They do not affect the values of the number when added.

The repeated division-by-8 system can be used to convert a decimal number to an octal number. This is illustrated in Figure 12–9, with the conversion of decimal number 500_{10} to its octal equivalent of 764_8.

In Figure 12–9, the number 500_{10} is divided by 8. The remainder, 4, is the least significant digit in the octal number system. The quotient of the first division, 62, is then divided by 8 with a remainder of 6, which is put in the 8^1 position of the octal number. The quotient of the second division, 7, is then divided by 8 with a remainder of 7, the most significant digit in the octal system for the decimal number 500. The 7 is in the 8^2 position of the octal number.

THE HEXADECIMAL (16) NUMBER SYSTEM

As microprocessors have become faster, the use of the octal number system has decreased and is being replaced by the hexadecimal number system. This allows for the processing of more bits at the same time. Because the hexadecimal system has a base of 16, it is composed of 16 digits and characters. Most digital systems process binary data in groups that are multiples of four bits. This makes the hexadecimal system convenient because each four-bit number represents one hexadecimal number.

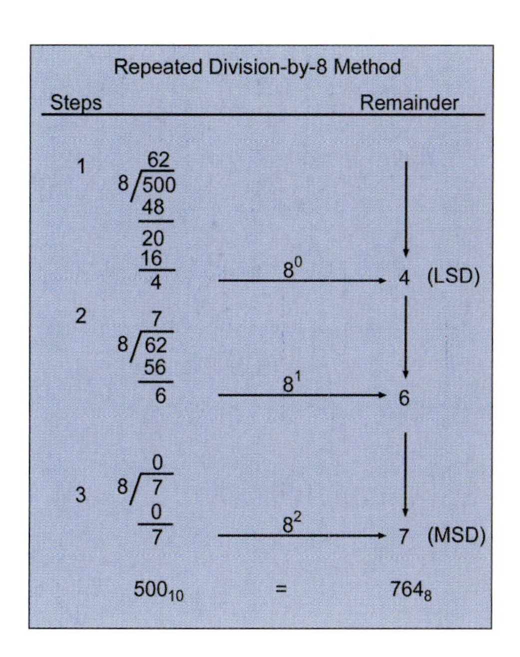

FIGURE 12–9 Decimal-to-octal conversion

The hexadecimal number system, also referred to as the alphanumeric hexadecimal number system, uses decimals 0 through 9 and alphabet letters A through F. The letters are used because it is necessary to represent 16_{10} different values with a single digit for each value. The letters A through F represent 10_{10} through 15_{10} respectively. Table 12–7 shows the corresponding values of the decimal, hexadecimal, and binary number systems.

Table 12–7 Decimal, Hexadecimal, and Binary Equivalents

Decimal	Hexadecimal	Binary
0	0	0000
1	1	0001
2	2	0010
3	3	0011
4	4	0100
5	5	0101
6	6	0110
7	7	0111
8	8	1000
9	9	1001
10	A	1010
11	B	1011
12	C	1100
13	D	1101
14	E	1110
15	F	1111

It may first seem awkward using letters for numbers, but after you become familiar with the notation, it becomes much easier. As with the previous number systems, each digit position of the hexadecimal number carries a positional weight. Positional values left of the hexadecimal point are

$$16^0 = 1_{10}$$
$$16^1 = 16_{10}$$
$$16^2 = 256_{10}$$
$$16^3 = 4{,}096_{10}$$
$$16^4 = 65{,}536_{10}$$
$$16^5 = 1{,}048{,}576_{10}$$
$$16^6 = 16{,}777{,}216_{10}$$

Positional values to the right of the hexadecimal point are

$$16^{-1} = \frac{1}{16} = 0.0625_{10}$$
$$16^{-2} = \frac{1}{16^2} = 0.00390625_{10}$$
$$16^{-3} = \frac{1}{16^3} = 0.000244140625_{10}$$
$$16^{-4} = \frac{1}{16^4} = 0.0000152587809625_{10}$$

A decimal number can be converted to a hexadecimal number by the repeated division-by-16 process. This is shown in Figure 12–10.

To convert 700_{10} to hexadecimal, you first divide by 16. The remainder of 12 from the first division is the least significant digit, rep-

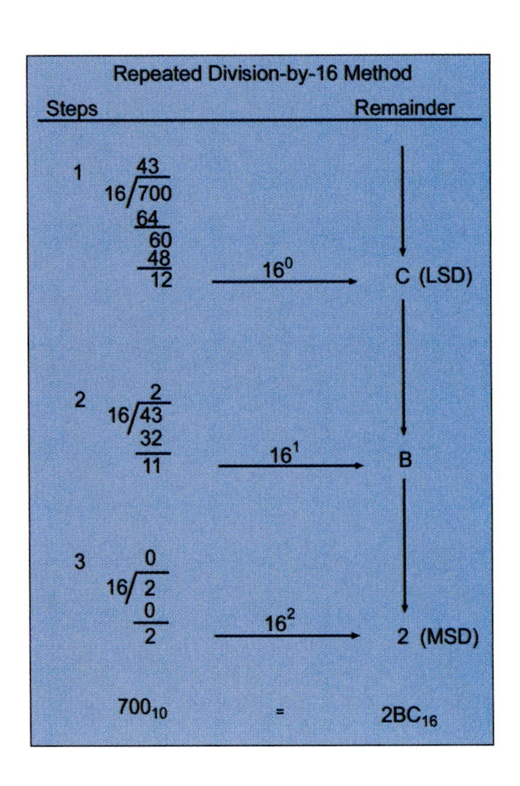

FIGURE 12–10 Decimal-to-hexadecimal conversion

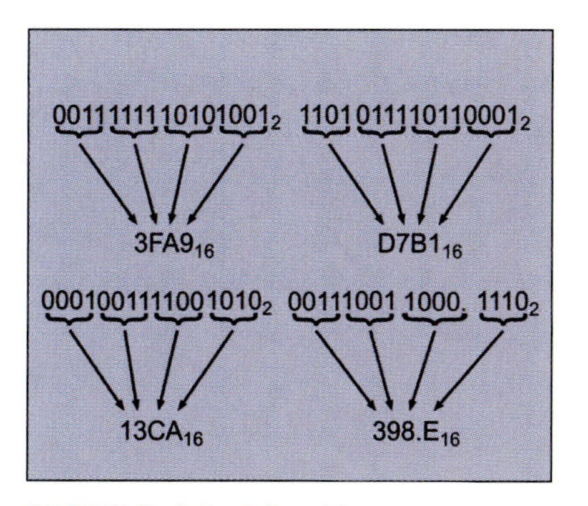

FIGURE 12–11 Binary-to-hexadecimal conversion

resented by C. The quotient of the first division, 43, is now divided by 16 with a remainder of 11. B represents 11 in the hexadecimal number system and is placed in the 16^1 position. The quotient 2 from the second division is now divided by 16 with a remainder of 2. The number 2 is placed in the most significant digit position, giving $2BC_{16}$ for the decimal number 700_{10}.

Binary numbers can easily be converted to hexadecimal values. The binary number is converted to groups of four, starting with the binary point. Zeros can be added to the most significant group so that four digits are available without changing the value of the binary number. Each of these groups is then replaced with the corresponding hexadecimal value (see Figure 12–11).

Hexadecimal numbers can be changed to binary numbers in a similar manner. Each of the hexadecimal numbers is changed to an equal-value four-digit binary number, starting at the hexadecimal point. Four examples are shown in Figure 12–12.

Counting with the hexadecimal system is similar to the decimal system.

0 1 2 3 4 5 6 7 8 9 A B C D E F 10 11 12 13 14 15 16 17 18 19 1A 1B 1C 1D 1E 1F

20 21 22 23 24 25 26 27 28 29 2A 2B 2C 2D 2E 2F 30 31 32 33 . . .

Two hexadecimal digits allow a count to FF_{16}, which is 255_{10}. A count beyond this decimal number requires additional hexadecimal digits. For example, 100_{16} equals 256_{10}, 101_{16} equals 257_{10}, and so on. The maximum three-digit hexadecimal number is FFF_{16}, or $4{,}095_{10}$. The maximum four-digit hexadecimal number, $FFFF_{16}$, is equal to $65{,}535_{10}$.

Occasionally, you may need to convert from hexadecimal to decimal. One way of making this conversion is to convert to binary and from binary to decimal. Conversion can be made directly to a decimal number by multiplying the number in each position by the power of 16 for that position and then adding the numbers together. An example of the process follows for the number $C53_{16}$.

$$C53_{16} = (12 \times 16^2) + (5 \times 16^1) + (3 \times 16^0) = 3{,}072 + 80 + 3 = 3{,}155_{10}$$

FIGURE 12–12 Hexadecimal-to-binary conversion

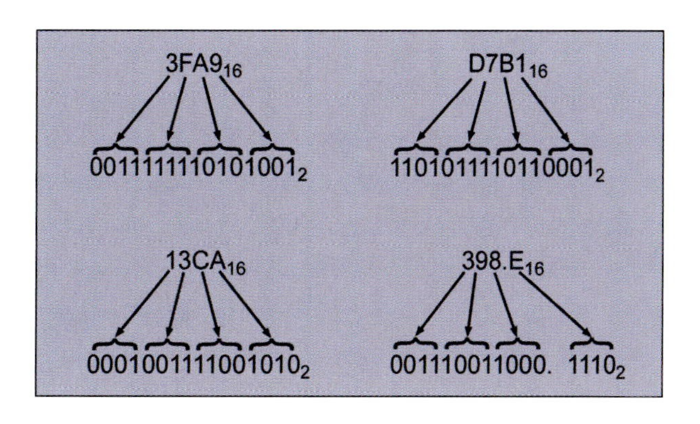

SUMMARY

The ten-digit decimal numbering system we use every day has limitations when dealing with modern digital and electrical systems. Basically, only two states exist with the digital numbering system, on or off. The requirement for the use of a variety of numbering methods, depending on the type of measurement or input needed, has led to the development of additional numbering systems for the digital world. The simplest form we have studied is the binary numbering system, containing only two digits. Because of the large number of digits needed by the binary system, the octal and hexadecimal system are often used instead. These allow larger numbers using smaller numbers of digits.

REVIEW QUESTIONS

1. Convert the number $353{,}652_{10}$ to
   - **a.** Binary
   - **b.** Octal
   - **c.** Hexadecimal
   - **d.** Egyptian
2. Consider the concept of the LSD and MSD. How does it affect computations such as multiplication, division, subtraction, and addition?
3. What is the octal point? The hexadecimal point? The binary point? How do they relate to the decimal point?
4. When would you use the *repeated division-by-2* method? When would you use the *repeated multiplication-by-2* method?
5. The most often used binary code for computer applications is called the ***A**merican **S**tandard **C**ode for **I**nformation **I**nterchange*, usually abbreviated as ***ASCII***. The most fundamental form for ASCII uses a seven-bit binary word. So-called extended ASCII uses an eight-bit binary word.
   - **a.** How many different characters or words can ASCII represent?
   - **b.** How many different characters or words can extended ASCII represent?
6. Use the following table for the indicated switch positions of Figure 12–6.
   - **a.** What binary and decimal numbers are represented?
   - **b.** Which, if any, of the positions are "not allowed?"

   Check your answers by converting from decimal to binary and back again.

Switch	↑	↑	↑	↑	↓	↑	↓	↑	↓	↓	↓	↓	↓	↑	↑	↑
Binary																
Decimal																

7. In some electrical system protection equipment, settings are applied by the use of a thirty-two-bit binary word. To shorten the amount of storage space, the binary word is usually abbreviated using an eight-bit hexadecimal word. What binary number is represented by $FC34A7A8_{16}$?
8. In each of the various number systems, it takes two bits to represent the radix. For example, in the binary system, the number $2 = 10_2$, in the octal system the number $8 = 10_8$, and so on. Explain why you always need two bits to express the radix of a given number system.
9. How many characters can ASCII (see question 5) represent? Express your answer in hexadecimal.
10. Convert the number $8{,}461.203_{10}$ to octal.

chapter 13

Computer Mathematics

OUTLINE

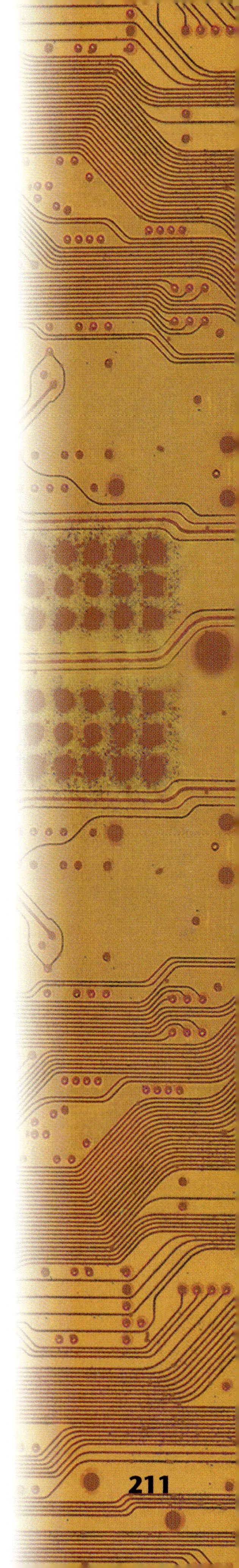

OVERVIEW

Virtually all operations within a number system can be simplified to addition and subtraction. Multiplication, division, exponentiation, and all other such mathematical operations are simply extensions of addition and subtraction.

Throughout your education, you have been taught to perform all of these operations in the decimal system; however, binary arithmetic is fundamental to all digital systems. To understand digital systems, you must learn the basics of binary arithmetic.

In this chapter, you will learn how to add, subtract, multiply, and divide using binary numbers. Decimal mathematics is reviewed as a guide to understanding the decimal system better.

OBJECTIVES

After completing this chapter, the student should be able to:

1. Add binary numbers.
2. Subtract binary numbers.
3. Multiply binary numbers.
4. Divide binary numbers.

GLOSSARY

Addend Any of a set of numbers to be added.[1]

Augend The first in a series of addends.

Dividend A quantity to be divided. In 45 / 3 = 15, 45 is the dividend.[6]

Divisor The quantity by which another quantity, the dividend, is to be divided. In 45 / 3 = 15, 3 is the divisor.[5]

Minuend The quantity from which another quantity, the subtrahend, is to be subtracted. In the equation 50 − 16 = 34, the minuend is 50.[2]

Multiplicand The number that is or is to be multiplied by another. In 8 × 32, the multiplicand is 8.

Multiplier The number by which another number is multiplied. In 8 × 32, the multiplier is 32.

Quotient The number obtained by dividing one quantity by another. In 45 / 3 = 15, 15 is the quotient.[4]

Subtrahend A quantity or number to be subtracted from another. In the equation 50 − 16 = 34, the subtrahend is 16.[3]

---

[1]Excerpted from *American Heritage Talking Dictionary.* Copyright © 1997 The Learning Company, Inc. All Rights Reserved.

[2]Excerpted from *American Heritage Talking Dictionary.* Copyright © 1997 The Learning Company, Inc. All Rights Reserved.

[3]Excerpted from *American Heritage Talking Dictionary.* Copyright © 1997 The Learning Company, Inc. All Rights Reserved.

[4]Excerpted from *American Heritage Talking Dictionary.* Copyright © 1997 The Learning Company, Inc. All Rights Reserved.

[5]Excerpted from *American Heritage Talking Dictionary.* Copyright © 1997 The Learning Company, Inc. All Rights Reserved.

[6]Excerpted from *American Heritage Talking Dictionary.* Copyright © 1997 The Learning Company, Inc. All Rights Reserved.

ADDITION

13.1 Decimal Addition

Addition is a manipulation of numbers that represent the combining of common physical quantities. For example, in the decimal system, 3 + 4 = 7 symbolizes combining /// with //// strokes to get /////// strokes.

Binary addition is much like decimal addition. Using two decimal numbers, if $54{,}678_{10}$ is added to $69{,}142_{10}$, a sum of $123{,}820_{10}$ is obtained. Table 13–1 shows how this decimal number is calculated.

EXAMPLE 1

Find the sum of the **addends** 54,678 and 69,142, using Table 13–1.

Solution:

First column (The one to the right, the 1s column or the 10^0 column)

Adding the rightmost column (10^0) gives (8 + 2 = 10). This is expressed in the sum as the digit 0 under 8 and 2 with a carry of 1 above the 7 of the **augend**.

Second column

The carry is then added in the second column, 1 + 7 + 4 equals 12. This is expressed as a 2 in the sum in the 10^1 column with a carry of 1 to be added in the third column.

Third column

Adding the third column gives 1 + 6 + 1 = 8. The 8 becomes part of the sum in the 10^2 column. There is no carry in this case, so a 0 is placed above the 4 in the 10^3 column.

Fourth column

Adding the fourth column gives 0 + 4 + 9 = 13. The 3 becomes part of the sum and the 1 is the carry to the 10^4 column above the 5.

Fifth column

Adding the fifth column gives 1 + 5 + 6 = 12. The 2 becomes part of the sum in the 10^4 column and the carry of 1 is added in the 10^5 column.

Table 13–1 Decimal Addition

Column#	6	5	4	3	2	1
Column Power	10^5	10^4	10^3	10^2	10^1	10^0
Carry	1	1	0	1	1	
Augend		5	4	6	7	8
Addend		6	9	1	4	2
Sum	1	2	3	8	2	0

13.2 Binary Addition

The same basic operations are performed with binary numbers, using the following four rules:

$0 + 0 = 0$

$0 + 1 = 1$

$1 + 0 = 1$

$1 + 1 = 10$ (0 with a carry of 1)

Note that the first three rules result in a single bit, 0 or 1. In the fourth case, the addition of two 1s results in a binary 2 (10).

EXAMPLE 2

Use Table 13–2 to add 011_2 and 001_2.

Solution:

First column

In the 2^0 column, adding $1 + 1$ results in a 0 in the sum of that column with a carry of 1.

Second column

Adding the 2^1 column gives $1 + 1 + 0 = 10$. The 0 remains in the 2^1 column and the 1 is a carried to the 2^2 column.

Third column

Adding the 2^2 column gives $1 + 0 + 0 = 1$. The sum therefore is 100_2, which is equal to 4_{10}. This checks because $011_2 = 3_{10}$ and $001_2 = 1_{10}$.

When there is a carry digit, there is a situation in which three bits are being added. The rules for handling each of the possible sums are:

Carry + Augend + Addend = Sum

$1 + 0 + 0 = 01$ (1 with a carry of 0)

$1 + 1 + 0 = 10$ (0 with a carry of 1)

$1 + 0 + 1 = 10$ (0 with a carry of 1)

$1 + 1 + 1 = 11$ (1 with a carry of 1)

Table 13–2 Binary Addition (All Digits Are to the Base 2)

Column#	3	2	1
Column Power	2^2	2^1	2^0
Carry	1	1	
Augend	0	1	1
Addend	0	0	1
Sum	1	0	0

EXAMPLE 3

To illustrate these possibilities, add 1011_2 to 1001_2, using Table 13–3.

In this binary addition example, 11_{10} is added to 9_{10} to obtain a sum of 20_{10}.

If four 1s are to be added, cumulatively add two numbers each time:

$$1 + 1 = 10$$
$$10 + 1 = 11$$
$$11 + 1 = 100$$

EXAMPLE 4

Add 111111_2 to 11001100_2, using Table 13–4.

Evaluate this addition in terms of base 10 numbers.

$$\begin{array}{r} 11001100_2 = 204_{10} \\ +00111111_2 = 63_{10} \\ \hline 100001011_2 = 267_{10} \end{array}$$

Table 13–3 Another Example of Binary Addition

Column#	5	4	3	2	1
Column Power	2^4	2^3	2^2	2^1	2^0
Carry	1	0	1	1	
Augend		1	0	1	1
Addend		1	0	0	1
Sum	1	0	1	0	0

Table 13–4 A Third Example of Binary Addition

Column#	9	8	7	6	5	4	3	2	1
Column Power	2^8	2^7	2^6	2^5	2^4	2^3	2^2	2^1	2^0
Carry	1	1	1	1	1	1	0	0	
Augend		1	1	0	0	1	1	0	0
Addend		0	0	1	1	1	1	1	1
Sum	1	0	0	0	0	1	0	1	1

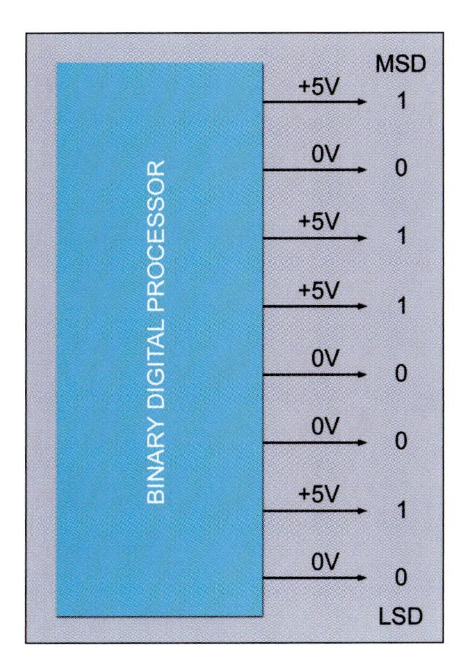

FIGURE 13–1 Parallel processing of the binary word 10110010

13.3 Computer Addition

There are several methods by which computers add numbers. One thing common to all computers is that numbers are added in pairs only. If there is a sum of three numbers, a pair will be added to obtain the first sum, and the remaining addend will then be added for the final sum.

Methods for addition are classified as either parallel or serial. The difference is the method in which the numbers are transmitted to and from the computer or processed within the computer.

Parallel Representation

In the parallel method, a numbers code is transmitted on a set of leads, one for each digit in the number code. The number 1001_2 would require four leads. The first and fourth lead would be high, and the second and third lead would be low to transmit the number 9_{10}. ($1001_2 = 9_{10}$) The parallel methods of addition will add two numbers by considering all positions in the numbers at once and producing a set of signals on a third set of leads that corresponds to the sum. Figure 13–1 shows parallel processing of the binary word 10110010. Note that there is one lead for each digit.

Serial Representation

With the serial method, numbers are transmitted to the machine in sequence, one digit at a time beginning with the MSD. As the sequence of signals is fed into the processing unit, the digits are combined progressively and the sum is given as a sequential output. Figure 13–2 shows the serial transmission of the binary word 10110010. Note that only a single transmission line is necessary because the bits are separated by time.

The parallel method is faster but requires more apparatus. The serial method requires less apparatus but generally requires more time for processing. Many factors go into the selection of one system over the other. New innovations in protocol, such as the Universal Serial Bus (USB) and IEEE 1394 protocol, have made serial transfer extremely fast.

FIGURE 13–2 Serial processing of the binary word 10110010

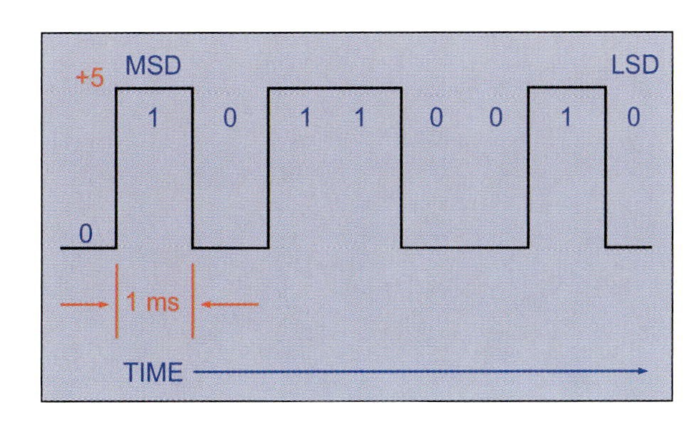

SUBTRACTION

13.4 Decimal Subtraction

Binary subtraction is performed in the same manner as decimal subtraction. Before attempting binary subtraction, review the decimal method of subtracting 2768_{10} from 7605_{10}. See Table 13–5.

Because the digit 8 in the **subtrahend** is greater than **minuend** 5, 1 is borrowed from the next higher order digit in the minuend. If the digit is 0 as in the example, 1 is borrowed from the next higher order that contains a number other than 0. That number is reduced, 6 to 5 in the example, and the digit skipped in the minuend becomes a 9. This is the equivalent of subtracting 1 from 20 with 19 being the difference. In the decimal system, the digit borrowed has the value of 10, so that the 5 in the 10^0 place is now 15, as shown in the top row of numbers. $15 - 8 = 7$.

Continuing: $9 - 6 = 3$ for the 10^1 position. Because 5 is less than 7 in the 10^2 position, 1 is borrowed to make the number 15, and the 7 in the 10^3 position is reduced to 6. $15 - 7 = 8$ and $6 - 2 = 4$ completes the problem.

13.5 Binary Subtraction

In the decimal system, the borrowed number has the value of the radix, or 10. In the binary number system, the borrowed number will also have the value of the base, or in this case, 2.

When one binary number is subtracted from another, the same method described for the decimal system is used.

The basic rules for binary subtraction are:

$0 - 0 = 0$

$1 - 1 = 0$

$1 - 0 = 1$

$10 - 1 = 1$ ($0 - 1$ with a borrow of 1)

The $10 - 1 = 1$ can be better understood when strokes are used.

// − / = /

To subtract larger binary numbers, subtract column by column from the 2^0 position to 2^n, where n is the power of the MSD. In Table 13–6, 101_2 is subtracted from 111_2.

Table 13–5 Decimal Subtraction

Column Power	10^3	10^2	10^1	10^0
Minuend after borrow	6	15	9	15
Minuend	7	6	0	5
Subtrahend	2	7	6	8
Difference	4	8	3	7

Table 13–6 Binary Subtraction

Column Power	2^2	2^1	2^0
Minuend after borrow			
Minuend	1	1	1
Subtrahend	1	0	1
Difference	0	1	0

In this case, it was not necessary to borrow a digit. The minuend is 7_{10}, and the subtrahend is 5_{10}. The difference is 2_{10} . This verifies the value of the difference obtained using the binary system.

Table 13–7 shows a second example.

In this example, it was necessary to borrow a digit in the 2^1 position. In that case, $10 - 1 = 1$. To confirm the results, the decimal solution to the problem is: $13 - 10 = 3$.

Like decimal numbers, binary numbers can also be negative. Subtracting 111_2 from 100_2 will result in a negative value, as shown in Table 13–8.

Note that, as with decimal numbers, we subtract the smaller number from the larger and then prefix the sign of the larger of the two numbers. In this case, the sign of the subtrahend.

The decimal values verify the difference obtained in the binary subtraction: $4 - 7 = -3$.

Two more examples of binary subtraction follow, with verification using decimal numbers.

Minuend after borrow:		0	1	1	1	10	1	10
Minuend:	1	1	0	0	0	1	0	0
Subtrahend:	–0	0	1	0	0	1	0	1
Difference:	1	0	0	1	1	1	1	1

Table 13–7 A Second Example of Binary Subtraction

Column Power	2^3	2^2	2^1	2^0
Minuend after borrow		0	10	
Minuend	1	1	0	1
Subtrahend	1	0	1	0
Difference	0	0	1	1

Table 13–8 A Negative Binary Number

Column Power	2^2	2^1	2^0
Minuend after borrow			
Minuend	1	0	0
Subtrahend	1	1	1
Difference	−0	1	1

The decimal solution for the preceding problem is: 196 − 37 = 159

Minuend after borrow:	0		10	10				
Minuend:	1	1	1	0	1	1	1	0
Subtrahend:	–1	0	1	1	1	0	1	0
Difference:	0	0	1	1	0	1	0	0

The decimal solution for the preceding problem is: 238 − 186 = 52

Note that when a borrow is required, 1 is obtained from the next high order bit that contains a 1. That bit becomes zero, as illustrated in the 2^6 position of both problems. All bits skipped that had 0 become 1, as best illustrated in the first problem.

For example, if 1 is subtracted from 1000_2, the result is 0111_2.

MULTIPLICATION

13.6 Decimal Multiplication

Multiplication is nothing more than a short method of addition. For example, the decimal multiplication problem 3 × 2 = 6 is the same as 3 + 3 = 6. 3 × 3 = 9 is the same as 3 + 3 + 3 = 9. The addition process is fine for small numbers, and most of us probably tried to remain with addition when we were first introduced to the multiplication process. But we soon found that this was futile when numbers such as 324 × 467 were introduced.

Multiplicand:	324
Multiplier:	× 467
First partial product:	2268
Second partial product:	1944
Third partial product:	1296
Carry:	02110
Final product:	151,308

To add 324 to itself 467 times would be quite a task. Using the short form of multiplication, the **multiplicand** is multiplied by the **multiplier** one digit at a time to obtain a partial product for each. The partial products are then added together. Note that the carries for each row of addition have been included in the example.

13.7 Binary Multiplication

Binary multiplication is much easier than decimal multiplication. In binary multiplication, we have to deal with only two digits, 0 and 1, and there are only four rules.

$0 \times 0 = 0$

$0 \times 1 = 0$

$1 \times 0 = 0$

$1 \times 1 = 1$

Numbers with several bits are processed just like decimal numbers. Partial products are obtained and they are added together. Look at the following examples. The results using binary multiplication methods are verified by substituting decimal numbers for the binary numbers.

Multiplicand:	11
Multiplier:	1
Final product:	11

The decimal equivalent of the binary multiplication is: $3 \times 1 = 3$

Multiplicand:	11
Multiplier:	11
First partial product:	11
Second partial product:	11
Carry:	10
Final product:	1001

The decimal equivalent of the binary multiplication is: $3 \times 3 = 9$

Multiplicand:	111
Multiplier:	101
First partial product:	111
Second partial product:	000
Third partial product:	111
Carry:	11100
Final product:	100011

The decimal equivalent for the preceding problem is: $7 \times 5 = 35$

Multiplicand:	1111
Multiplier:	1010
First partial product:	0000
Second partial product:	1111
Third partial product:	0000
Fourth partial product:	1111
Carry:	1111
Final product:	10010110

The decimal equivalent for the preceding problem is: $15 \times 10 = 150$.

Keep in mind that, just as in decimal multiplication, you must keep track of any zeros by setting a zero product under the 0 bit in the multiplier. This is very important when the zero occupies the LSD.

DIVISION

13.8 Decimal Division

Division is the reverse of multiplication. It is the process of determining how many times one number can be subtracted from another. In the decimal number system, most students are probably familiar with short and long division. In short division, only a single digit is the divisor, whereas, in long division, two or more digits are in the divisor. See the examples that follow.

EXAMPLE 1

Short division

$$8\overline{)1616}\quad = 202$$

Where:

8 is called the divisor
1616 is called the dividend
202 is called the quotient

EXAMPLE 2

Long division

```
    457
19)8683
  -76
   108
   -95
    133
   -133
    000
```

Where:

19 is the divisor
8683 is the dividend
457 is the quotient
000 is the remainder

In division, the most significant digit in the **dividend** is examined to determine if the **divisor** is smaller or greater in value. In our short division example using radix 10, 8 is greater than 1, so the **quotient** is 0 for this position. In these cases where the divisor is a single digit, the two most significant digits are selected. Eight is divided into 16 to give a quotient of 2 in the 10^2 position. The third MSD is 1 and smaller than the divisor of 8. The quotient in the 10^1 position is 0 with the one carrying for the next round. 8 into 16 is 2.

The same process is used for the long division problem. The divisor 19 is larger than the MSD of 8, giving a quotient of 0 in the 10^3 position. The 8 carries, so in the second round, 19 is divided into 86 with a quotient of 4 in the 10^2 position. Four is multiplied by 19 to give a product of 76, which is subtracted from 86. The remainder is 10. The 8 in the 10^1 position is brought down to provide a remainder of 108. One hundred and eight divided by 19 equals 5 with a remainder of 13. The 3 in the 10^0 position is brought down to provide a remainder of 133. One hundred and thirty-three divided by 19 equals 7 with no remainder.

13.9 Binary Division

Binary division is performed much like decimal division. Binary division, however, is a simpler process because only two numbers are used rather than 10, as in the decimal number system. See the examples of binary division that follow.

EXAMPLE 3

Long division

```
                 110     Quotient
Divisor    10)1100      Dividend
              -10
               10       Remainder
              -10
               00

                   111   Quotient
Divisor    101)100011   Dividend
              -101
                 111    Remainder
                -101
                  101   Remainder
                 -101
                  000   Remainder
```

When using long division, the dividend is examined beginning with the MSD to determine the number of bits required to exceed the value of the divisor. In the first example, this occurs above the 2^2 position of the dividend. A 1 is placed in this position and multiplied times the divisor. The value of the divisor is subtracted from the dividend. This results in a remainder of 1. Bring down the 0 in the 2^1 position to the remainder. Place a 1 above the 2^1 position of the dividend and multiply times the divisor. Subtract the product from the remainder. This leaves zero remainder. Put a 0 about the 2^0 position.

The results of the first binary division can be confirmed using the decimal number system $1100_2 = 12_{10}$, and $10_2 = 2_{10}$.

$$\frac{1100_2}{10_2} = \frac{12_{10}}{2_{10}} = 6_{10} = 110_2$$

The same approach is taken in the second example. The dividend is examined beginning with the MSD to determine how many places are necessary to just exceed or equal the value of the divisor. This is true at the 2^2 position of the dividend. A 1 is placed above this position and multiplied times the divisor. Subtract 101 from 1000 in the dividend. This results in a remainder of 11, which is less than 101 of the divisor. Like decimal mathematics, this remainder must be less than the divisor or an error has been committed. Bring down 1 from the dividend to make the remainder greater than the divisor. Place 1 in the quotient above the 2^1 position of the dividend and multiply times the divisor. Subtract 101 from 111, leaving a remainder of 10. Bring down the 1 from the dividend and again divide by the divisor. Place a 1 in the quotient above the 2^0 position and multiply times the divisor. Subtract the product from the remainder. The result is zero.

Confirming with decimal arithmetic yields

$$\frac{100011_2}{101_2} = \frac{35_{10}}{5_{10}} = 7_{10} = 111_2$$

SUMMARY

Binary arithmetic is much like decimal arithmetic. The basic mathematic operations of adding, subtracting, multiplying, and dividing can be performed using binary numbers. These processes are easier than the decimal process because only two numbers, 0 and 1, are used. Results of the binary processes can be confirmed using the decimal number system.

PRACTICE PROBLEMS

1. Addition
   - a. 101110 + 11011 = ?
   - b. 1111 + 1011 = ?
   - c. 001110001 + 100011 = ?
   - d. 100001 + 101101 = ?
2. Subtraction
   - a. 100100 − 10110 = ?
   - b. 101101 − 0101 = ?
   - c. 101 − 011 = ?
   - d. 111111 − 100001 = ?
3. Multiplication
   - a. 1001 × 11 = ?
   - b. 0110 × 100 = ?
   - c. 1100 × 101 = ?
   - d. 11111 × 101110 = ?
4. Division
   - a. 1001 / 11 = ?
   - b. 110011011 / 111010 = ?
   - c. 1110011 / 111 = ?
   - d. 11011011 / 1011 = ?

Differential and Operational Amplifiers

OUTLINE

OVERVIEW

In this chapter, you will learn about two very commonly used amplifier circuits: differential amplifiers and operational amplifiers.

Both of these amplifier circuits are very versatile and are widely used in linear applications. Their popularity is partially due to the fact that they are available as integrated circuits (ICs) at a very low cost. Both of these types of amplifiers are extremely easy to use and allow the electrician to build useful circuits without worrying about the complex internal circuitry.

OBJECTIVES

After completing this chapter, the student should be able to:

1. Draw the schematic symbols of differential and operational amplifiers.
2. Describe the operation of differential and operational amplifiers.
3. Describe the basic operations of circuits using differential and operational amplifiers.

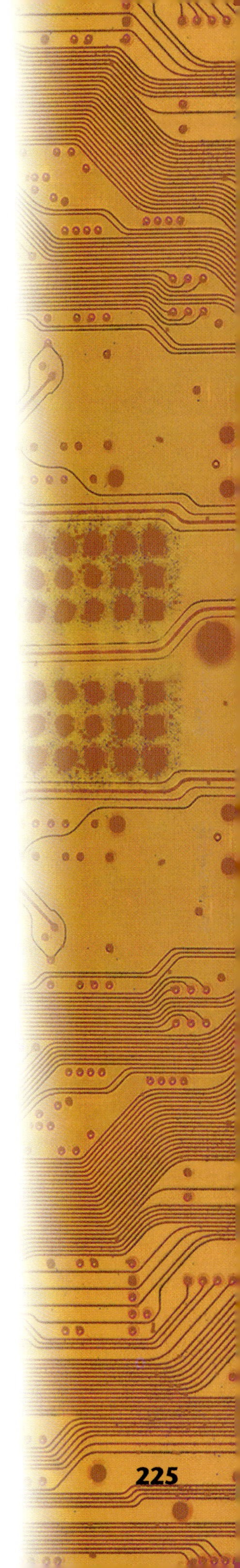

DIFFERENTIAL AMPLIFIERS

14.1 Basic Circuit

The differential amplifier gets its name from the fact that it produces an output that is proportional to the difference between its two inputs. Figure 14–1 shows the basic arrangement of a differential amplifier. The transistors Q_1 and Q_2 must be as close to identical as possible. Ideally, they are exactly alike. R_{C_1} and R_{C_2} are also identical to keep the symmetry in the circuit. The input to transistor Q_1 is called the inverting input (I), and the input to transistor Q_2 is called the noninverting input (NI).

The circuit requires two power supplies: $+V_{CC}$ connected to the collector circuit and $-V_{EE}$ connected to the emitter circuit. Figure 14–2 shows a bipower supply that can be used for this purpose.

When a BJT is used in an amplifier, the collector-base junction must be reverse biased and the emitter-base junction must be forward biased. The transistors used in Figure 14–1 are NPN transistors, and the voltages $+V_{CC}$ and $-V_{EE}$ establish the required biasing conditions in the amplifier.

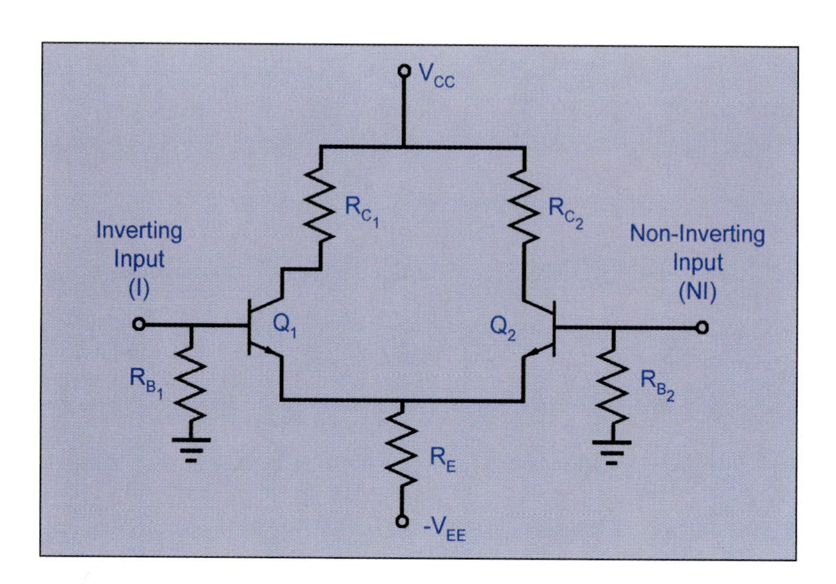

FIGURE 14–1 A differential amplifier

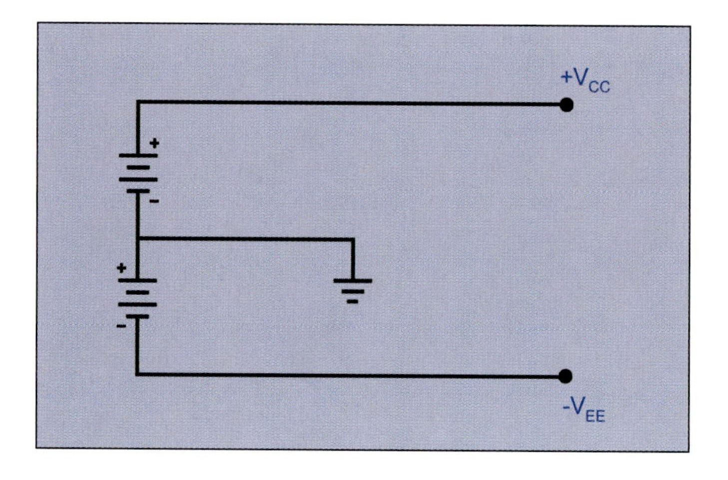

FIGURE 14–2 A bipolar DC power supply

14.2 Modes of Operation

There are three basic modes of operation for a differential amplifier, each based on the way that input is fed into the amplifier.

1. The single-ended mode
2. The differential mode
3. The common mode

Single-Ended Mode

A differential amplifier is operating in the single-ended mode when an active signal is connected to only one of its inputs. The inactive input is normally connected to ground (directly or through a resistor). The amplifier is classified as an inverting amplifier or a noninverting amplifier depending on which input is active.

Figure 14–3 shows a differential amplifier driven at the inverting input (Q_1). If the input increases in the positive direction, Q_1 (NPN transistor) will increase in conduction. This will cause the collector current (I_{C_1}) to increase, which in turn causes the voltage drop across the collector resistor $R_{C_1}(I_{C_1} \times R_{C_1})$ to increase. The voltage at the collector of Q_1 ($V_{CC} - I_{C_1} \times R_{C_1}$) decreases. Thus, an increase in the input to Q_1 (inverting input) causes a decrease in the collector voltage of Q_1. This means that the output of Q_1 is inverted or 180° out-of-phase with its input.

Now consider what is happening to the collector of Q_2. When Q_1 is driven to conduct by the positive going input, its emitter current increases, thereby increasing the voltage drop across R_E. The emitter of Q_2 (which is also connected to R_E) also experiences a higher voltage. The base of Q_2 (noninverting input) is connected to ground, so its base-emitter PN junction experiences a smaller forward bias and conducts

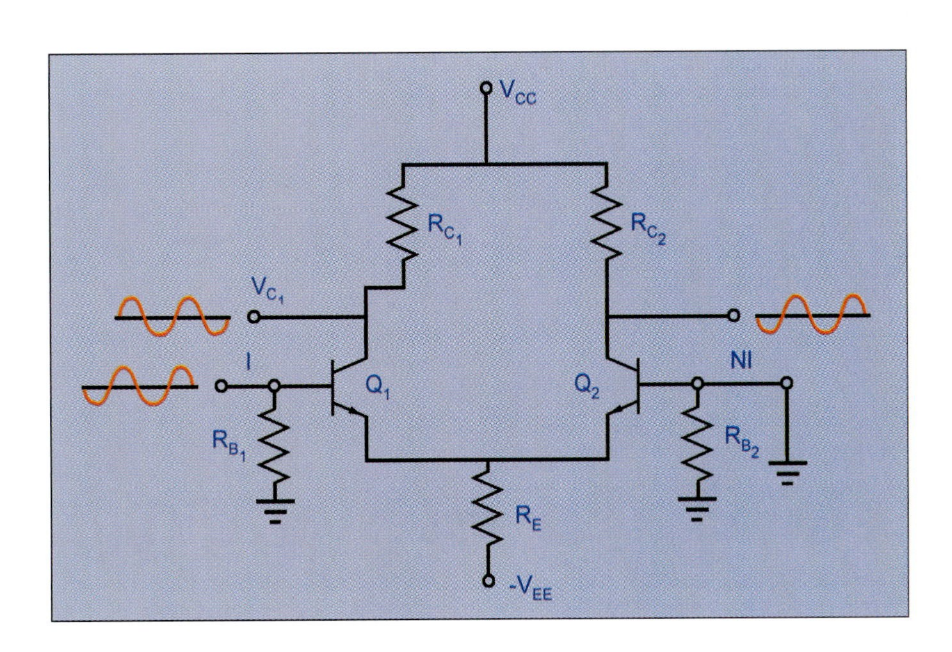

FIGURE 14–3 Differential amplifier operating in the single-ended mode

less current. In fact, the magnitude of this voltage is almost equal to the magnitude of the input signal. The only difference is the voltage drops of the PN junctions. This causes the collector current (I_{C_2}) to be small, which in turn results in a smaller voltage drop across resistor R_{C_2}. The voltage at the collector of Q_2 ($V_{CC} - I_{C_2} \times R_{C_2}$) will increase in the positive direction. Consequently, the collector of Q_2 increases in the positive direction as the input increases, thus causing a noninverted output at the collector of Q_2.

This shows that the circuit in Figure 14–3 can be used to obtain an inverted (out-of-phase) and a noninverted (in-phase) output. The output is said to be single ended if it is measured between the collector of Q_1 or Q_2 and ground. The output is said to be differential if it is measured between the collector of Q_2 and the collector of Q_1. The differential output has twice the swing of the individual outputs.

As an example, consider an input signal of 1 $V_{p\text{-}p}$ applied to the inverting input, while the noninverting terminal is connected to ground. This creates an amplified inverted signal being produced at the collector of Q_1, while an amplified noninverted output is produced in the collector of Q_2. If the transistors in the circuit amplify by a factor of 3, the output (single ended) at Q_1 will be 3 $V_{p\text{-}p}$ (inverted) and the output at Q_2 will be 3 $V_{p\text{-}p}$ (noninverted). This is shown in Figure 14–4.

When the collector of Q_1 goes to its negative maximum (−1.5 V), the collector of Q_2 goes to its positive maximum (+1.5 V). The differential output measured between Q_2 and Q_1 is (+1.5 V) − (−1.5 V) = +3 V. Also, when the collector of Q_1 goes to its positive maximum (+1.5 V), the collector of Q_2 goes to its negative maximum (−1.5 V).

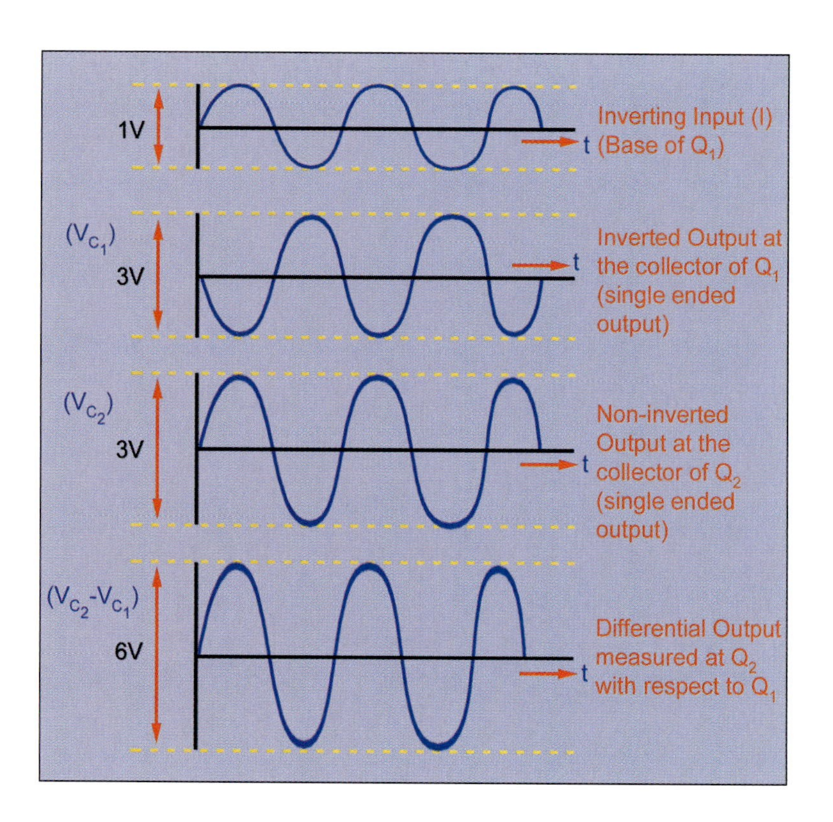

FIGURE 14–4 Single-ended and differential outputs

The differential output measured between Q_2 and Q_1 is (−1.5 V) − (+1.5 V) = −3 V. Thus, we can see that the differential output in Figure 14–4 has a swing from +3 V to −3 V for a total peak-to-peak voltage of 6 V. This is twice the swing of the individual single-ended outputs.

It must be noted that in the previous example, the difference between the two inputs is 1 $V_{p\text{-}p}$ − 0 V (GND) = 1 $V_{p\text{-}p}$. The single-ended output amplified this difference three times, while the differential output amplified the difference six times.

Differential Mode

In the differential mode of operation, two separate signals are applied to both the amplifier inputs. The magnitude of the output voltage is the difference between the two individual inputs. Figure 14–5 shows a differential amplifier driven in the differential mode.

Assume the following conditions for Figure 14–5:

1. The inverting input (Q_1) is fed with 1 $V_{p\text{-}p}$
2. The noninverting input (Q_2) is fed with 3 $V_{p\text{-}p}$.
3. Both input voltages are in phase and the same frequency.
4. Each of the two transistor circuits has a voltage gain of 3.

Both the transistors, being identical, will conduct proportional to their inputs. This means the following:

1. The collector of Q_1 (which has an input voltage of 1 $V_{p\text{-}p}$) will have an output voltage of 3 $V_{p\text{-}p}$.
2. The collector of Q_2 (which has an input of 3 $V_{p\text{-}p}$) will have an output voltage of 9 $V_{p\text{-}p}$.
3. Each transistor's output will be inverted from its input.

The outputs at the collectors of Q_1 and Q_2 are shown in Figure 14–6. Note that both the outputs with respect to ground are out of phase with their corresponding inputs with respect to ground.

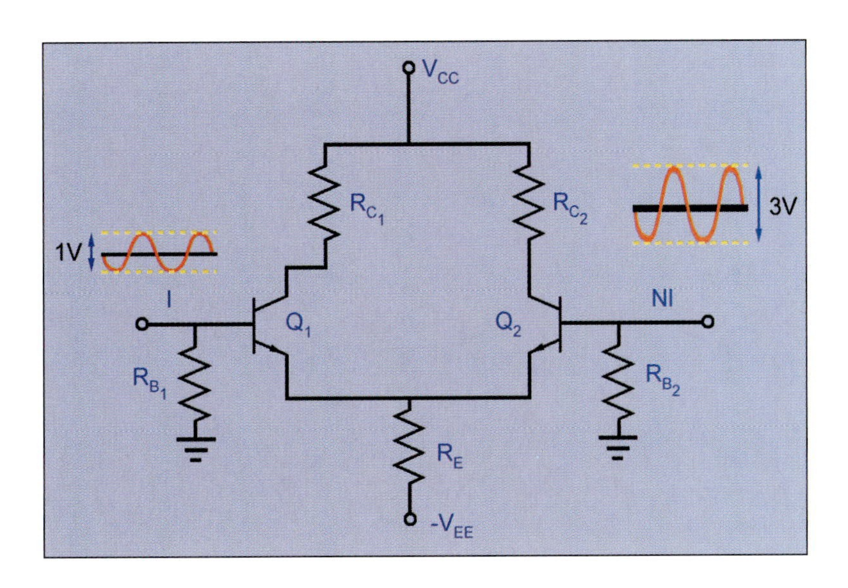

FIGURE 14–5 Differential amplifier operating in the differential mode

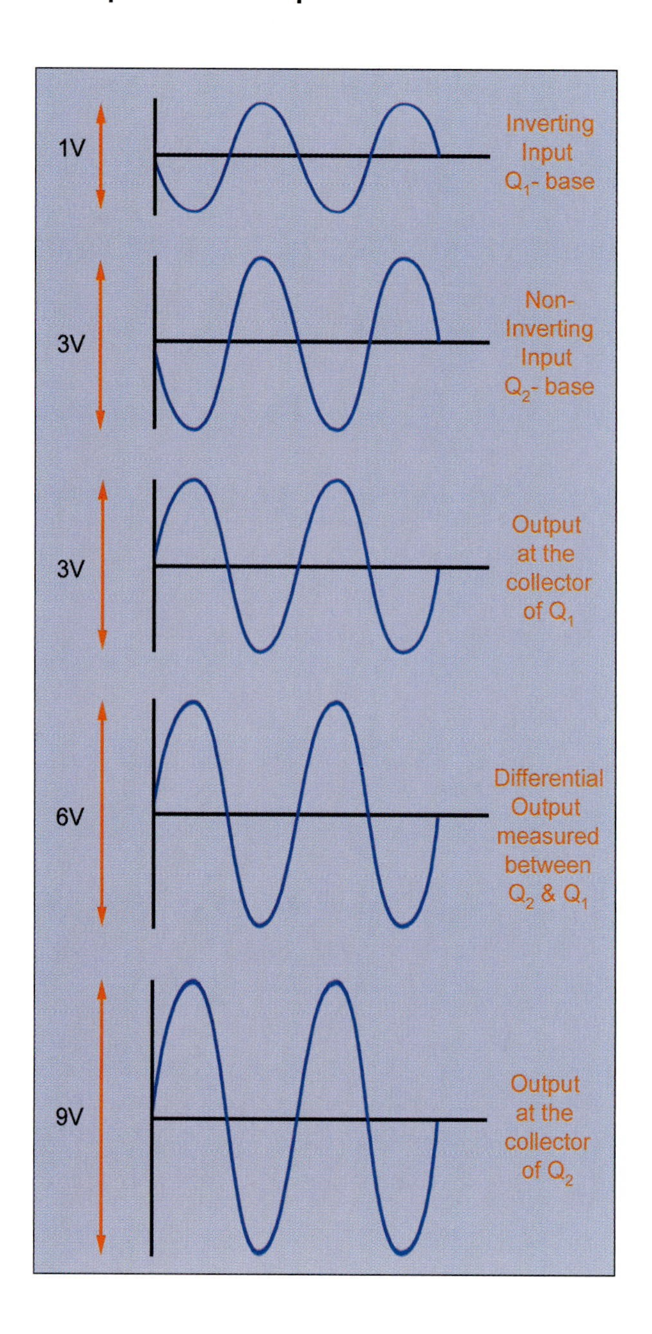

FIGURE 14–6 Input and output waveforms of differential amplifier operating in the differential mode

The differential output measured at Q_2 with respect to Q_1 can be calculated as either the difference between the two outputs:

$$V_o = V_{Q_2} - V_{Q_1} = 9\ V_{p\text{-}p} - 3\ V_{p\text{-}p} = 6\ V_{p\text{-}p}$$

or the product of the gain and the difference of the two inputs:

$$V_o = A_V \times (V_{iQ_2} - V_{iQ_1}) = 3 \times (3\ V_{p\text{-}p} - 1\ V_{p\text{-}p}) = 6\ V_{p\text{-}p}$$

Note that when the inputs are made equal, the differential output becomes zero. Also note that if either of the inputs is phase-shifted by 180°, the inputs effectively add ($V_o = 9\ V_{p\text{-}p} - (-3\ V_{p\text{-}p}) = 12\ V_{p\text{-}p}$).

Common Mode

In the common mode of operation, identical signals are applied simultaneously to both the inverting and the noninverting inputs. Ideally, identical transistors will produce identical outputs with identical inputs. The collector voltages being identical throughout the cycle produce a differential output of 0 V. The advantage of this mode of operation is that any noise or undesired signal that appears simultaneously at the two inputs does not generate an output from the circuit.

Power-line hum is a commonly occurring noise signal that interferes with high-gain amplifiers. The 60 Hz frequency of the power line radiates signals that may be picked up by electronic circuits. Such hum is usually sensed in common mode. That is, it affects both wires of a signal circuit equally.

In Figure 14–7 the power-line noise is being impressed on lines a and b equally with respect to ground. The desired signal, however, is being applied across lines a and b.

If the input signal line is connected to a differential amplifier, as shown in Figure 14–8, the 60 Hz noise is being connected equally to both the inverting and noninverting inputs. As you saw earlier, this means that the hum will be cancelled at the output terminals.

The desired signal, on the other hand, is changing positive at one input while it changes negative at the other. This means that it will be amplified at the output. In other words, hum and noise constitute the common-mode signal and are cancelled.

The inputs to the inverting and the noninverting inputs after the hum or noise is picked up are shown in Figure 14–9. The hum is of the same phase in both the inputs, and so the differential output is unaffected by it. Identical transistors and resistors produce identical collector voltages, and hence the differential output measured between Q_1 and Q_2 is zero for the noise. In other words, the differential amplifier has rejected the common-mode signals (which is noise); conversely, the input signal is fed in differential mode and is amplified.

FIGURE 14–7 Common mode

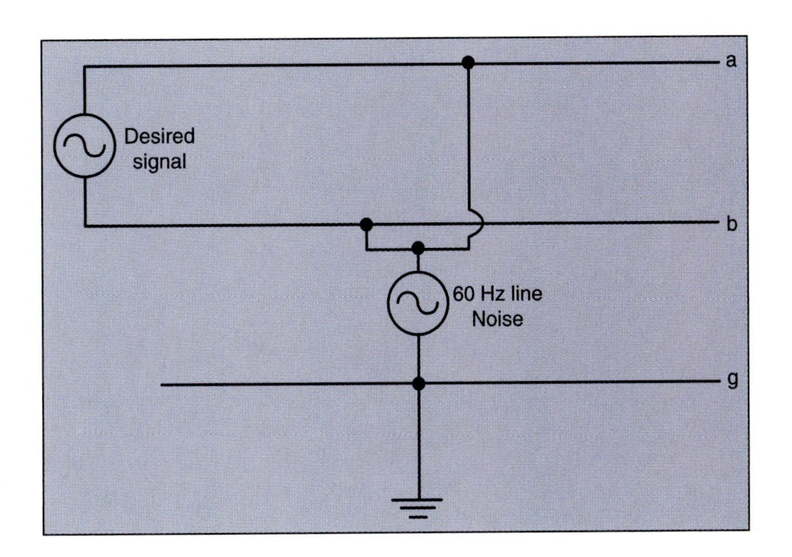

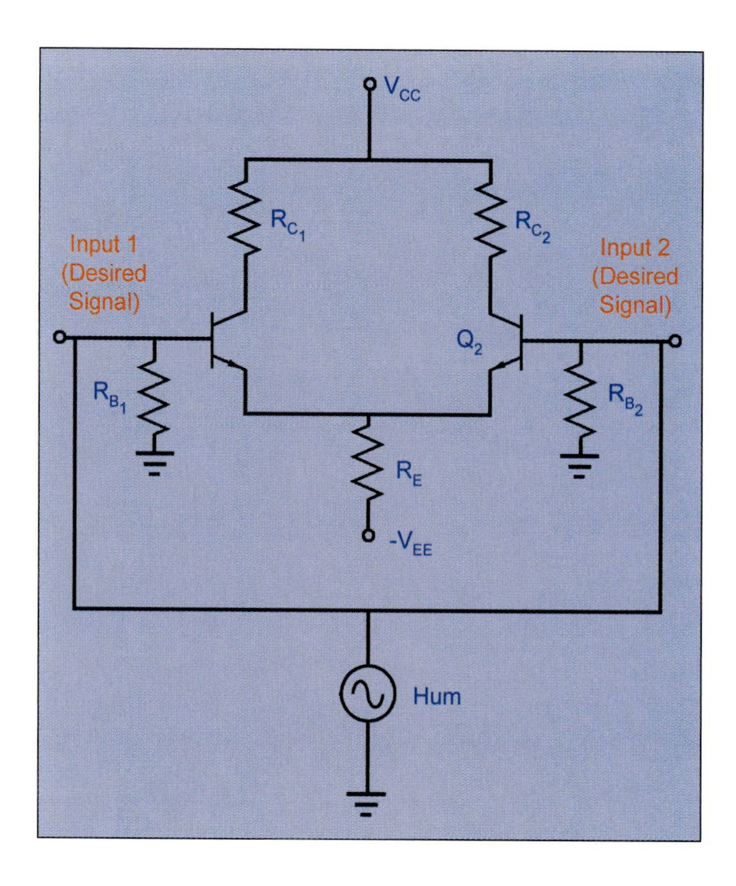

FIGURE 14–8 Differential amplifier with hum fed in common mode and desired signal in differential mode

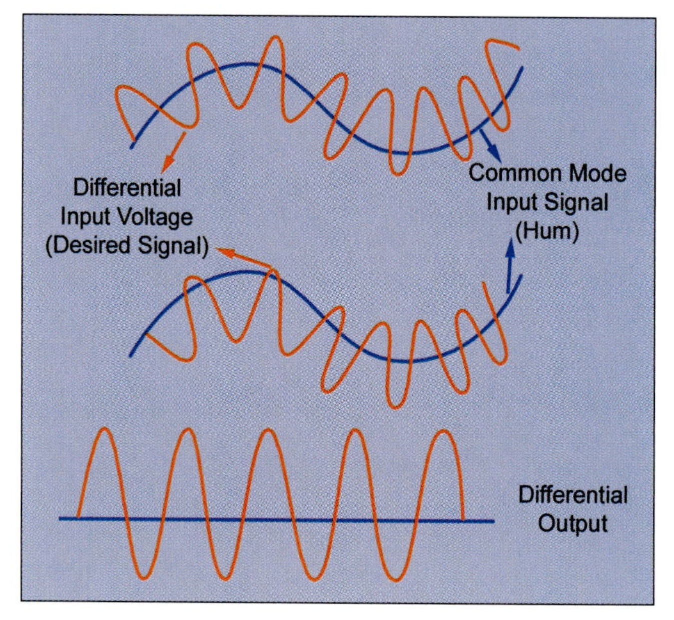

FIGURE 14–9 Input and output waveforms demonstrating noise rejection

14.3 Common-Mode Rejection Ratio (CMRR)

Common-mode signal rejection occurs only if the two transistors and their associated resistors are perfectly matched. In reality, the differential amplifier will not have perfect balance, and one transistor may have a higher gain than the other. This causes some common-mode signal to appear at the output. The ability of a circuit to reject common-mode signals is called common-mode rejection ratio or CMRR.

$$\text{CMRR} = \frac{A_{V(DIFF)}}{A_{V(COMM)}}$$

where:

$A_{V(DIFF)}$ = voltage gain of the amplifier for differential signals
$A_{V(COMM)}$ = voltage gain of the amplifier for common-mode signals

If $A_{V(DIFF)}$ and $A_{V(COMM)}$ are expressed in dB, then

$$\text{CMRR}_{dB} = A_{V(DIFF)dB} - A_{V(COMM)dB}$$

EXAMPLE 1

A differential amplifier has the following input and output signals.

Common-mode input = 1 V
Common-mode output = 0.02 V
Differential input = 0.1 V
Differential output = 10 V

Solution:

Step 1

$$A_{V(DIFF)} = \frac{\text{Differential output}}{\text{Differential input}} = \frac{10\text{ V}}{0.1\text{ V}} = 100$$

Step 2

$$A_{V(DIFF)dB} = 20 \times \log_{10} A_{V(DIFF)} = 20 \times \log 100 = 40\text{ dB}$$

Step 3

$$V_{V(COMM)} = \frac{\text{Common} - \text{mode output}}{\text{Common} - \text{mode input}} = \frac{0.02\text{ V}}{1\text{ V}} = 0.02$$

Step 4

$$A_{V(COMM)dB} = 20 \times \log_{10} A_{V(COMM)} = 20 \times \log_{10} 0.02 = -33.98\text{ dB}$$

Step 5

$$\text{CMRR} = \frac{A_{V(DIFF)}}{A_{V(COMM)}} = \frac{100}{0.02} = 5{,}000$$

Step 6

$$\text{CMRR}_{dB} = A_{V(DIFF)dB} - A_{V(COMM)dB} = 40\text{ dB} - (-33.98\text{ dB}) = 73.98\text{ dB}$$

or

$$\text{CMRR}_{dB} = 20 \times \log_{10}(\text{CMRR}) = 20 \times \log_{10} 5{,}000 = 73.98\text{ dB}$$

14.4 Analysis

The characteristics and properties of differential amplifiers can be demonstrated by performing DC and AC analysis on the circuit.

DC Analysis

Figure 14–10 shows a differential amplifier operation with +10 V(V_{CC}) and −10 V(V_{EE}). In this circuit, do not consider any AC inputs applied because you are going to strictly analyze DC currents and voltages.

To ease the analysis, redraw the circuit as shown in Figure 14–11. Here the emitter resistor (R_E) has been changed into two parallel resistors of 4.4 kΩ each.

Because the differential amplifier is balanced, you can now separate the two circuits as shown in Figure 14–12. Any calculations done on one half of the circuit can be applied symmetrically to the other half.

FIGURE 14–10 A differential amplifier circuit

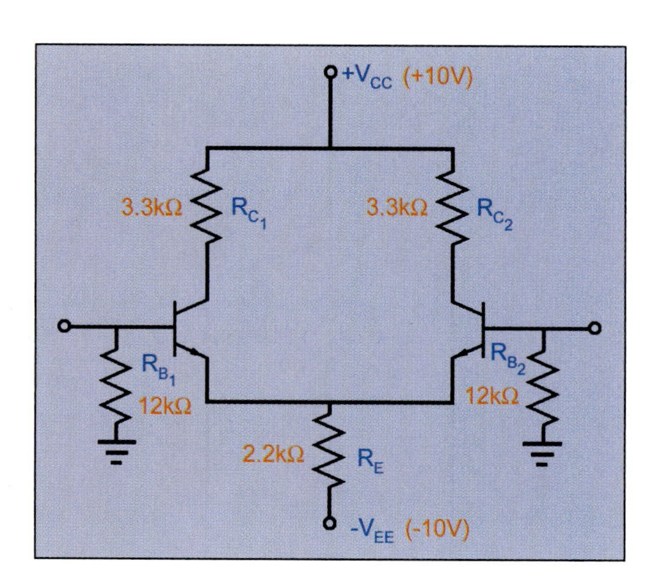

FIGURE 14–11 The circuit of Figure 14–10 with R_E split into two equal resistors in parallel

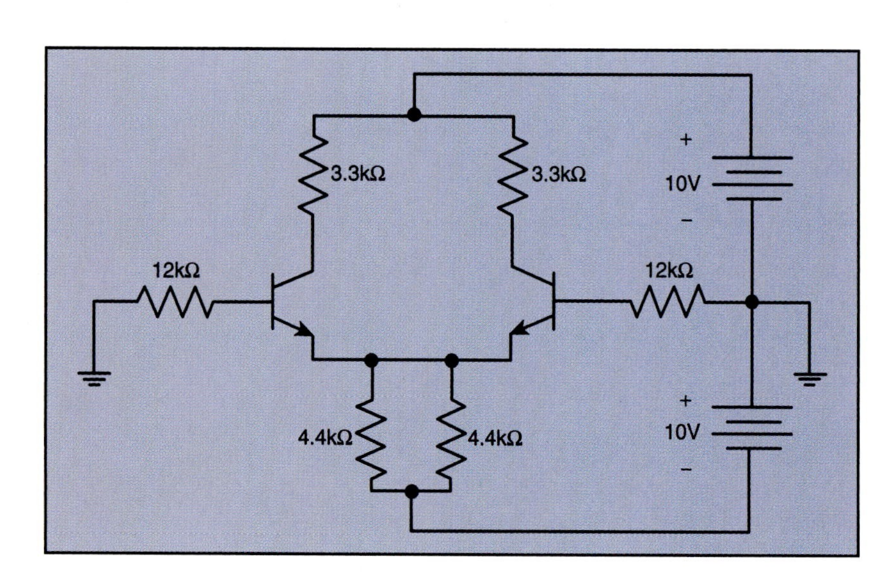

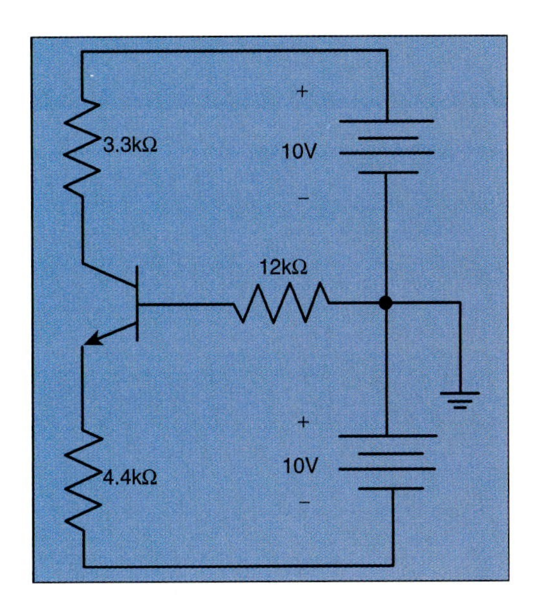

FIGURE 14–12 The right half of Figure 14–11 used to calculate the DC quiescent values

EXAMPLE 2

The following steps use the circuit of Figure 14–12.

Solution:

Step 1

Sum the voltages around the base-emitter circuit.

$$-10\text{ V} + 12\text{ k}\Omega \times i_b + V_{BE} + 4.4\text{ k}\Omega \times (1 + \beta) \times i_b = 0$$

Substituting values and assuming that $\beta = 200$ gives

$$-10\text{ V} + 12{,}000\ \Omega \times i_b + 0.7\text{ V} + 4{,}400\ \Omega \times (1 + 200) \times i_b = 0$$

Simplifying and collecting terms gives

$$896{,}400\ i_b = 9.3\text{ V} \Rightarrow i_b = \frac{9.3\text{ V}}{896{,}400\ \Omega} = 10.4\ \mu\text{A}$$

Step 2

Current through the emitter resistor is (Figure 14–12)

$$I_{4.4\text{ k}\Omega} = (1 + \beta) \times i_b = 201 \times 10.4\ \mu\text{A} = 2.09\text{ mA}$$

Step 3

The collector current is

$$\beta \times i_b = 200 \times 10.4\ \mu\text{A} = 2.08\text{ mA}$$

Step 4

Summing the voltages around the collector-emitter circuit gives

$$-20\text{ V} + 3.3\text{ k}\Omega \times 2.08\text{ mA} + V_{CE} + 4.4\text{ k}\Omega \times 2.09\text{ mA} = 0$$

Simplifying and collecting terms gives

$$V_{CE} = 20\text{ V} - 16.06\text{ V} = 3.94\text{ V}$$

AC Analysis

When calculating values for transistor circuits, AC values are normally expressed in lower case (e.g., r_E) and DC values are expressed in upper case (e.g., R_E).

AC analysis is used to determine the values of $A_{V(DIFF)}$ and $A_{V(COM)}$ using the values of resistors used in the circuit. In order to determine the gains, first determine the value of the transistor's AC emitter resistance r_E. Note that the emitter voltage drop for BJTs is usually considered to be relatively constant for small-signal work. In this example, use 50 mV as the value that is typical of the types of BJTs used for this application.

$$r_E = \frac{50\text{ mV}}{2.09\text{ mA}} = 23.92\ \Omega$$

The AC base resistance is calculated using

$$r_B = \frac{R_B}{\beta} = \frac{12\text{ k}\Omega}{200} = 60\ \Omega$$

The value of $A_{V(DIFF)}$ is calculated with the formula

$$A_{V(DIFF)} = \frac{R_C}{(2 \times r_E) + r_B} = \frac{3.3\text{ k}\Omega}{(2 \times 23.92\ \Omega) + 60\ \Omega} = 30.6$$

The value of $A_{V(COM)}$ is calculated with the formula

$$A_{V(COM)} = \frac{R_C}{(2 \times R_E)} = \frac{3.3\text{ k}\Omega}{(2 \times 2.2\text{ k}\Omega)} = 0.75$$

OPERATIONAL AMPLIFIERS (OP-AMPS)

14.5 Basic Characteristics

Operational amplifiers are circuits that use differential amplifiers and exhibit characteristics that make them very easy to use in electronic circuits. Some of the characteristics are:

- High gain
- High input impedance
- Low output impedance
- Common-mode rejection
- Ease of design by using external circuit values

The word 'operational' in operational amplifiers (op-amps) is derived from the fact that they were originally designed to perform mathematical operations such as addition, subtraction, division, multiplication, and even differentiation and integration. Although modern-day digital computers with their speed and accuracy have taken the place of op-amps when it comes to mathematical operations or computations, op-amps are still used in a multitude of applications today. Some of the present-day uses include signal conditioning, process control, and communications.

14.6 Stages of an Op-Amp

Characteristics of an op-amp such as high gain, high input impedance, and low output impedance cannot be attained by a single transistor. An op-amp is an integrated combination of several stages of amplifiers. Figure 14–13 shows the different stages of an op-amp.

- Stage 1 (input stage) is a differential amplifier that provides a high input impedance and high CMRR.
- Stage 2 is another differential amplifier whose inputs are derived from the outputs of the first stage. This stage serves to improve on the differential voltage gain and common-mode rejection provided by the first stage.
- Stage 3 (output stage) is a common-collector (emitter-follower) stage. This stage has very low output impedance. The output is a single terminal and is referred to as a single-ended output.

Note in Figure 14–13 that the op-amp does not provide a differential output. The single-ended output shows only one phase with respect to ground. When the input is applied to the noninverting terminal, the output and input are in phase. When the input is applied to the inverting terminal, the output is 180° out of phase with the input. Operational amplifiers are normally represented by a triangle with a (+) to indicate a noninverting input and a (−) to indicate an inverting input. Figure 14–14 shows a simplified block diagram of an op-amp.

14.7 A Typical Op-Amp

Circuitry and Schematic Symbols

The op-amp stages are packaged in the form of a single integrated circuit. Figure 14–15 shows the schematic diagram of an integrated circuit

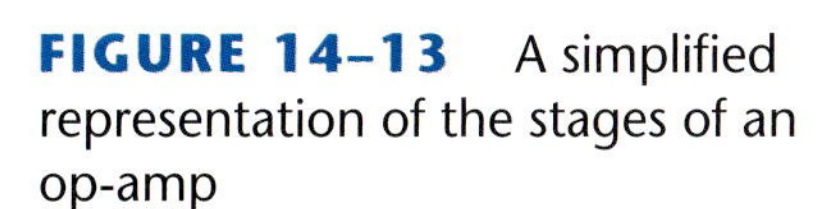

FIGURE 14–13 A simplified representation of the stages of an op-amp

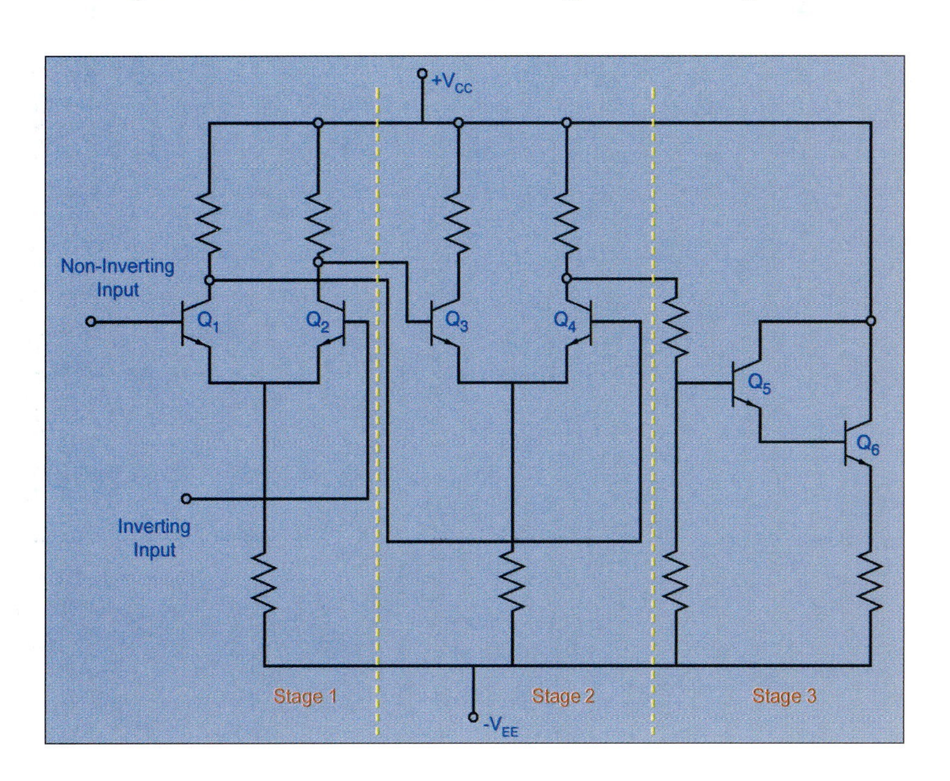

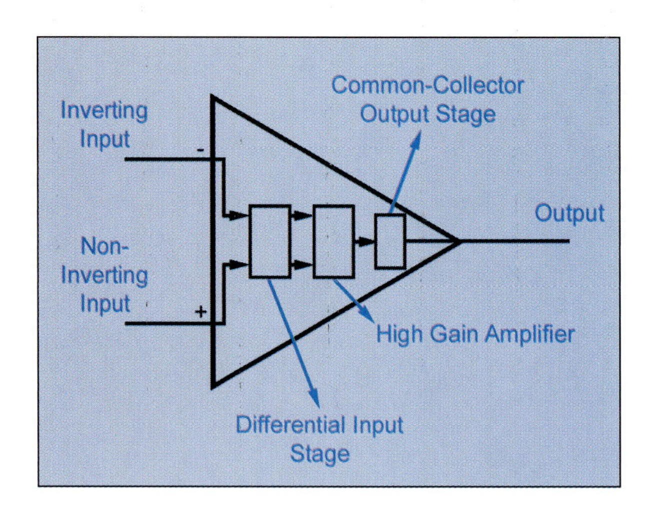

FIGURE 14–14 Simplified block diagram of an op-amp

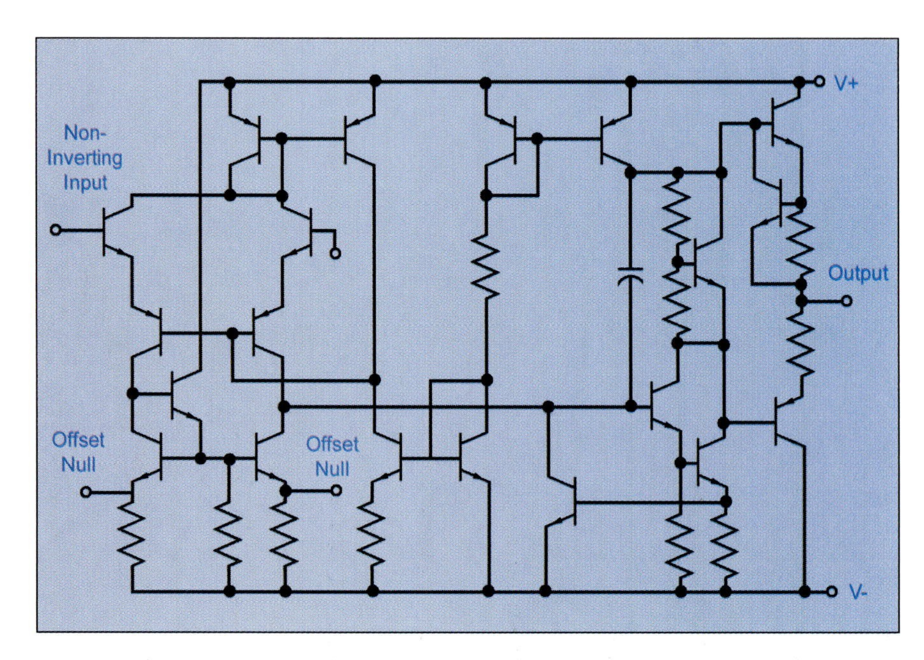

FIGURE 14–15 Schematic diagram of an integrated circuit op-amp

op-amp. The offset null terminals are used to take care of DC-offset error. As has been already discussed, it is not possible to perfectly match the transistors and resistors of the op-amp. This mismatch creates an error in the output and can be corrected by using the nulling terminals.

A typical op-amp is represented by the symbols shown in Figures 14–16a and b. Figure 14–16a shows the inputs, outputs, and the biasing supply connections. Figure 14–16b shows only the inputs and outputs and is the way that the op-amp is usually drawn in a circuit.

External Nulling Circuitry

Figure 14–17 shows an op-amp nulling circuit. When no input voltage is applied, the output voltage of the op-amp should be zero with respect to the ground. But if an offset exists, the output voltage is likely to be different from zero. This can be easily corrected by adjusting the potentiometer setting in the nulling circuit until the output voltage becomes zero.

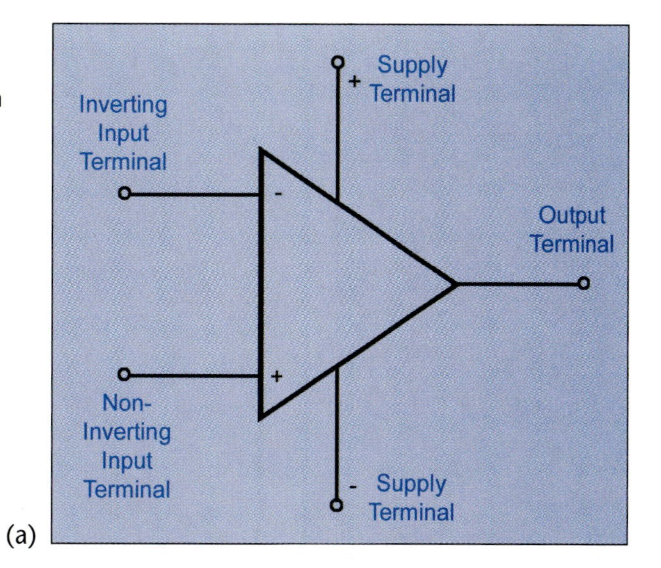

FIGURE 14–16 Schematic symbols of an op-amp; a) showing power connections, b) as usually seen in circuit diagrams

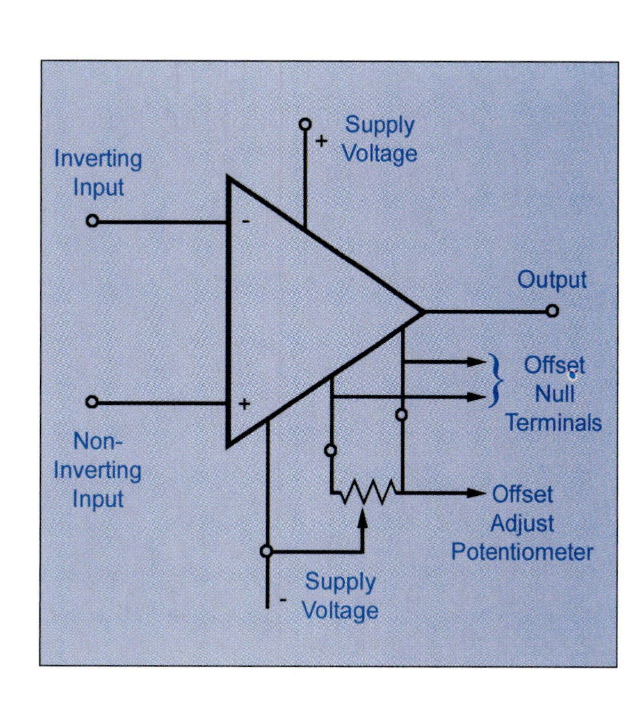

FIGURE 14–17 An op-amp nulling circuit

Slew Rate

Slew rate is defined as the maximum rate of change of the output voltage of an op-amp. The output voltage cannot make an instantaneous change. If the slew rate is 0.04 V/ms, it means that the output voltage changes over a range of 40 mV in 1 ms. A high value of slew rate means that the output of the op-amp is not able to slew (change) fast enough with the input. This produces slew-rate distortion. Slew-rate distortion is an important factor to consider at high frequencies.

Figure 14–18 shows the op-amp's response to an instantaneous change in the op-amp's input.

Figure 14–19 shows slew-rate distortion. Note that when a sinusoidal input signal is applied, the output signal becomes triangular because of the op-amp's inability to follow the rate of change of input. This deviation of the output waveform from the input constitutes slew-rate distortion.

Slew rate also limits the output of the op-amp from producing its full output swing. We can use the following equation to predict the value of the maximum signal frequency that can be applied to an op-amp without distortion being produced at the output.

$$f_{\text{max}} = \frac{\text{slew rate}}{6.28 \times V_{\text{p}}}$$

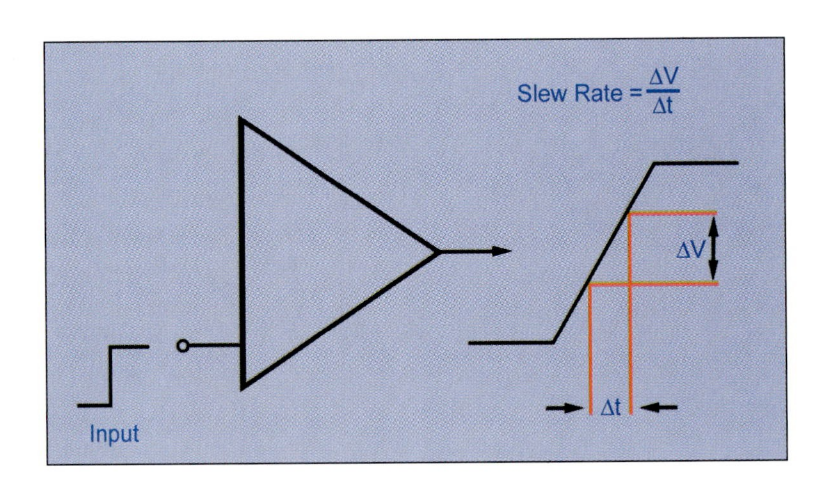

FIGURE 14–18 Op-amp's response to an instantaneous change in input

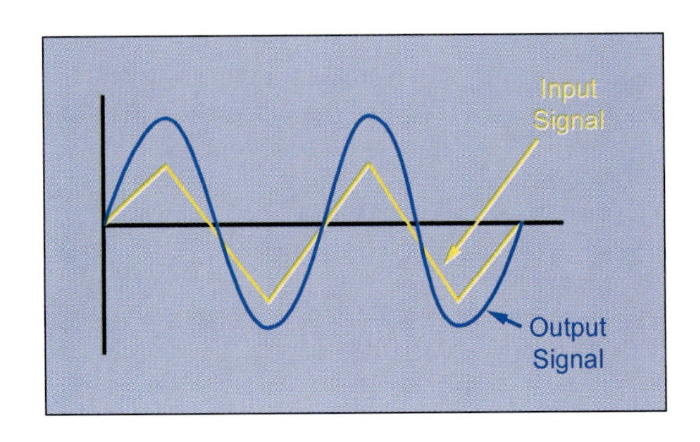

FIGURE 14–19 Slew-rate distortion

Where:

f_{max} = the maximum frequency that can be applied
V_p = peak output voltage

If an op-amp can produce a maximum output swing of 13 V when using a 15 V power supply and the slew rate is $0.4\frac{V}{\mu s}$, the maximum operating frequency of the op-amp can be determined as follows:

$$f_{max} = \frac{0.4\frac{V}{\mu s}}{6.28 \times 13\ V} = 4.89\ kHz$$

Some typical operating values for op-amps like LM 741C include the following:

- Open-loop voltage gain (A_{OL}), which is the value of the voltage gain without any feedback, is very large and typically about 200,000 (106 dB).
- Output impedance is typically small and has a value of 75 Ω.
- Input impedance is typically large and has a value of 2 MΩ.
- CMRR is very large and is 90 dB.
- Offset adjustment range is ±15 mV.
- Small-signal bandwidth is 1 MHz.
- Output voltage swing is ±13 V.
- Slew rate is 0.5 V/μs.

14.8 Open-Loop Operation of the Op-Amp

The open-loop operation of the op-amp involves circuits that have no feedback; that is, there is no connection between the output and the input. The op-amp's input terminals are called differential input terminals because the output voltage (V_O) depends on the difference in voltage (E_d) between them and the open-loop gain of the amplifier (A_{OL}).

$$V_O = E_d \times A_{OL}$$

Where:

E_d = (voltage at the (+) input) − (voltage at the (−) input)

Figure 14–20a and 14–20b show op-amp circuits that operate in the open-loop mode. In this mode, they essentially function as comparators, where they compare the inputs to the inverting (−) and noninverting (+) terminal and amplify the difference. Keep in mind that A_{OL} is extremely large, on the order of 200,000 or more. Typical operating voltages of the op-amp are ±15 V, which limits the output voltages of the op-amp to $\pm V_{sat}$ (±13 V or ±14 V). Even with an E_d as low as ±65 μV, the output will go to $\pm V_{sat}$.

$$V_O = \pm 65\ \mu V \times 200{,}000 = \pm 13\ V = \pm V_{sat}$$

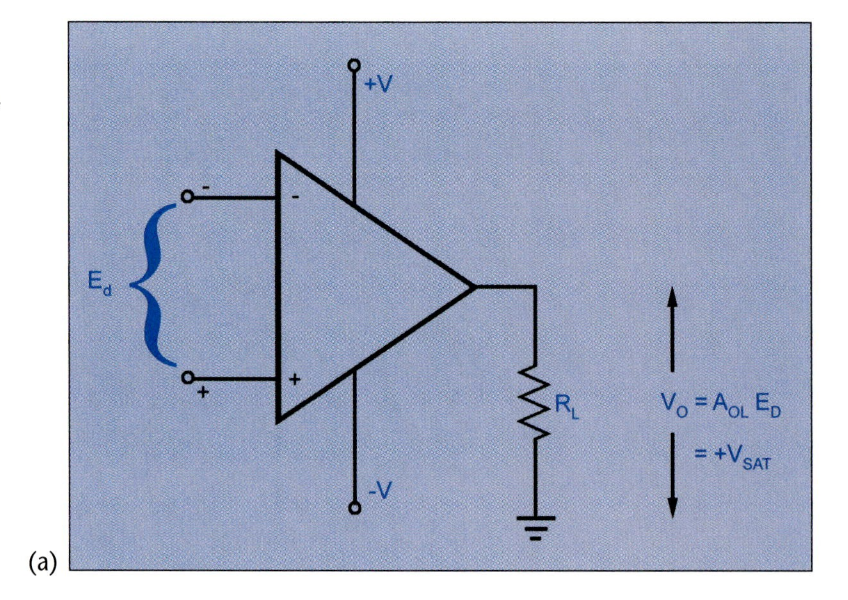

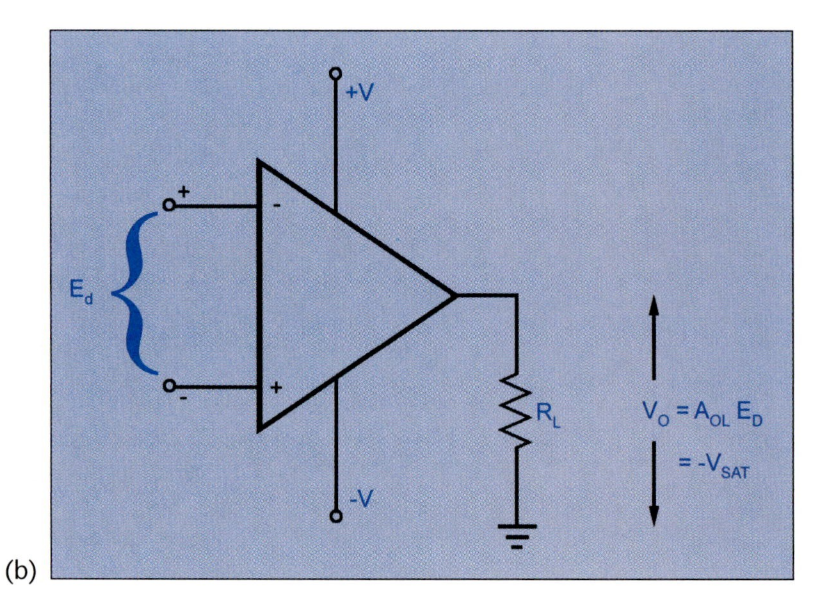

FIGURE 14–20 Op-amp input and output characteristics; a) noninverting input more positive, b) inverting input made more positive

Clearly, this high-gain amplifier forces the output V_O to stay between $+V_{sat}$ and $-V_{sat}$, depending on whether E_d is positive or negative. In Figure 14–20a, the (+) terminal is made more positive with respect to the (−) terminal, and this causes the output V_O to be positive ($+V_{sat}$). In Figure 14–20b, the (+) input is made negative with respect to the (−) input, and this causes the output V_O to be negative ($-V_{sat}$). Table 14–1 shows the data for three examples of open-loop operation.

14.9 Closed-Loop Operation of the Op-Amp

When a portion of the output signal is fed back to the input, the circuit is called a closed-loop circuit. Op-amps are generally operated with feedback in the closed-loop mode. There are two types of feedback used in op-amp circuits:

- Negative feedback
- Positive feedback

Table 14–1 Example Problem

Voltage at (+) input	Voltage at (−) input	$E_d = (+V) - (-V)$	$V_O = A_{OL} \times E_d$	Actual V_O
2 V	0.5 V	+1.5 V	300,000 V	$+V_{sat} = +13$ V
1 V	2 V	−1 V	−200,000 V	$-V_{sat} = -13$ V
30 mV	0 V	30 mV	6,000 V	$+V_{sat} = +13$ V

Negative Feedback Circuits

In these circuits, a portion of the output is fed back to the inverting (−) input of the op-amp. Negative feedback is used to reduce the gain and increase the bandwidth of an amplifier. Amplifier circuits that are constructed using op-amps use negative feedback.

The explanations of operation that follow depend on two important points:

1. The voltage between the (+) terminal and (−) terminal is zero. This is easily proven by looking at Figure 14–13.
   a. Notice that the emitters of Q_1 and Q_2 are connected together and are therefore at the same potential.
   b. Both Q_1 and Q_2 are forward biased; consequently, the base voltages for both transistors are given by the formula $V_B = V_E + V_{BE}$, where V_{BE} is the 0.7 volts for the base-emitter junction.
2. The input impedance of the op-amp is extremely high. (The LM 741C has a 2 MΩ input impedance.) This means that there is negligible current drawn into the op-amp's input terminals.

Keep these two points in mind as you read the following sections.

Inverting amplifiers Figure 14–21 is an inverting amplifier circuit. The resistor R_f connected between the output and (−) or inverting

FIGURE 14–21 Inverting amplifier

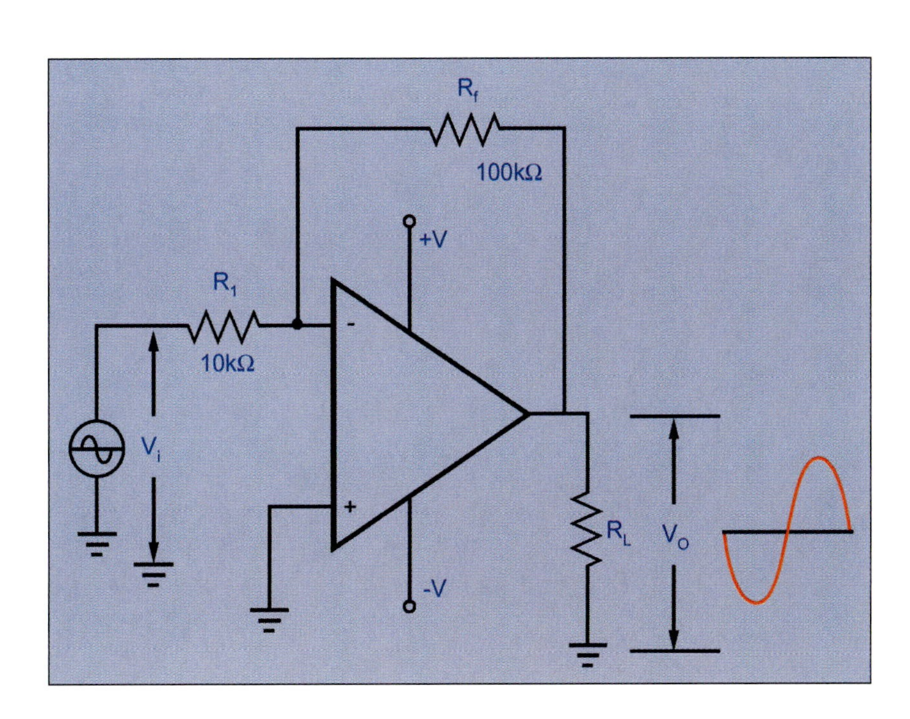

terminal provides negative feedback. The input V_i to the amplifier is connected to the inverting terminal via the resistor R_1, and the noninverting terminal is grounded. Because of the first point that the two terminals are at the same potential, the inverting input (−) is also at ground potential and is known as virtual ground. The output of the amplifier is 180° out of phase with the input. This means that it is inverted with respect to the input.

Derivation of the voltage gain (A_V) equation is as follows:

1. The existence of virtual ground at the (−) input causes the right end of R_1 to be grounded; therefore, I_1 (I_{R_1}) can be calculated from Ohm's law as:

$$I_1 = \frac{V_i}{R_1}$$

2. The output voltage V_O causes a current I_2 to flow through R_f.

$$I_2 = -\frac{V_O}{R_f}$$

3. There is almost no current flow through the input terminals of the op-amp (from the second point previously discussed), therefore $I_1 = I_2$, and:

$$-\frac{V_O}{R_f} = \frac{V_i}{R_1}$$

4. Rearranging terms yields the closed-loop voltage gain $A_{V(CL)}$:

$$A_{V(CL)} = \frac{V_O}{V_i} = -\frac{R_f}{R_1}$$

Notice the simple elegance of the op-amp as an amplifier. The gain of the amplifier is dependent only on the input and feedback resistor values. The negative sign in the gain indicates that there is a phase inversion present. For the circuit shown in Figure 14–21 the voltage gain is calculated as:

$$A_{V(CL)} = -\frac{R_f}{R_1} = -\frac{100\text{ k}\Omega}{10\text{ k}\Omega} = -10$$

A voltage gain of −10 suggests that if an input of 1 V is applied to the circuit shown in Figure 14–21, the output voltage is −10 V. On the other hand if an input voltage of −1 V is applied, the output voltage is +10 V.

Noninverting amplifiers Figure 14–22 illustrates a noninverting amplifier. In this amplifier, the input (V_i) is applied to the (+) input. Once again note the presence of negative feedback, in the form of the resistor R_f connected between the output and input. In this amplifier, the output voltage is in phase with the input.

The voltage gain is slightly different in this connection. Based on the first point that the two input terminals are at the same potential,

FIGURE 14–22 Noninverting amplifier

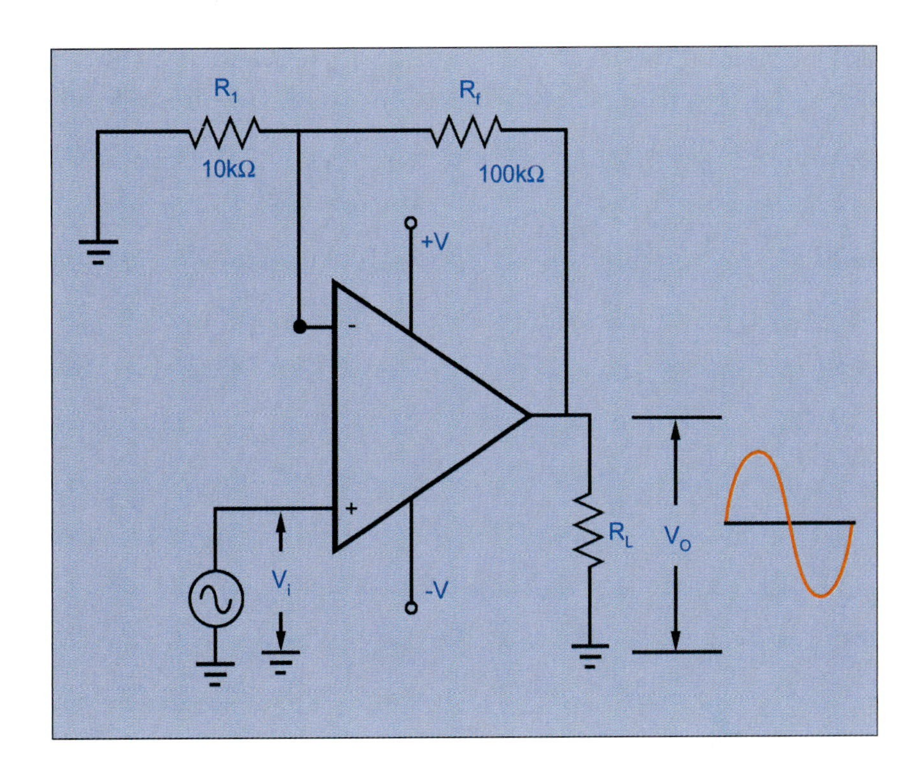

the inverting terminal is at a potential V_i. It can be seen that V_O is divided between the resistors R_f and R_L. The input voltage is calculated using the voltage divider rule and given by:

$$V_i = V_O \frac{R_1}{R_1 + R_f}$$

Rearranging terms gives the voltage gain $A_{V(CL)}$

$$\frac{V_O}{V_i} = A_{V(CL)} = \frac{R_1 + R_f}{R_1}$$

The gain of the circuit in Figure 14–22 is calculated as:

$$A_{V(CL)} = \frac{R_1 + R_f}{R_1} = \frac{10\text{ k}\Omega + 100\text{ k}\Omega}{10\text{ k}\Omega} = 11$$

A voltage gain of +11 suggests that when an input voltage of 1 V is applied, the output voltage is +11 V and when the input voltage applied is −1 V, the output voltage is −11 V.

Voltage followers This is a special case of a noninverting amplifier, with a $A_{V(CL)} = 1$. Figure 14–23 illustrates a voltage follower. Negative feedback is still present in the circuit, but here the resistors R_f and R_1 are absent. Because the two input terminals are at the same potential, the voltage V_i applied to the noninverting (+) terminal also appears at the inverting (−) terminal. The output is connected to the inverting (−) input by a wire, and hence the output voltage V_O follows the input voltage V_i.

FIGURE 14–23 Voltage follower

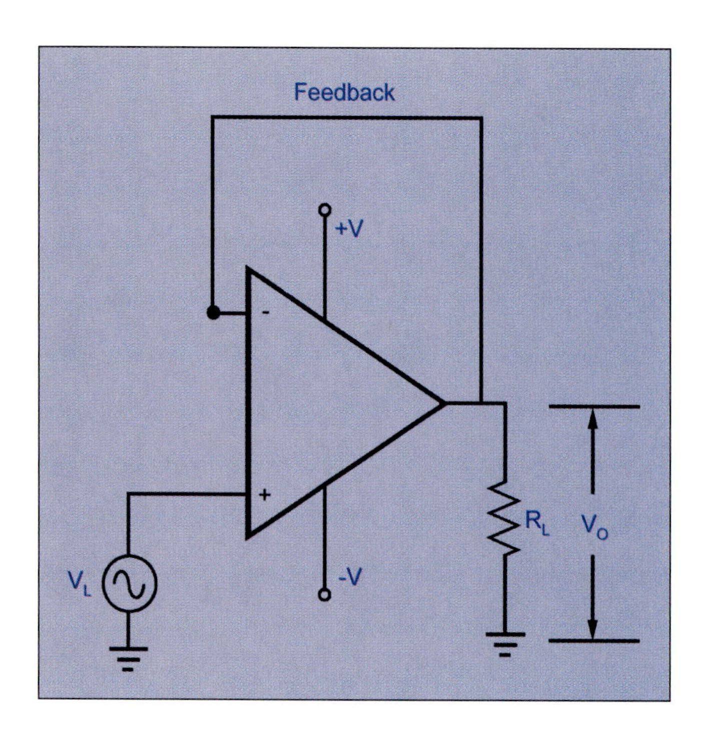

The voltage gain of this amplifier is 1. You may wonder what the use of this amplifier with a gain of 1 is. Although the amplifier has no gain, it serves as a buffer, which offers high input impedance and low output impedance and can be used for the purposes of impedance matching.

Summing amplifiers An inverting amplifier connected with two or more inputs is a summing amplifier or mixer. Figure 14–24 shows a summing amplifier with two inputs, V_1 and V_2, applied to the inverting input. The output voltage is the inverted sum of the input voltages. These summing amplifiers are used to add or mix AC or DC voltages.

The output voltage V_0 for the summing amplifier of Figure 14–24 is given by the following equation:

$$V_O = -\frac{R_f}{R_1}V_1 - \frac{R_f}{R_1}V_2$$

If V_1 is 3 V, V_2 is 5 V, and the resistors R_1, R_2, and R_f are each of 10 kΩ, the output voltage for the summing amplifier shown in Figure 14–24 is given by:

$$V_O = -\frac{10\ \text{k}\Omega}{10\ \text{k}\Omega}3\ \text{V} - \frac{10\ \text{k}\Omega}{10\ \text{k}\Omega}5\ \text{V} = -8\ \text{V}$$

The inputs V_1 and V_2 can be scaled by any factor by changing the values of R_1 and R_2. If R_1 is made 5 kΩ and R_2 and R_f remain the same (10 kΩ), the output voltage V_O is given by

$$V_O = -\frac{10\ \text{k}\Omega}{5\ \text{k}\Omega}3\ \text{V} - \frac{10\ \text{k}\Omega}{10\ \text{k}\Omega}5\ \text{V} = -11\ \text{V}$$

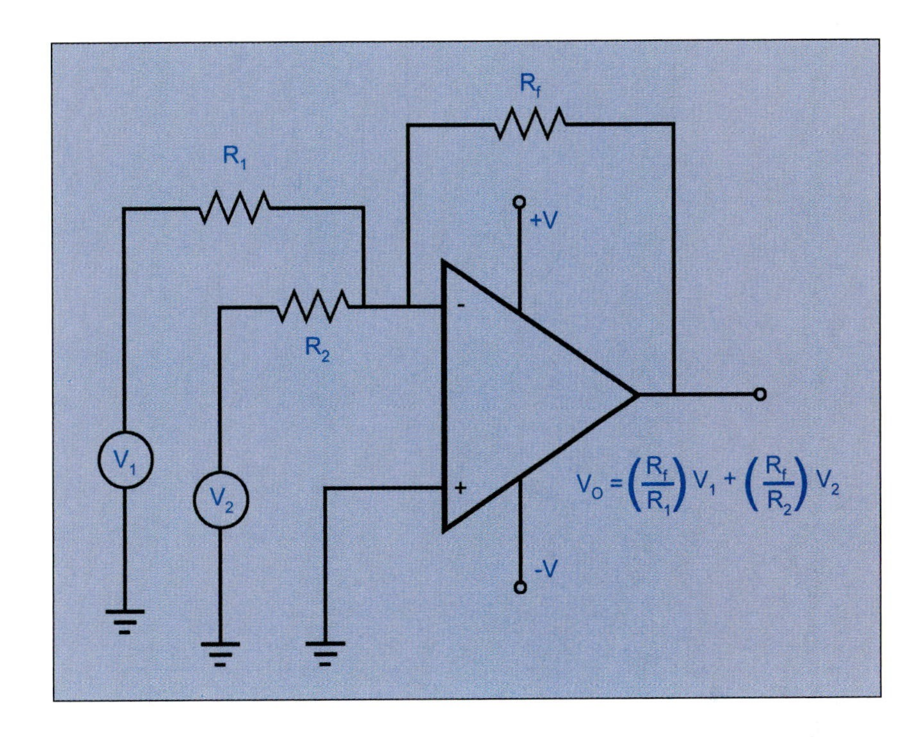

FIGURE 14–24 Summing amplifier

The number of inputs of the summing amplifier/mixer can be extended. By scaling the gains of the different inputs, the inputs can be mixed in different proportions. A summing amplifier can thus be used in an audio mixer to mix the inputs from different microphones. The virtual ground on the inverting input provides isolation between the individual microphones, and there is no interaction between inputs. By scaling the gains of different inputs, a very low voice can be made more audible than a loud guitar. Figure 14–25 shows an audio mixer.

Positive Feedback Circuits

In positive feedback circuits, signal from the output of the op-amp is applied to the noninverting (+) input of the op-amp. Figure 14–26 shows an op-amp circuit with positive feedback. Positive feedback increases the gain, so the output of the op-amp circuit changes between $+V_{sat}$ to $-V_{sat}$. Positive feedback serves to reinforce comparator action, and in addition, it makes the circuit immune to noise voltages.

The op-amp circuit of Figure 14–26 is called Schmitt trigger. The output of this circuit changes between $+V_{sat}$ and $-V_{sat}$. Resistors R_1 and R_2 divide the output voltage and apply a portion to the noninverting input. When the output is at its positive maximum ($+V_{sat}$), the voltage that is fed back to the (+) input is V_{UTP}, or upper threshold point, and is given by applying the voltage divider rule:

$$V_{UTP} = V_{sat} \times \frac{R_1}{R_1 + R_2}$$

When the output is at its negative maximum ($-V_{sat}$), the voltage fed back to the (+) input is called V_{LTP}, or lower threshold point.

$$V_{LTP} = -V_{sat} \times \frac{R_1}{R_1 + R_2}$$

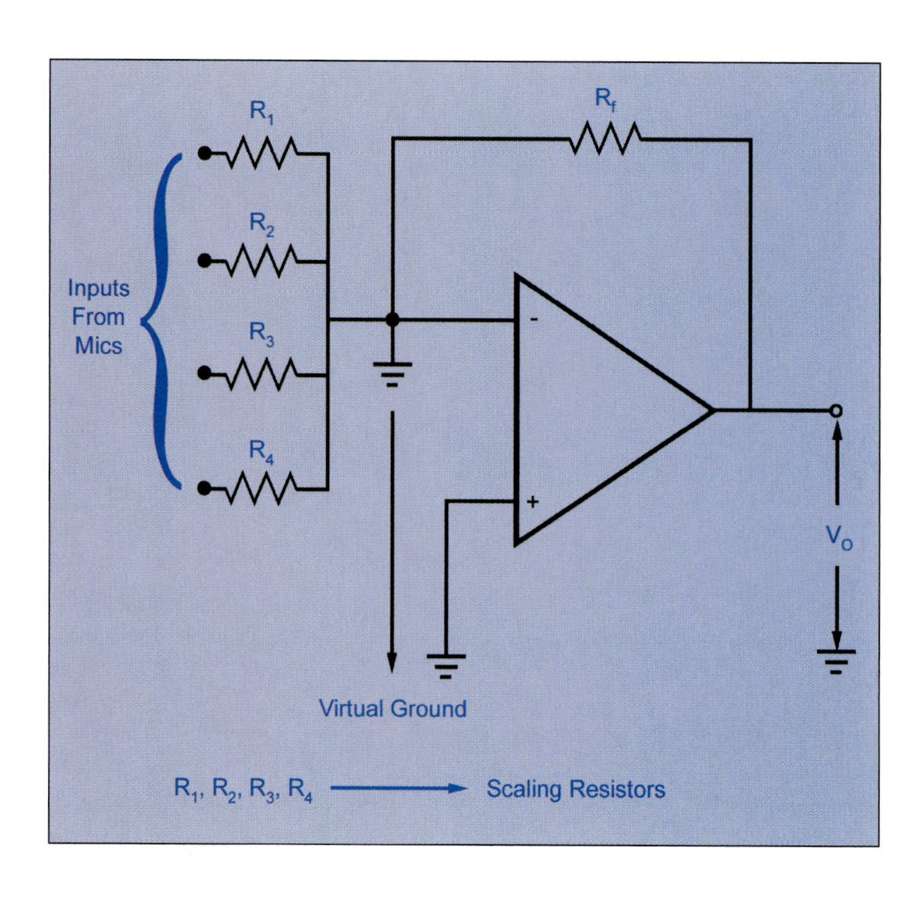

FIGURE 14–25 An audio mixer

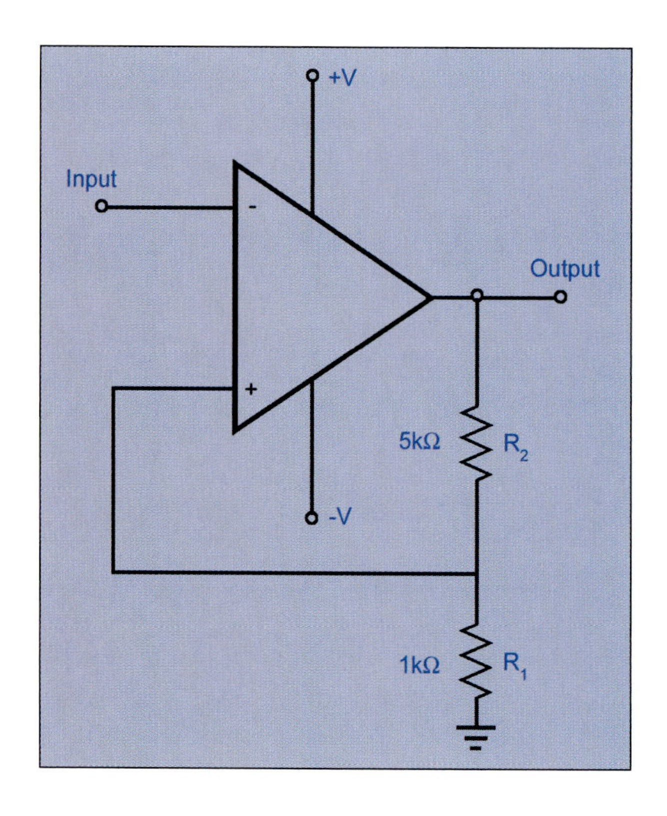

FIGURE 14–26 A Schmitt trigger

EXAMPLE 1

Suppose that the Schmitt trigger of Figure 14–26 operates at $\pm V_{sat} = \pm 15$ V. R_1 and R_2 are 5 kΩ and 1 kΩ. The values of V_{UTP} and V_{LTP} can be calculated as

$$V_{UTP} = +15 \times \frac{5\ \text{k}\Omega}{5\ \text{k}\Omega + 1\ \text{k}\Omega} = 2.5\ \text{V}$$

$$V_{LTP} = -15 \times \frac{5\ \text{k}\Omega}{5\ \text{k}\Omega + 1\ \text{k}\Omega} = -2.5\ \text{V}$$

Figure 14–27 shows the Schmitt trigger in operation. The input to the Schmitt trigger is a sinusoidal signal. First suppose that the output is at $+V_{sat}$. The voltage applied to the noninverting terminal is V_{UTP}. As the input goes positive, it crosses the upper threshold point at 2.5 V. This causes the inverting input to be more positive than the noninverting input, and so the output switches to $-V_{sat}$. Now, the voltage that is fed to the noninverting terminal is V_{LTP}. As the input voltage crosses the lower threshold (−2.5 V), the output switches to $+V_{sat}$.

The advantage of positive feedback is that it makes the circuit immune to noise signals. In normal comparators, there is a chance of false triggering due to noise. However, when the noisy signal passes through the Schmitt trigger's upper threshold point and lower threshold point, false triggering does not occur. Figure 14–28 compares the operation of the Schmitt trigger to that of regular comparators.

14.10 Active Filters

Circuits that pass certain frequencies through a circuit, while preventing certain other frequencies from passing through, are called filters.

FIGURE 14–27 Operation of Schmitt trigger

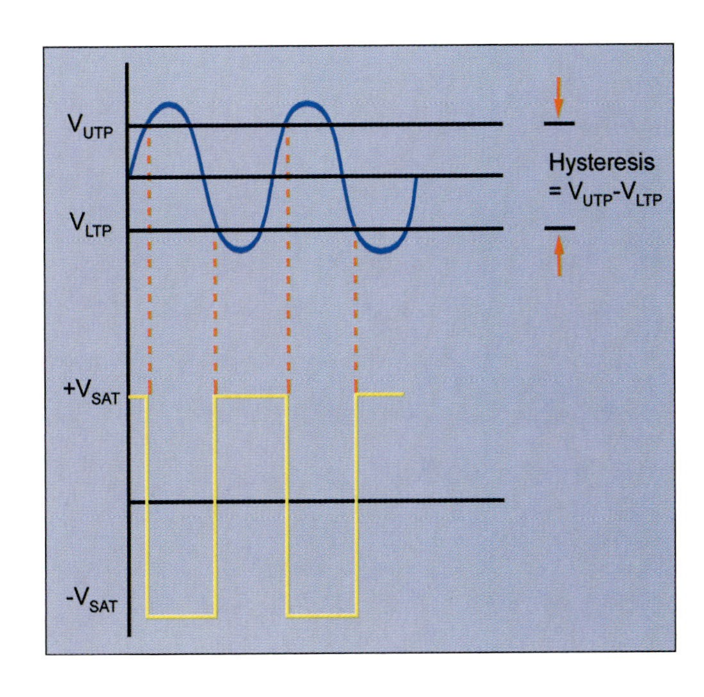

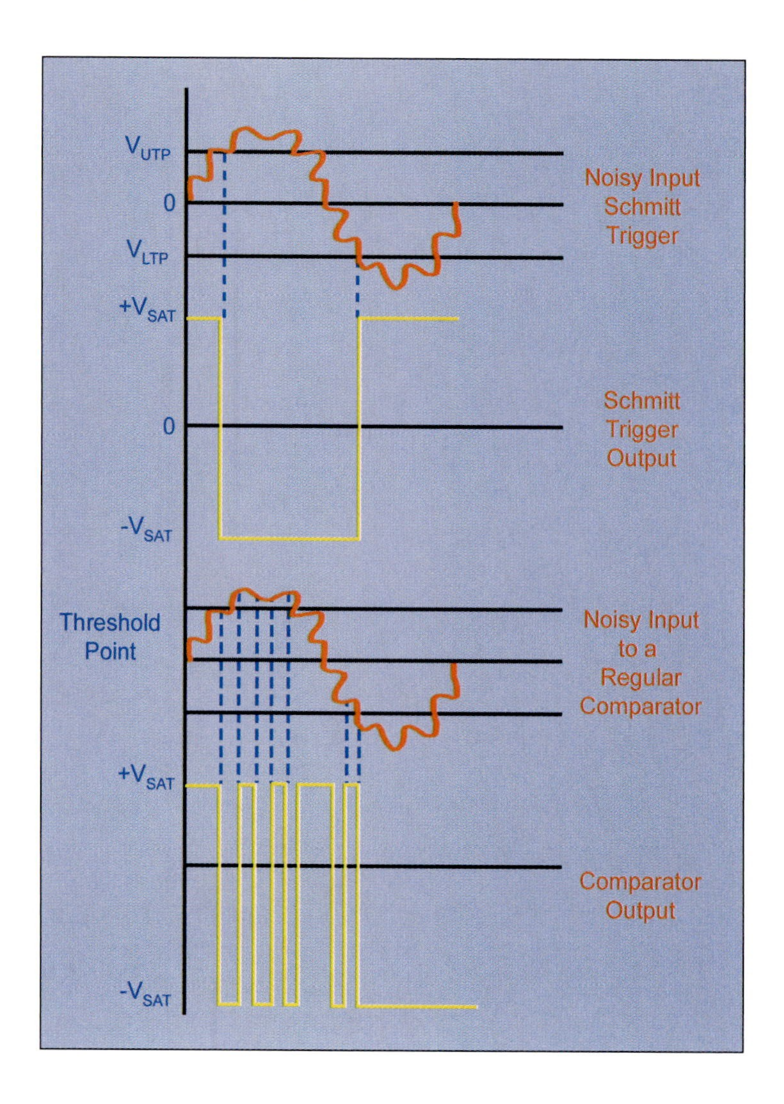

FIGURE 14–28 Elimination of false triggering using Schmitt trigger

Filters that are constructed using resistors, inductors, and capacitors are called passive filters. Passive filters offer no gain in the circuit. However, filters that are constructed using op-amps are called active filters. Active filters offer high gain (or amplification) to frequencies that it passes, while it attenuates frequencies that it blocks.

Active filters constructed using op-amps employ both positive and negative feedback. These filters are constructed using resistors and capacitors in the feedback paths. The capacitor in the feedback path has a lagging effect on the current and simulates an inductor. Active filters can be classified as follows:

- Active low-pass filter
- Active high-pass filter
- Active band-pass filter
- Active band-stop filter

Active Low-Pass Filter

An active low-pass filter passes low frequencies, while it attenuates high frequencies. Figure 14–29 shows the characteristics of an active

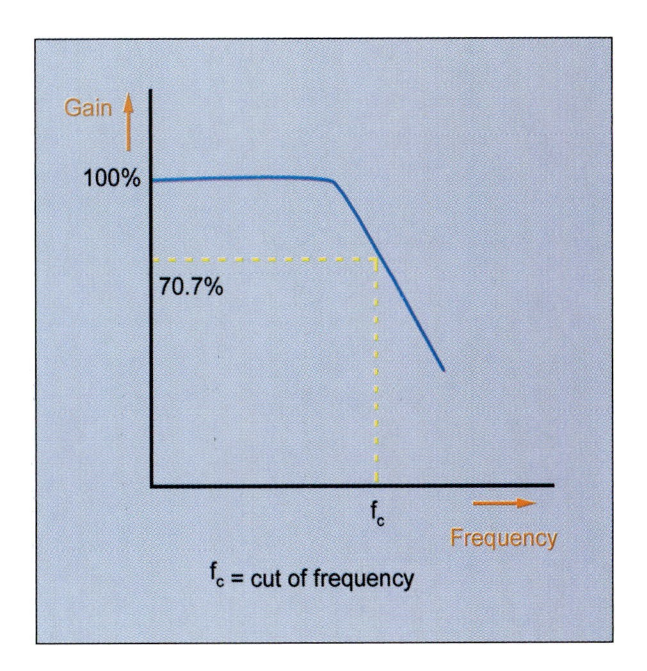

FIGURE 14–29 Active low-pass filter characteristics

low-pass filter. It indicates that gain is maximum at lower frequencies, and as frequency increases, gain starts dropping. The cutoff frequency is determined as the point at which the gain drops to 70.7 percent of the maximum value.

Figure 14–30 illustrates an op-amp circuit that functions as a low-pass filter. Note the presence of negative feedback, in the form of resistor R_f connected between the output and (−) terminal. Positive feedback is present in the form of capacitor C_1 connected between the output and (+) input.

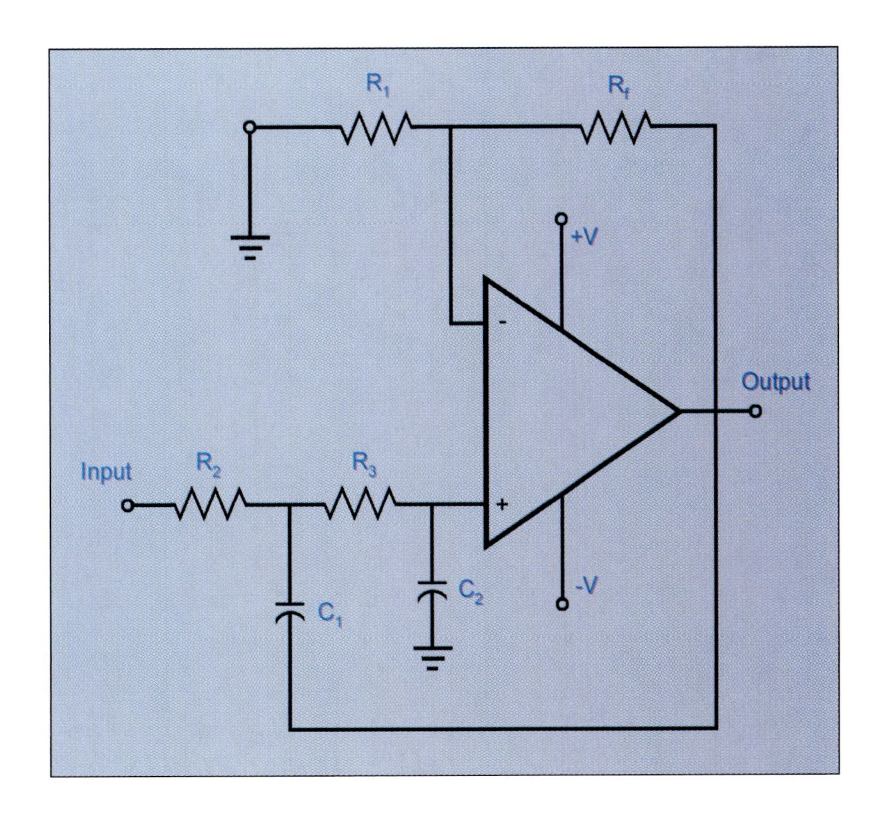

FIGURE 14–30 Active low-pass filter

Active High-Pass Filter

An active high-pass filter attenuates low frequencies, while it passes or amplifies high frequencies. Figure 14–31 shows the characteristics of an active high-pass filter. The graph indicates that the gain is low at lower frequencies and starts increasing as the frequency increases.

Figure 14–32 illustrates an op-amp circuit that functions as a high-pass filter.

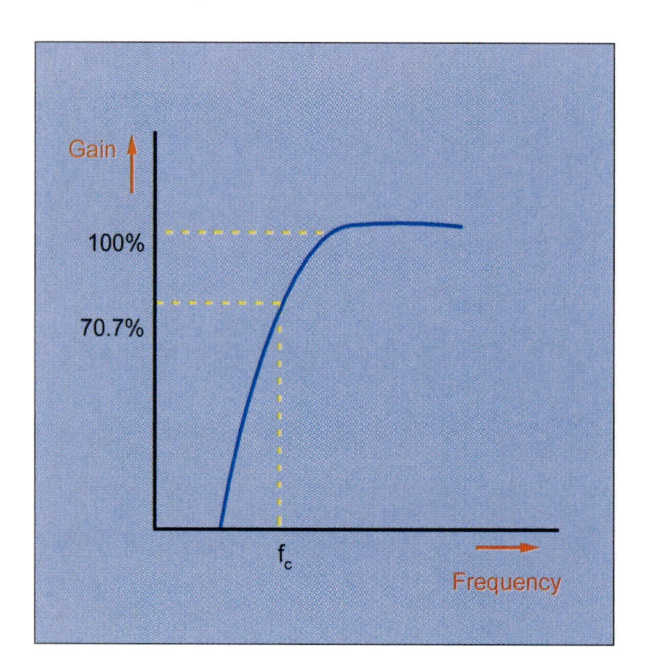

FIGURE 14–31 Active high-pass filter characteristics

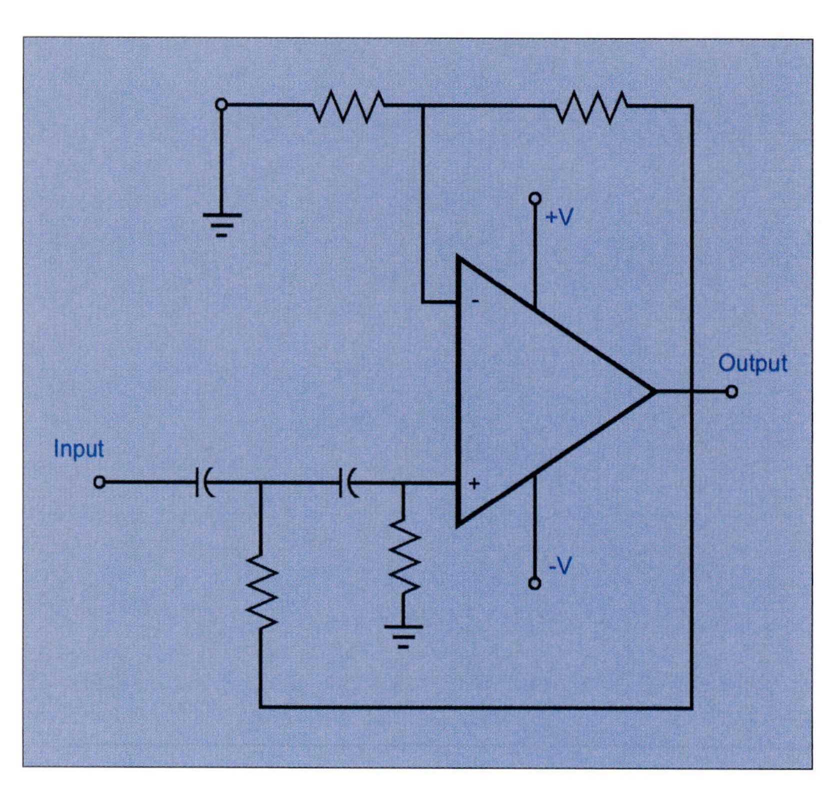

FIGURE 14–32 Active high-pass filter

Active Band-Pass Filter

An active band-pass filter passes a band of frequencies. Figure 14–33 shows the characteristics of a band-pass filter. The graph indicates that the gain is high at the band of frequencies between f_{C_2} and f_{C_1}. At frequencies less than f_{C_1} and greater than f_{C_2}, the gain is significantly lower.

Figure 14–34 illustrates an op-amp circuit that functions as a band-pass filter.

FIGURE 14–33 Active band-pass filter characteristics

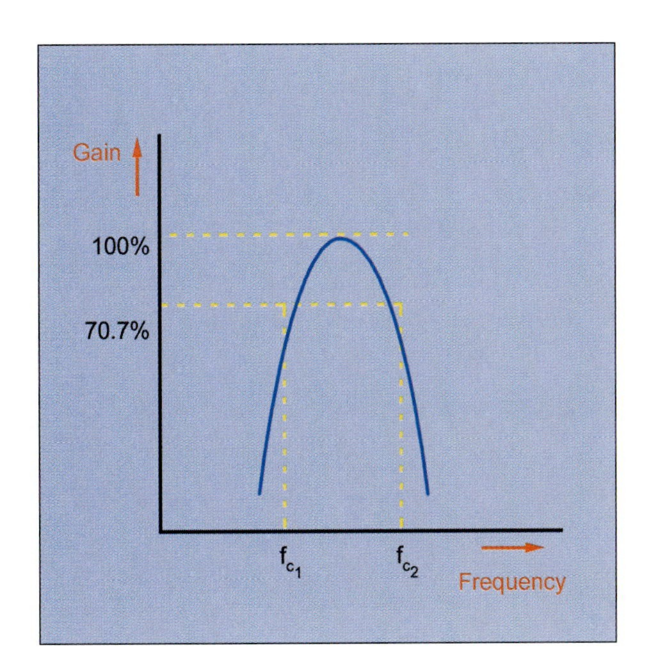

FIGURE 14–34 Active band-pass filter

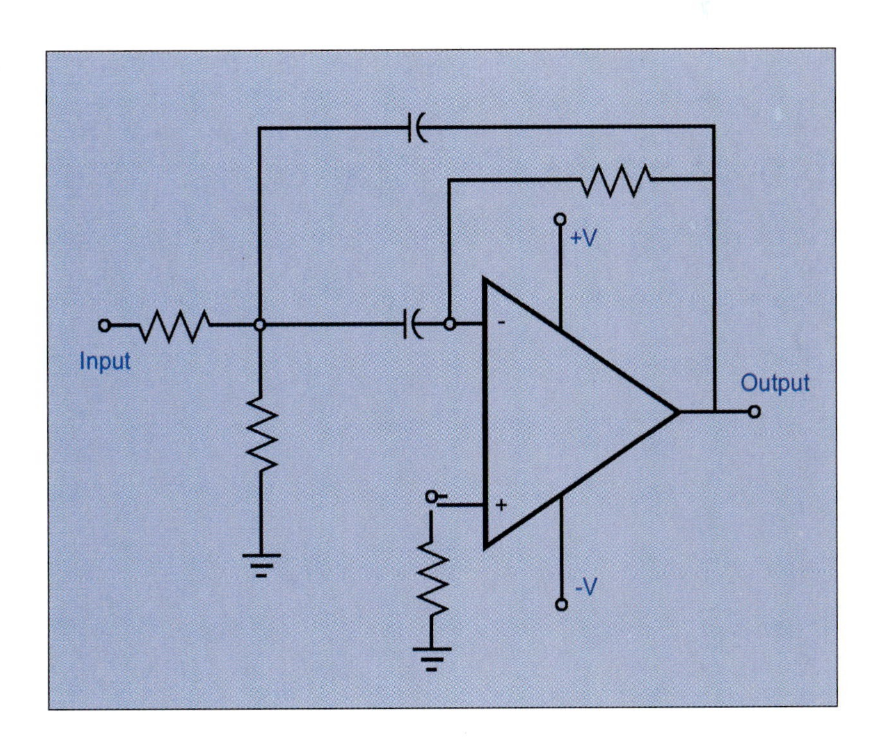

Active Band-Stop Filter

An active band-stop filter stops or attenuates a range of frequencies. The band-stop filter is also called a notch filter. Figure 14–35 shows the characteristics of an active band-stop filter. The graph shows that the gain is low at the band of frequencies between f_{C_2} and f_{C_1}. At frequencies less than f_{C_1} and frequencies greater than f_{C_2}, the gain is significantly higher.

Figure 14–36 illustrates an op-amp circuit that functions as a band-stop filter.

FIGURE 14–35 Active band-stop filter characteristics

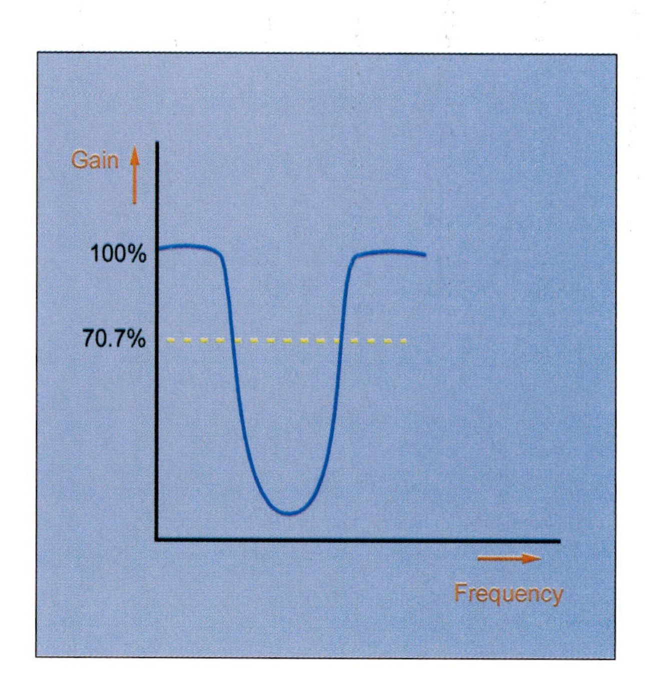

FIGURE 14–36 Active band-stop filter

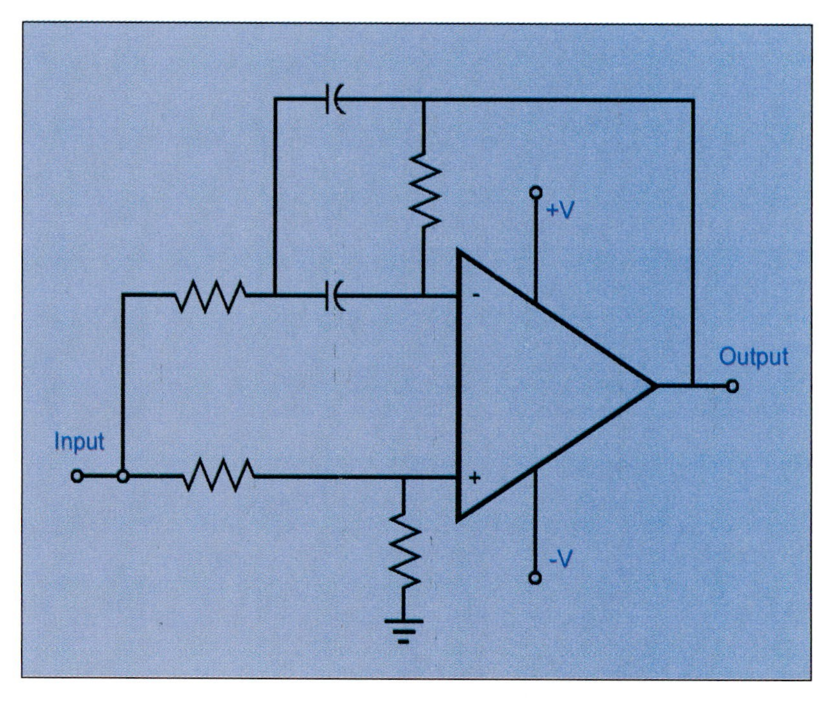

SUMMARY

Differential amplifiers can be operated or fed in three different ways.

1. Single-ended mode: The input signal is fed to only one of the transistor base circuits. In this mode, the differential amplifier works very much like a single transistor amplifier. The amplifier may be either inverting or noninverting, depending on which of the two transistors is driven.
2. Differential mode: In this mode, the input is applied between the two bases so that when the signal sends one base positive, it drives the other base negative. The gain in this condition is quite high and is a function of the difference between the two signals.
3. Common mode: When the same signal is applied to both bases, it is cancelled in the output. This is because one of the transistors creates a negative signal and the other creates a positive. So the two cancel in the output circuit.

Differential amplifiers are used whenever high amplification, input isolation from ground, and removal of common-mode noise is desired. Differential amplifiers are also used as the first two stages of an operational amplifier.

Operational amplifiers got their name from their early use in analog computers for performing mathematical operations such as addition, subtraction, and so on. In this application, they were often employed in analog computers. With the advent and improvement of digital computers, analog computers have all but disappeared.

Op-amps are now used in a wide variety of applications, including linear high-gain amplifiers, integrators for high-power walk-in circuits, trigger circuits, summing amplifiers, active filters, and many other such applications.

REVIEW QUESTIONS

1. The differential amplifier uses a positive and negative voltage supply with the two connected in the middle at common. Discuss this structure and try to determine reasons for such a method.
2. Describe the three modes of operation for a differential amplifier. List some possible applications for each mode.
3. One input of a differential amplifier is called the inverting input. Why? What is the difference between this input and the other one?
4. Look at the differential amplifier and its waveforms shown in Figures 14–5 and 14–6. What would the output be if the inputs were changed to 3 V and 3 V? 8 V and 5 V? 3 V and −3 V?
5. A certain differential amplifier has a differential mode voltage gain of 200 and a common mode voltage gain of 0.06. What is the CMRR?
6. Look at the circuit of Figure 14–10. How would the quiescent point be changed if R_E is changed to 5.6 kΩ? What if R_E is left alone and both collector resistors are changed to 5 kΩ?
7. Discuss the concept of offset null. What is it? Why is it important?
8. What is slew rate and how does it affect the frequency response of an op-amp?
9. Describe negative feedback and positive feedback. What are their uses in operational amplifier circuits?
10. You are designing an audio stage for a communications receiver. This particular receiver needs a low-pass filter that will allow voice signals but no extraneous noise above approximately 2.5 kHz. How would you provide this feature?

OPERATING PRINCIPLES

15.1 Feedback Principles

Consider the wrecking ball of a construction crane, as in Figure 15–1a. Originally, the ball is hanging straight down; however, when the crane arm moves, the ball begins to swing back and forth. Assume that the ball first swings to the right—we call this the positive direction. Eventually, the ball will expend all of its energy and will start to swing the other direction. The ball will pass back through the starting point and swing out to the left. Again, it will reach the end of its swing and start back to the right. This type of motion is called **oscillation**.

This swinging action is shown graphically in Figure 15–1b. Note the angles shown on Figure 15–1b, as indicated by the degree symbol (°). Although the pendulum does not actually swing to plus or minus 90°, we can refer to one full oscillation as 360°; therefore, one-fourth of the total swing is 90°.

Two important points need to be understood about this swinging action:

1. The ball will swing at a definite frequency that is determined primarily by the length of the cable holding it.

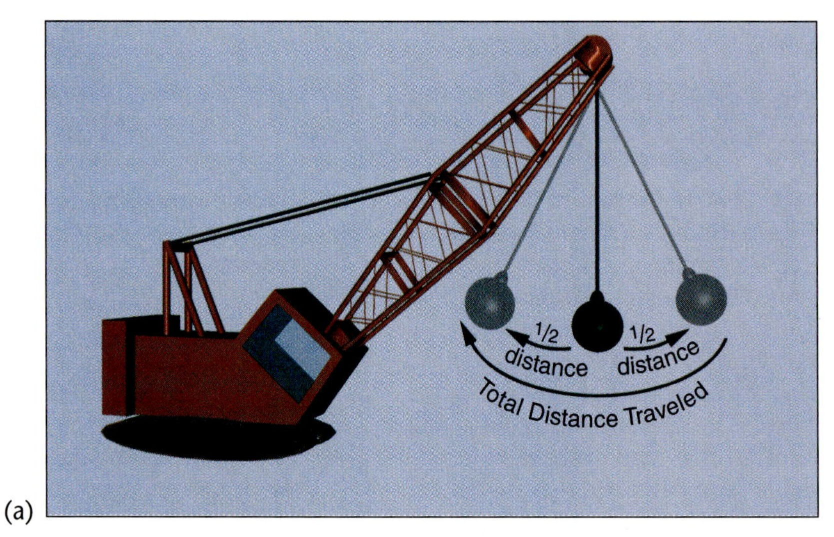

(a)

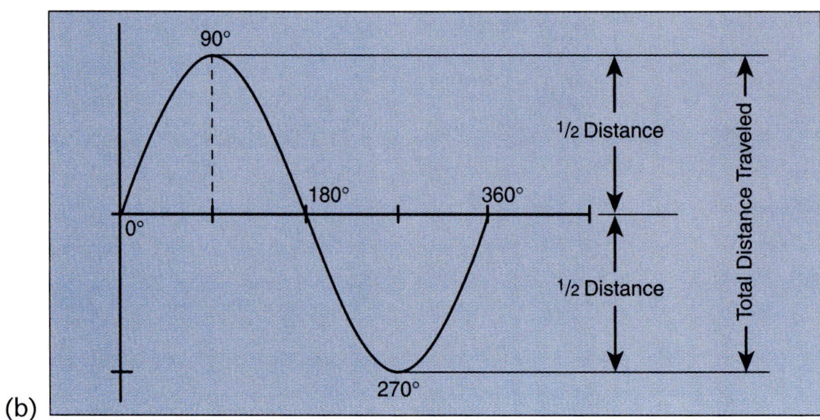

(b)

FIGURE 15–1 Oscillations and sine waves; a) the physical motion of a wrecking ball, b) time versus distance graph of the wrecking ball

2. Friction and air resistance will gradually cause the swinging action to reduce, until it finally comes to rest again.

The gradual reduction of the length of the swing is called **damping**. Note that even though the distance traveled with each swing is reduced, the time that each swing takes remains the same for a simple pendulum. This is because the frequency of the swing depends primarily on the length of the cable.

To offset this damping effect, we need to add energy to the motion of the ball, thus replacing the energy taken out by friction and air resistance. This added energy in an oscillator circuit is called **positive feedback**. Positive feedback must be in phase with the original sine wave so that it adds the needed energy to overcome and maintain signal loss. Another name for this type of feedback is **regenerative feedback** (see Figure 15–2).

The needed positive feedback effect can be accomplished only if the **feedback** is greater than the signal loss and if the feedback is in phase.

15.2 The Barkhausen Criterion

This criterion (a standard on which judgment or decision may be made) expresses the relationship between the circuit feedback factor and the voltage gain of the circuit for proper oscillator operation.

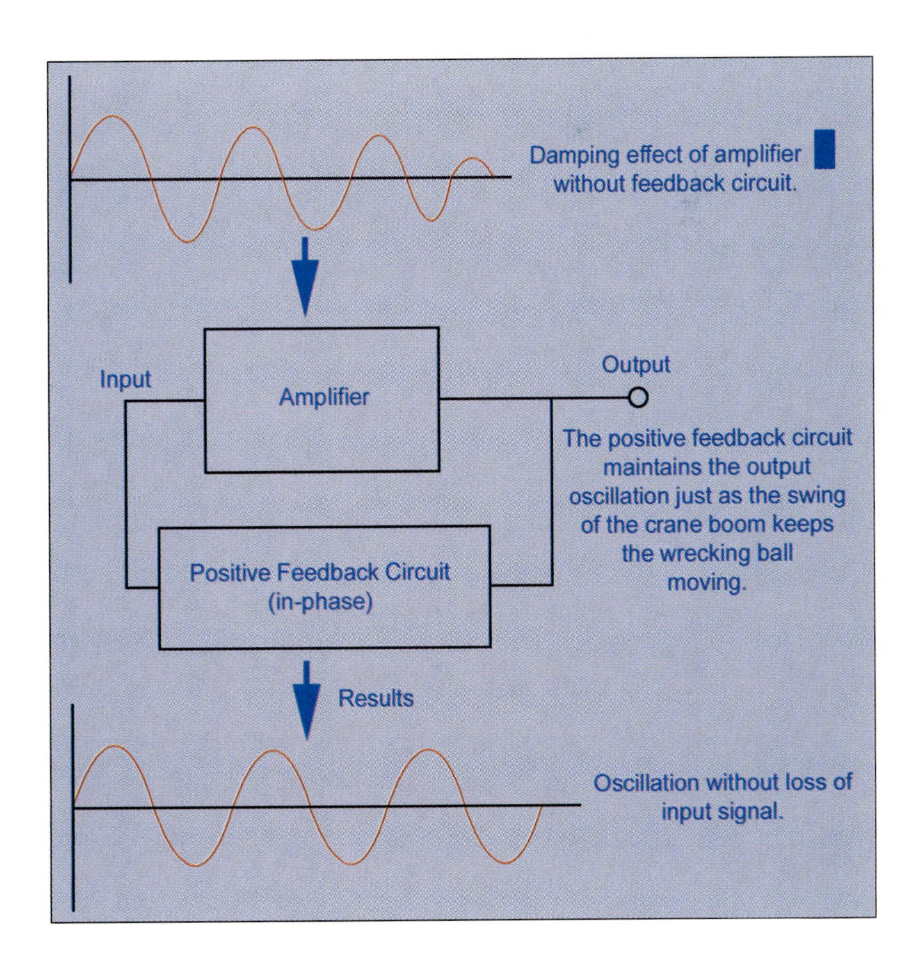

FIGURE 15–2 Damped oscillations and positive feedback

The Barkhausen criterion is expressed as

$$\alpha_V \times A_V = 1$$

Where:

α_V = fraction of the output signal that is fed back to the input
A_V = output voltage gain

This says that the product of the feedback and the gain must be exactly equal to 1 in order to maintain the oscillation at a constant level. If $\alpha_V \times A_V < 1$, then the oscillations will fade within cycles. If $\alpha_V \times A_V > 1$, then the oscillations increase each cycle and will eventually drive the oscillator into clipping and saturation.

The effects of this criterion can be seen in Tables 15–1, 15–2, and 15–3. Each table is used to analyze an amplifier with a forward voltage gain of $A_V = 100$. In each case, we assume a starting point of 0.1 V peak. The inputs and outputs of subsequent cycles are given in the tables.

In Table 15–1, the fraction of the output that is fed back with each cycle is set to 0.0075 of the output. Thus, the Barkhausen criterion is 0.75 → less than 1. The four rows are analyzed as follows:

1. The starting point is assumed to be 0.1 V peak to peak. With a gain of 100, the output will be 10 V. The fraction fed back is equal to 0.0075 or, in this case, .075 volts.
2. With 0.075 volts now at the input, the amplifier produces an output of 7.5 V. The feedback is now 0.0075 × 7.5 V = 0.0563 V.
3. The 0.0563 volts at the input produces 5.63 volts at the input. The feedback is given by .0075 × 5.63 V = .042 V.
4. In the fourth oscillation, the output is 100 × .042 V = 4.2 V. The feedback is .0075 × 4.2 = .0315 V.

Table 15–1 Negative Relationship between Feedback and Gain

Sine Wave Cycle	V_{in}	V_{out} (A_V)	Feedback (α_V)
1st	.1 V-pk	10 V-pk	.075 V-pk
2nd	.075 V-pk	7.5 V-pk	0.0563 V-pk
3rd	0.0563 V-pk	5.63 V-pk	.042 V-pk
4th	0.42 V-pk	4.2 V-pk	.0315 V-pk

Table 15–2 Positive Relationship between Feedback and Gain

Sine Wave Cycle	V_{in}	V_{out} (A_V)	Feedback (α_V)
1st	.1 V-pk	10 V-pk	.5 V-pk
2nd	.5 V-pk	50 V-pk	2.5 V-pk
3rd	2.5 V-pk	250 V-pk	12.5 V-pk
4th	12.5 V-pk	1250 V-pk	62.5 V-pk

Obviously, the output is dying off and will eventually reach zero. Thus, you see that with a Barkhausen criterion of less than 1, the oscillator will not sustain its output.

Table 15–2 shows the same oscillator with a Barkhausen criterion of 5; therefore, the feedback fraction is .05.

1. With the same starting point, the output will again be 10 V. The amount of feedback is $.05 \times 10 = 0.5$ V.
2. With 0.5 V at the input, the output is now 50 V, and the feedback is $50 \times .05 = 2.5$ V.
3. With 2.5 V at the input, the output is 250 V, and so on.

Clearly, with this activity, the oscillator will eventually reach saturation.

Table 15–3 analyzes the circuit with a Barkhausen criterion of $100 \times .01 = 1$. Note that in each case the feedback is exactly what is needed to sustain the input and the output at their starting points. Such an oscillator will be stable and continue its output.

15.3 Oscillation Frequency

The basic oscillation-producing components of an oscillator are the capacitor and inductor. This LC circuit is the familiar "tank" circuit. Recall that these components, connected in parallel, provide alternating charge and discharge outputs. Refer to Figure 15–3. There are two formulas that you need to recall. The first is for the resonant frequency of an RC circuit. The second is for the resonant frequency of an LC "tank" circuit.

For an RC circuit:

$$f_r = \frac{1}{2\pi RC}$$

For an LC circuit:

$$f_r = \frac{1}{2\pi\sqrt{LC}}$$

The oscillator can have varying frequency ranges by varying the values of the RC or LC components. Oscillators that have this capability are called variable-frequency oscillators (VFO). One method of frequency control is the lead-lag network. Figure 15–4 shows how a series-parallel RC network can be configured.

Table 15–3 Oscillator Feedback and Gain Equal 1

Sine Wave Cycle	V_{in}	V_{out} (A_V)	Feedback (α_V)
1st	.1 V-pk	10 V-pk	.1 V-pk
2nd	.1 V-pk	10 V-pk	.1 V-pk
3rd	.1 V-pk	10 V-pk	.1 V-pk
4th	.1 V-pk	10 V-pk	.1 V-pk

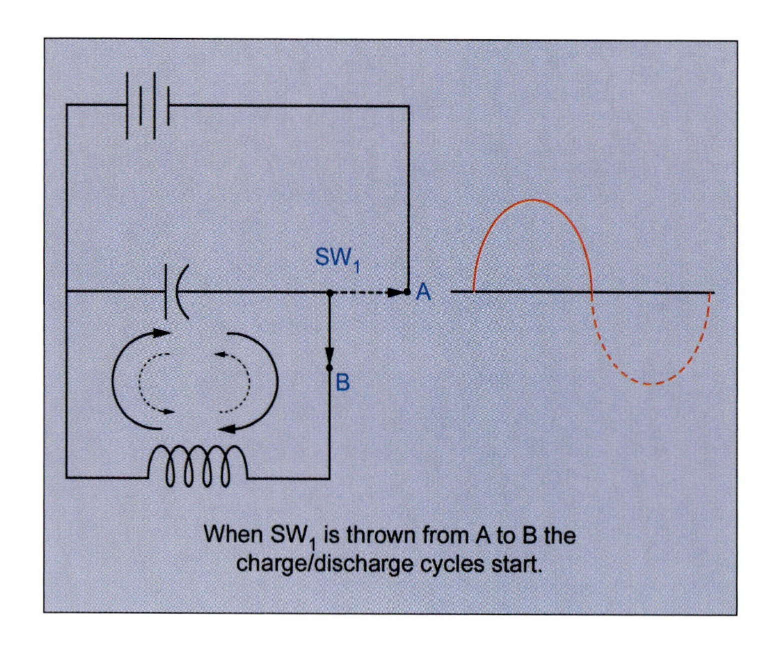

FIGURE 15–3 LC tank oscillations

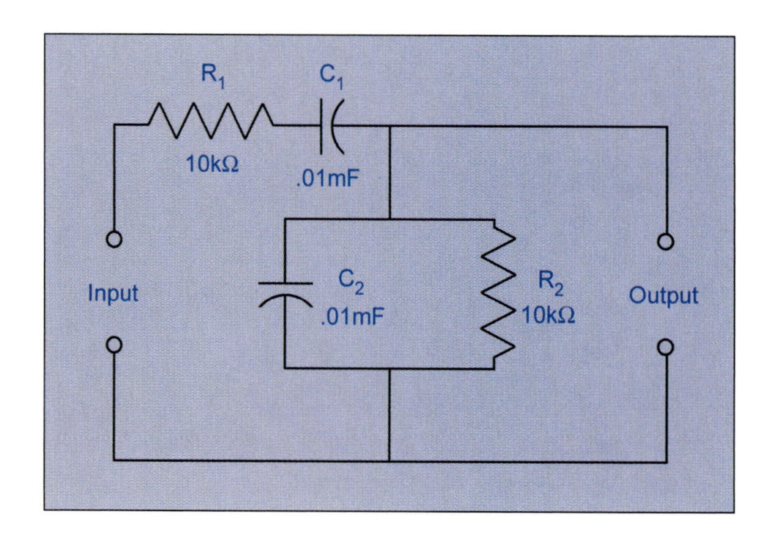

FIGURE 15–4 Lead-lag RC networks

EXAMPLE 1

Solve for the resonant frequency (f_r) of the circuit in Figure 15–4.

$$f_r = \frac{1}{2\pi RC} = \frac{1}{2 \times \pi \times 10{,}000 \times .00000001} = 1{,}592 \text{ Hz}$$

By redrawing the circuit in Figure 15–4 and replacing the resistor with an inductor (refer to Figure 15–5), we can calculate the resonant frequency of the "tank" circuit.

In this case, the resonant frequency is calculated as:

$$f_r = \frac{1}{2\pi\sqrt{LC}} = \frac{1}{2 \times \pi \times \sqrt{.0005 \times .00000001}} = 71{,}176 \text{ Hz}$$

FIGURE 15–5 LC tank oscillator circuit

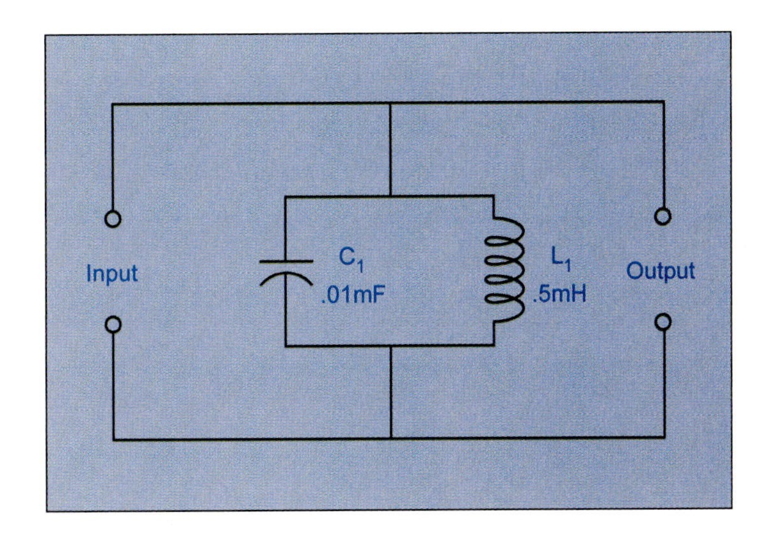

By reducing the LC component values, the oscillator can be made to create an output signal in the high megahertz range. For example, substitute the values of 10 pF and 5 μH for the capacitor and the inductor respectively.

$$f_r = \frac{1}{2\pi\sqrt{LC}} = \frac{1}{2 \times \pi \times \sqrt{5\ \mu H \times 10\ pF}} = 22.51\ MHz$$

VFOs are sometimes built using such circuits with continuously variable inductors or capacitors to adjust the frequency as required.

15.4 Oscillator Characteristics

Oscillators can be sensitive to stray or line capacitance. This is capacitance that comes from the skin effect of wires and interelectrode capacitances of transistor junctions. These excess capacitance sources can cause false readings or increase lag time at very high frequencies.

Recall the ELI and ICE sayings from your basic electricity lessons. Current will lag voltage in inductor circuits and will lead voltage in capacitive circuits. This is still true when dealing with the needed phase shifts of an oscillator. Unknown feedback paths can be developed in complicated circuits because of the unwanted capacitance and inductance.

Another impact on oscillator stability is the power supply voltage. An increase or decrease in power supply voltage can cause the biasing of the circuit to become unbalanced. For example, cutoff points could be reached earlier or saturation levels exceeded. Other effects are lowering or increasing the need for feedback from the output. Or, in a crystal oscillator, the increased voltage will vary the amplitude of the frequency.

OSCILLATOR TYPES

15.5 Phase-Shift Oscillators

The RC-Coupled Oscillators

RC oscillators are generally used only below a frequency of 1 MHz. RC circuits are used to produce the required phase shift for an oscillator to function. The RC oscillator is made of three basic sections: the amplifier, the phase-shift network, and the feedback loop. Figure 15–6 shows a typical common-emitter amplifier with the phase-shift network and the feedback loop.

RC phase-shift oscillators are built from three RC circuit combinations. In Figure 15–6, note the circled areas labeled A, B, and C. Each RC combination provides a phase shift of approximately 60°. The variable resistor (R_{V1}) provides for fine-tuning the oscillator's phase shifting to ensure 180° feedback to the input of the amplifier. The 180° phase shift is necessary because the output signal of a common-emitter amplifier is 180° out-of-phase with the input. The 180° phase shift of the output becomes in-phase with the input and thus provides signal stability, providing that $A_V \alpha_V = 1$ at the resonant frequency of the RC network.

EXAMPLE 1

In Figure 15–6, each capacitor is equal to .062 μf, and R_1, R_2, and R_3 are equal to 6.2 kΩ. At what frequency will a 180° phase shift occur for the proper operation of the oscillator?

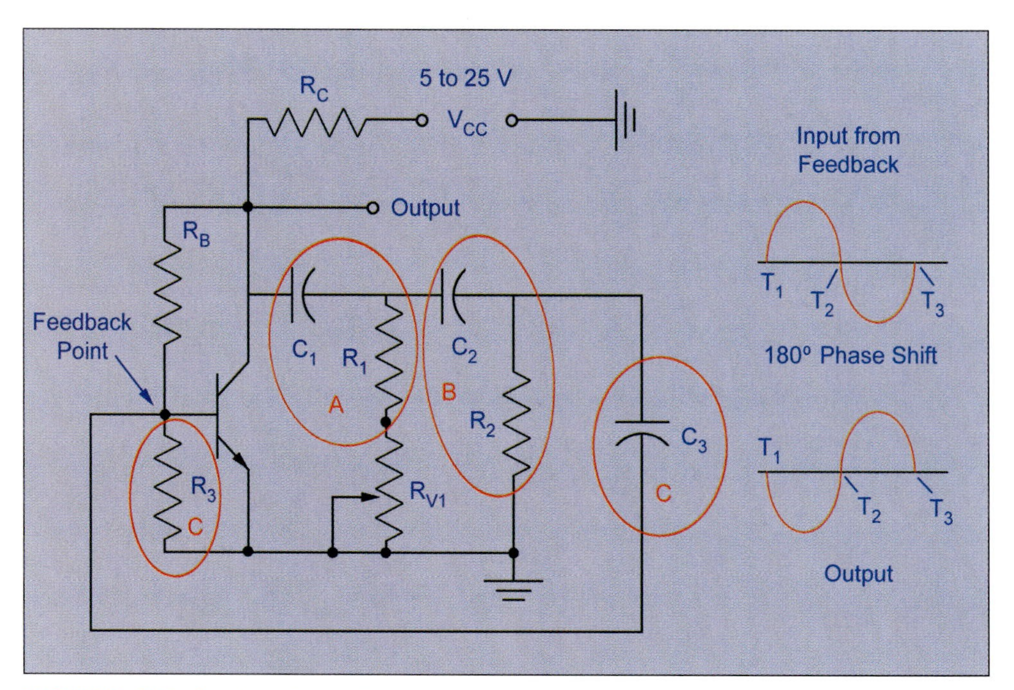

FIGURE 15–6 RC phase-shift common-emitter oscillator

Note: A new formula for calculating the combined three RC networks for the 180° phase shift is

$$f = \frac{1}{15.39 \times RC}$$

Using this new formula, solve for the frequency.

$$f = \frac{1}{15.39 \times 0.062\ \mu F \times 6.2\ k\Omega} = 169\ Hz$$

Although the phase-shift oscillator is one of the easiest to understand, it is rarely used because it is not stable and it is difficult to tune.

Phase-Shift Bridged-T Oscillators

This is a variation on the phase-shift oscillator that allows for very good frequency stability. What is sacrificed in this oscillator circuit is the range of frequencies at which the oscillator operates. Figure 15–7 shows the circuit for this type of oscillator.

C_1, C_2, and R_3 provide a high-pass filter to the network. This is because capacitive reactance decreases as frequency increases. Notice that these three components can be drawn in a 'T' shape—thus the name. R_1 and R_2 connected as a 'T' with C_3 create a low-pass filter to the network. This is because C_3 acts as a shunt to ground at high frequencies.

The circuit values are selected so that the frequency of oscillation is between the low-pass cutoff and high-pass cutoff of the two bridge circuits.

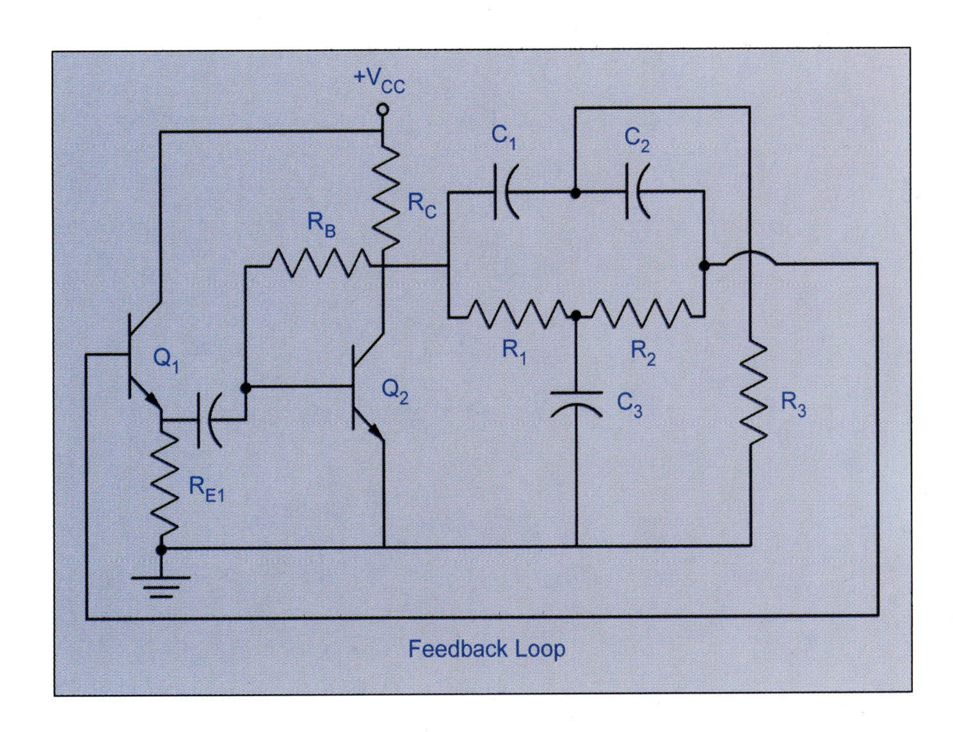

FIGURE 15–7 Phase-shift bridged-T oscillator

Wien-Bridge Oscillators

The Wien-bridge oscillator is a common low-frequency RC oscillator. This oscillator provides no phase shift at resonant frequency. The result of this "non-phase shifting" characteristic is that neither the amplifier nor the feedback network produces a phase shift.

This oscillator uses two feedback circuits—one positive and one negative. Figure 15–8 shows a schematic of the Wien-bridge oscillator. The positive feedback circuit is used to control the operating frequency of the oscillator. The **negative feedback** circuit is used to control the gain of the oscillator.

Positive feedback circuit Note that $R_1C_1 = R_7C_2$. R_2 and R_6 are added trimming pots to ensure "fine-tuning" of the positive feedback circuit. Given these equivalent values, the resonant frequency is equal to $f_r = \frac{1}{2\pi RC}$.

Negative feedback circuit The negative feedback path is a closed loop consisting of R_4, R_3, and two diodes in parallel with R_3. R_4 is a potentiometer used to control the gain of the circuit.

The closed-loop voltage gain of a noninverting amplifier is

$$A_V = \frac{R_f}{R_i} + 1$$

$R_i = R_5$ and $R_f = R_4 + R_3$ as long as the diodes are not conducting. If the oscillator output tries to go above the voltage drop of $R_4 + R_5$ by

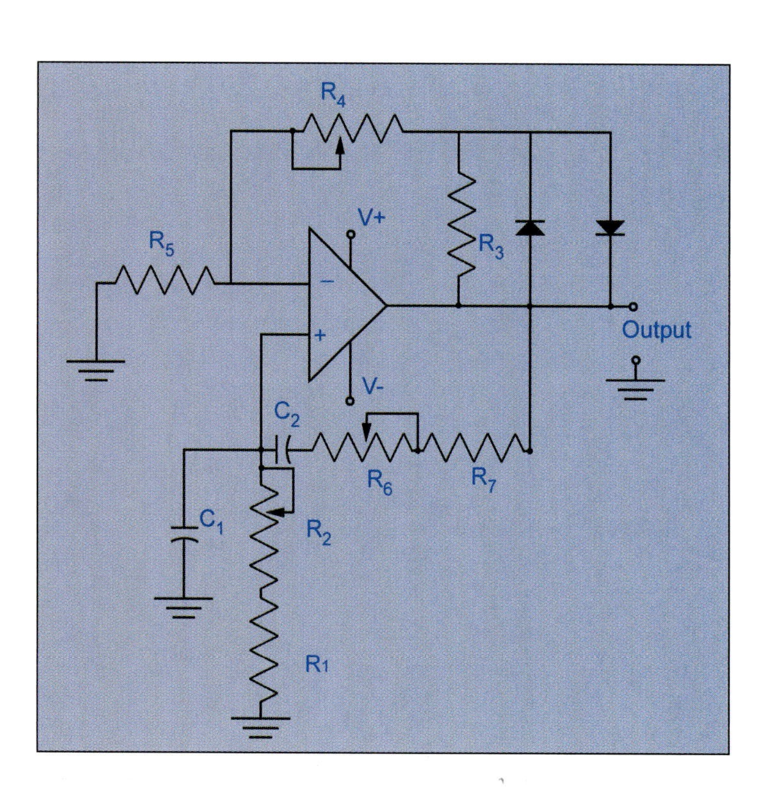

FIGURE 15–8 Wien-bridge oscillator

more than .7 volts, one of the diodes will "turn on" and effectively short out R_3, reducing the voltage gain to

$$A_V = \frac{R_4}{R_5} + 1$$

15.6 Hartley and Colpitts Oscillators

The Hartley and Colpitts oscillators are very similar in design. The major difference is how the feedback circuit is produced. Figures 15–9 and 15–10 show a comparison between the Hartley and Colpitts oscillators. Note that the feedback circuit from the Hartley oscillator is taken from the tapped coil of the inductor or autotransformer (L_1) and that the feedback from the Colpitts oscillator is taken from the tap between C_1 and C_2.

In the Hartley oscillator, Figure 15–9, C_1 prevents the RF choke (RFC) and L_1 from shunting V_{CC} to ground. The value of C_1 is made very

FIGURE 15–9 Hartley oscillator

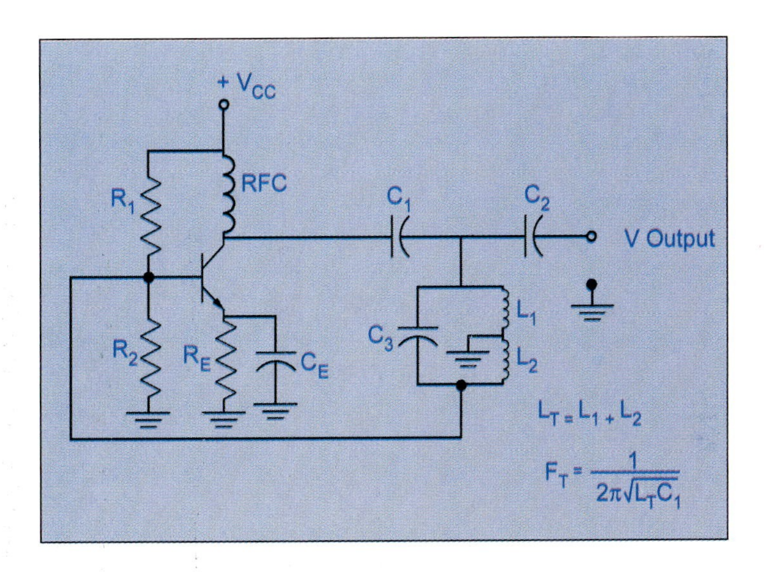

FIGURE 15–10 Colpitts oscillator

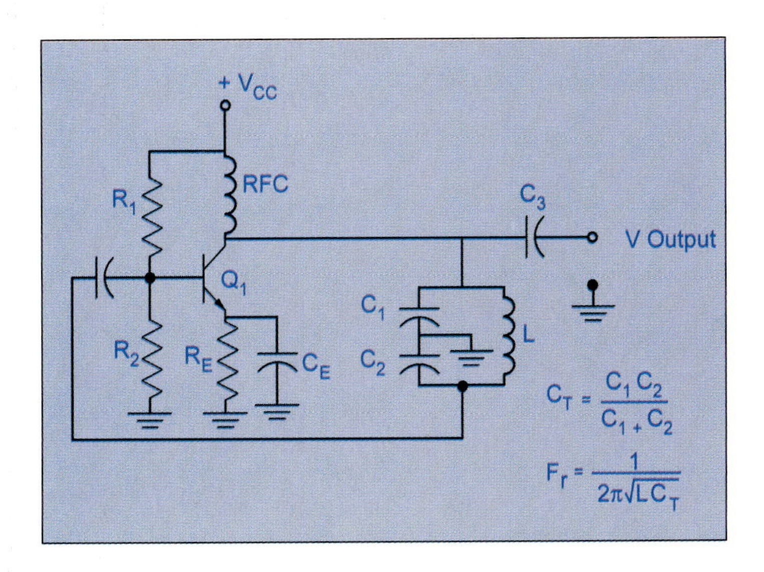

high to prevent it from affecting the resonant frequency circuit calculations. Note that the output voltage is developed across L_1 and that the feedback voltage is developed across L_2. The ratio of L_1 to L_2 is very important because the amount of feedback percentage is determined by where the tap is placed.

In a similar manner, the output of the Colpitts oscillator is taken from C_1, and the feedback is developed across C_2.

Table 15–4 lists the gain and the resonant frequency for each of the oscillators.

Any parallel resonant tank circuit will lose efficiency when loaded. A transformer is used to couple the output of the oscillators to the load. This prevents false readings and loss of oscillator efficiency. Figures 15–11 and 15–12 show Hartley and Colpitts oscillators using transformers to couple their outputs to the load.

15.7 Crystal Oscillators

The most stable of the oscillators is the crystal-controlled oscillator. Crystal oscillators have a quartz crystal that is used to control the operational frequency.

Certain types of crystals exhibit what is called the **piezoelectric** effect. This means that under compression the crystal produces an elec-

Table 15–4 Gain and Resonant Frequency of the Hartley and Colpitts Oscillators

Oscillator Type	Gain (A_V)	Resonant Frequency (f_r)
Hartley	$\frac{L_2}{L_1}$	$f_r = \frac{1}{2\pi\sqrt{(L_1 + L_2)C_1}}$
Colpitts	$\frac{C_2}{C_1}$	$f_r = \frac{1}{2\pi\sqrt{L\left(\frac{C_1 \times C_2}{C_1 + C_2}\right)}}$

FIGURE 15–11 Transformer coupling for a Hartley oscillator

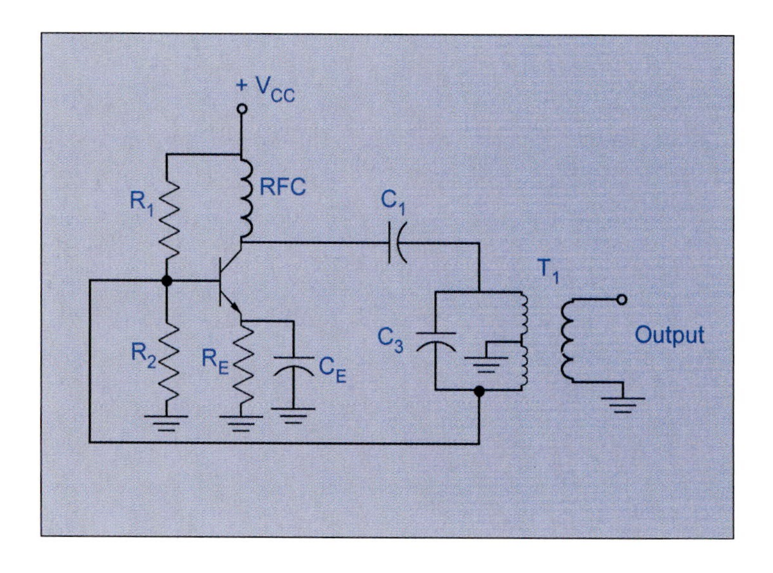

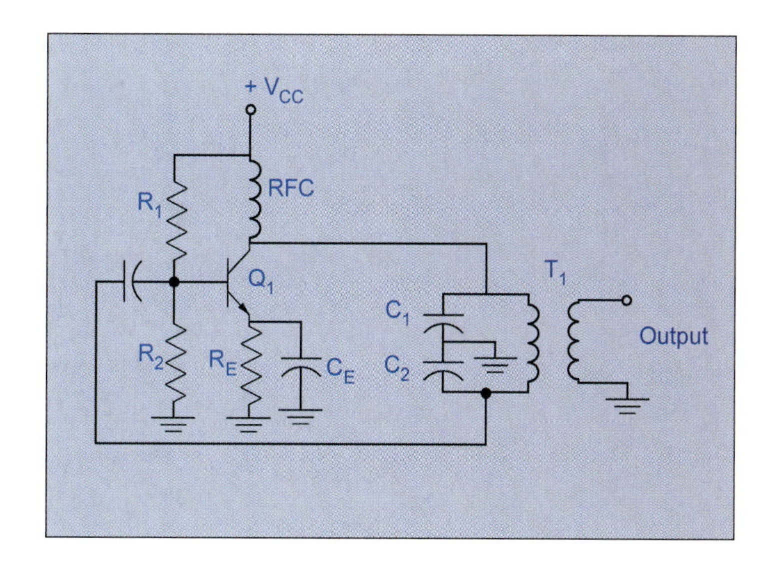

FIGURE 15–12 Transformer coupling for a Colpitts oscillator

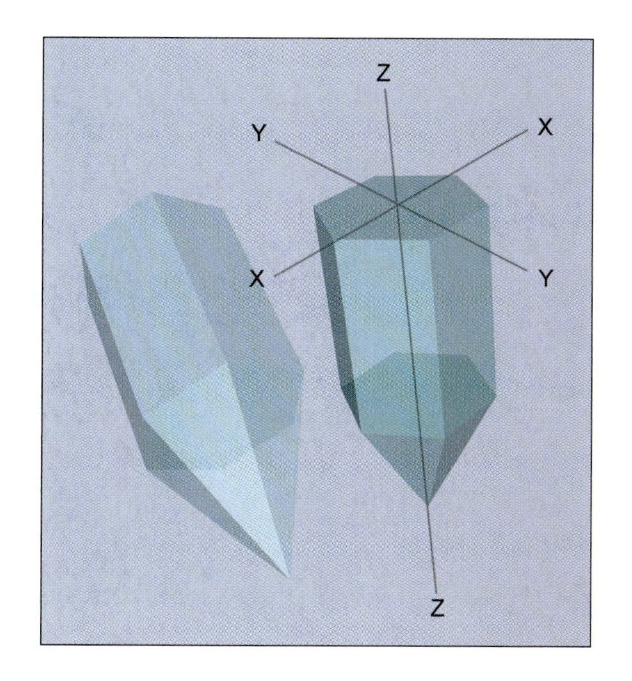

FIGURE 15–13 Quartz crystal

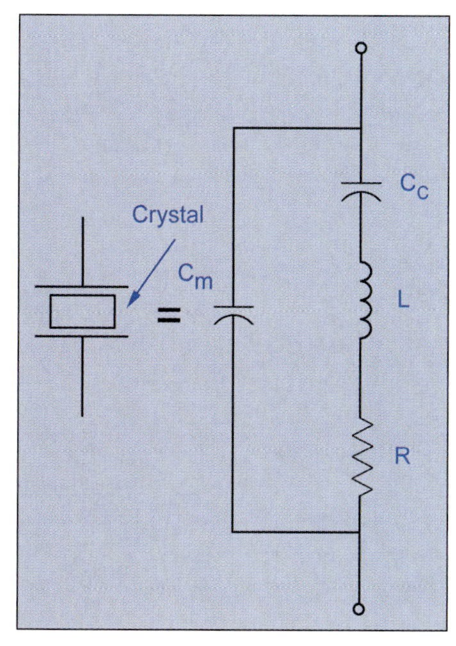

FIGURE 15–14 Crystal equivalent circuit

tric charge on its surface. The reverse is also true; when an electric charge is put across a crystal, it expands and contracts. The frequency of this expansion and contraction depends on the physical dimensions of the crystal. You can produce different frequencies by cutting crystals to different dimensions.

There are three common crystals used in oscillators: quartz, tourmaline, and Rochelle salt. The Rochelle salt is the best from a performance standpoint, but it breaks easiest. The tourmaline is the most rugged, but it is the most unstable. Quartz is the best all-around crystal for oscillator work, and it is most used. Quartz crystals are made of silicon dioxide, SiO_2. This is the same material used for the insulating layer of the MOSFET gate. Figure 15–13 is a picture of a typical quartz crystal.

Figure 15–14 shows the equivalent circuit for a crystal. Note that the crystal has all four components of an oscillator circuit.

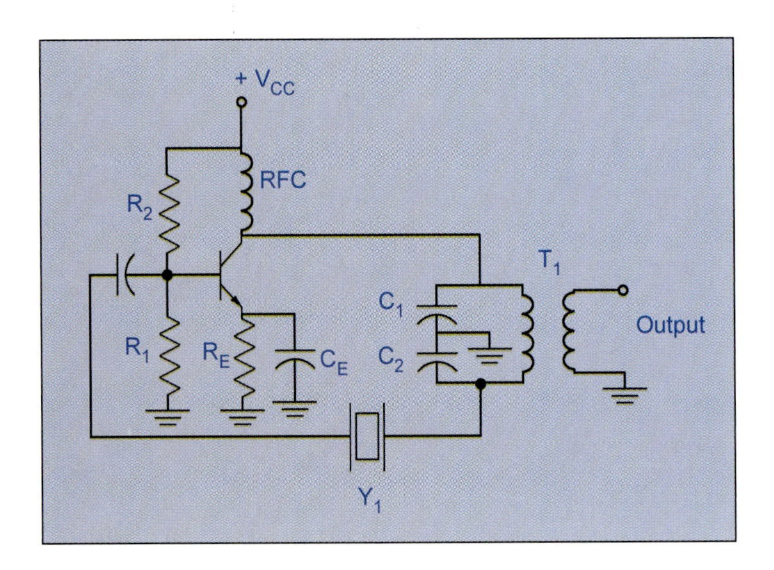

FIGURE 15–15 Crystal-controlled Colpitts oscillator

C_C = crystal capacitance

C_M = mounting capacitance

L = crystal inductance

R = crystal resistance

The "resonant" frequency of a crystal is determined by its thickness; the thinner the crystal, the higher the frequency.

Crystals also have a very high Q because of their natural frequency of vibration. This high Q is important because the higher the Q, the higher the frequency stability. Or put another way, with a high Q, the oscillator will operate at resonance only within a very narrow frequency band.

Crystal oscillators can operate in parallel or series. Figure 15–15 shows a crystal-controlled Colpitts oscillator (CCO). Note that the crystal (Y_1) is in series with the "tank" circuit. At resonance, the series crystal will have a minimum resistance and allow full feedback from the tank circuit. At frequencies other than resonance, the crystal will act as a band-stop circuit and prevent feedback from returning to the transistor's base.

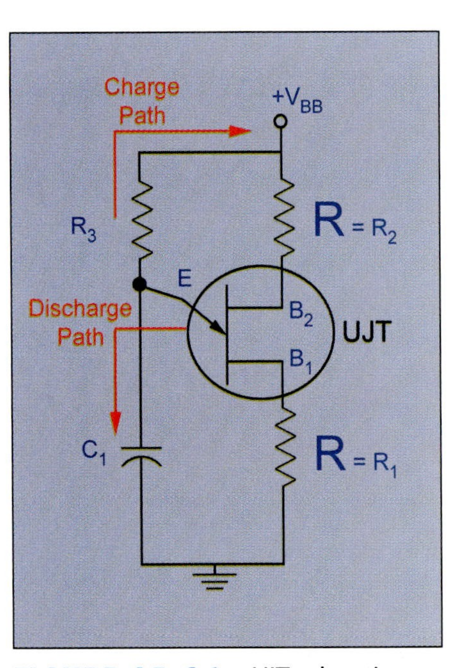

FIGURE 15–16 UJT relaxation oscillator

15.8 Relaxation Oscillators

The UJT, or unijunction transistor, can be used in an oscillator circuit called a relaxation oscillator. The relaxation oscillator is a circuit that uses the charge/discharge cycle of a capacitor or inductor to produce a pulse output. The output pulse is normally used to fire or trigger an SCR or a triac. The output pulse waveform is either sawtooth or rectangular. Figure 15–16 shows a simple relaxation oscillator circuit.

Figure 15–17 is the equivalent circuit for the UJT and is used for the following discussion. Recall from an earlier lesson that the equivalent circuit of a UJT is a diode connected to a voltage divider. For the UJT to trigger or conduct, the emitter-base junction must be .7 V more

FIGURE 15–17 UJT equivalent circuit

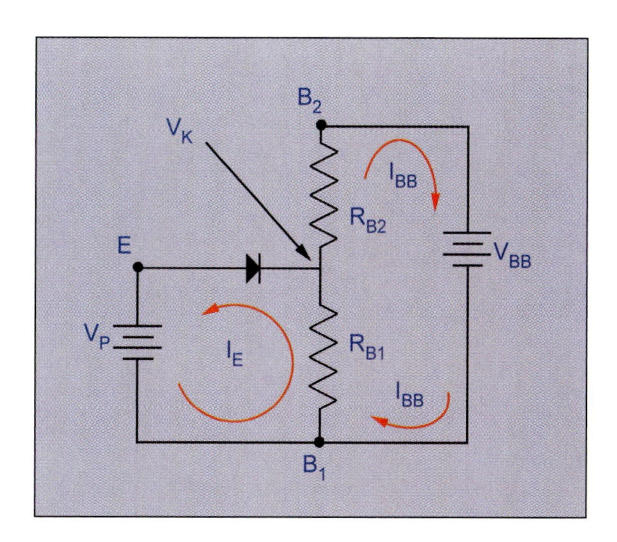

FIGURE 15–18 Relaxation oscillator charge/discharge waveforms

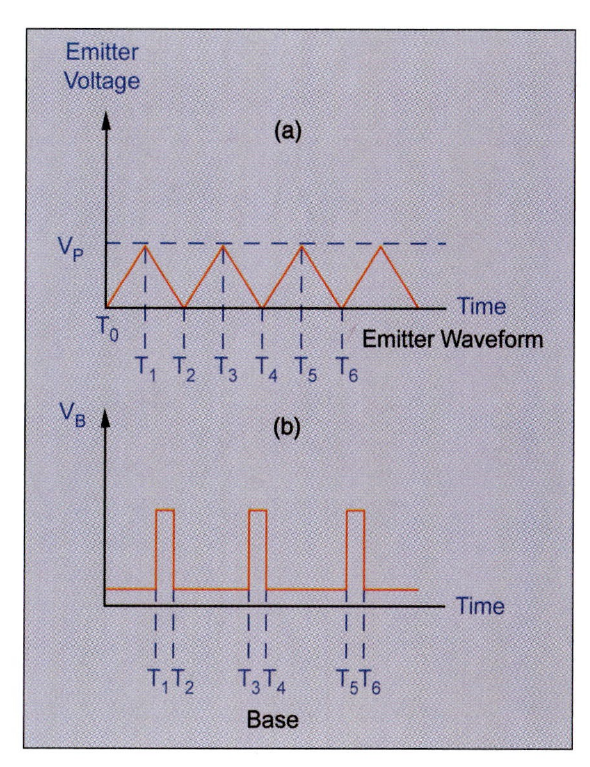

positive than its cathode. The cathode potential is determined by a combination of V_{BB}, R_{B1}, and R_{B2}. The normal voltage divider equation $\left(\frac{R_{B1}}{R_{B1} + R_{B2}}\right)$ is called the intrinsic standoff ratio, η.

So

$$V_K = \eta V_{BB}$$

And the triggering voltage is

$$V_P = \eta V_{BB} + 0.7 \text{ V}$$

Figure 15–18 shows the waveforms for the operating relaxation oscillator. As the capacitor charges, the voltage across the emitter builds until it is equal to V_P. At V_P, the negative resistance of B_1 goes to almost 0 and the capacitor rapidly discharges. This rapid discharge produces a waveform similar to the one shown in Figure 15–18b.

Both of the waveforms in Figure 15–18 can be used as an output, depending upon where the output is taken. The frequency of the oscillator is controlled by the RC time constant. Referring to Figure 15–16, you see that R_3 and C_1 provide the time constant ($T = R_3C_1$) for the charge-discharge cycle. Recall that the frequency is equal to the inverse of the time constant. Thus, the frequency is equal to $\frac{1}{R_3C_1}$. Note that this is slightly different than the previous calculations because we are dealing with pulses per second. Also, recall the formula to charge the capacitor to 63 percent is $T = RC$.

For example, assume that $R_3 = 20\ \text{k}\Omega$ and that $C_1 = 15\ \mu\text{F}$. The frequency for the oscillator would be

$$f = \frac{1}{RC} = \frac{1}{(20\ \text{k}\Omega)(15\ \mu\text{F})} = 3.3\ \text{Hz}$$

Assume that the desired oscillator frequency is 200 Hz and the value of R_3 is 30 kΩ. What would be the required value of C_1? Start by rearranging the frequency equation and then insert the given values.

$$C = \frac{1}{fR} = \frac{1}{30\ \text{k}\Omega(200)} = 16.7\ \mu\text{F}$$

15.9 Direct Digital Synthesis

When many operational frequencies are needed, there are several options that can be used. Two are the phase-locked loop (PLL) frequency synthesizer and the direct digital synthesizer (DDS). The synthesizer of choice is the DDS. The advantage of a DDS system over a crystal is that the DDS can be programmed for a high number of narrow-band frequencies.

Another name for the DDS oscillator is the numerically controlled oscillator. The DDS system has the following components (see Figure 15–19).

- Frequency tuning word—The input number to the phase accumulator for the value of the next phase increment.
- Phase accumulator—Produces (with the tuning word and clock) the binary digit for the sine lookup table.
- Sine lookup table—A set of sine wave voltage values that are sent, one value at a time, to the digital-to-analog (D/A) converter.
- Digital-to-analog converter (DAC)—Produces an approximation of the sine wave voltage, at a specified phase value, for the input voltage from the sine lookup table.
- Clock—Provides a constant reference square pulse frequency for the phase accumulator and the DAC.
- Low-pass filter—Provides the filtering function for sine wave cleanup generated from the DAC.

FIGURE 15–19 DDS oscillator

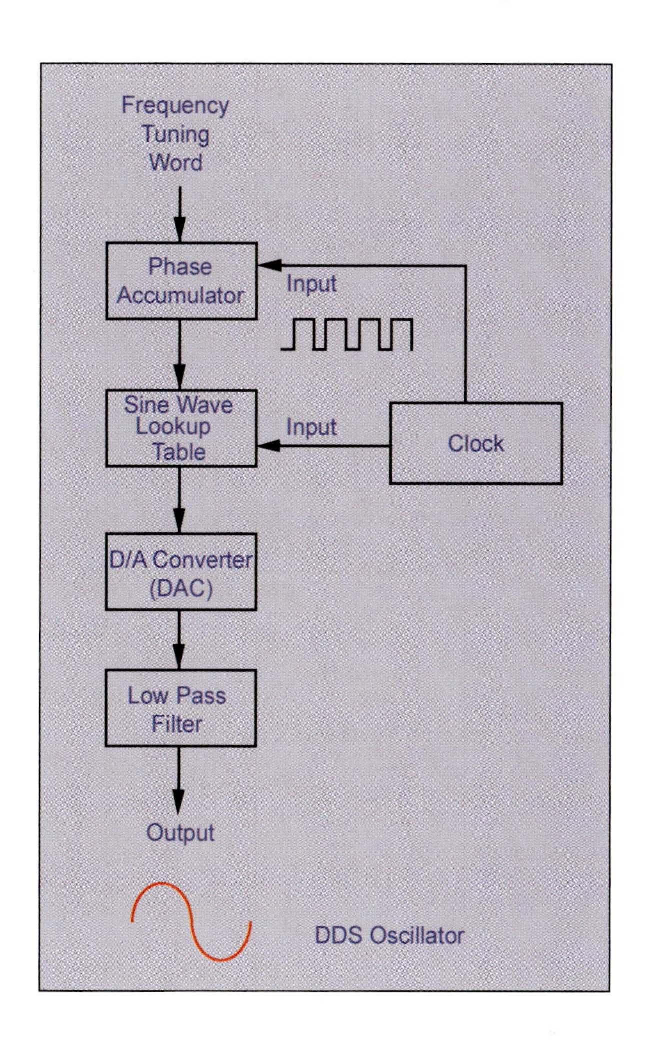

SUMMARY

The oscillator is an electronic circuit that uses positive feedback to sustain its output. A small signal, usually created by random noises in the input circuit, is amplified by the circuit. A small portion of the output signal is fed back to the input in such a way that the input is reinforced or strengthened. This is called positive feedback.

For the oscillator to operate, the gain of the amplifier multiplied times the percentage of the output that is fed back must be greater than 1. In this way, the input signal will be kept at its original level or higher. This product is called the Barkhausen criterion and is calculated using the following equation:

$$BC = \alpha_V A_V$$

Where:

α_V = the percent of the output that is feedback to the input

A_V = the forward gain of the amplifier

Note that if the feedback is negative (i.e., 180° out of phase), the feedback will not occur.

The frequency of an oscillation is determined by the passive elements in the circuit. Some oscillators use inductors and capacitors to determine their resonant frequency. Others, especially those that use integrated circuits or operational amplifiers, use resistors and capacitors for their frequency control.

The most stable of all the standard oscillator types uses a quartz crystal for the frequency control element. This type of circuit takes advantage of the piezoelectric properties of a properly cut crystal. In junction with an RC or LC circuit, the crystal-controlled oscillator provides an extremely stable output.

The relaxation oscillator uses the charging of a capacitor through a resistor to produce a pulse type of output. The capacitor charges until the junction of a UJT is forward biased. When that happens, the capacitor discharges through the UJT. When the capacitor is completely discharged, the UJT is reset and the capacitor begins its charge again. The frequency of the relaxation oscillator is controlled by the RC time constant of the capacitor input circuit.

Modern digital circuits use a variety of circuits to produce stable, controllable output waveforms. The frequency of output is controlled by a clock that controls the input of a digital series of pulses to a sine wave lookup table and a DAC converter. The output is smoothed to a sine wave by use of a low-pass filter.

REVIEW QUESTIONS

1. Figure 15–20 shows one-half of a sine wave that is being generated by a rotating circle. As the circle rotates, the distance from the point on the circumference to the horizontal diameter is measured. These distances are then plotted on a horizontal axis. What will the height of the sine wave be when the circle has rotated 45°, 135°, and 254° respectively?
2. Look at Table 15–3. What would happen if the gain of the amplifier were reduced to 80? Increased to 120?
3. A certain oscillator has a gain of 50. What will happen if 2 percent of the output is fed back to the input? 1 percent? 3 percent?
4. In Figure 15–6, which component provides the feedback path from output to input? Which components determine the frequency of the output?
5. In Figure 15–16, what will happen if the size of resistor R_1 is increased? Decreased?

FIGURE 15–20 Creating a sine wave with a rolling circle

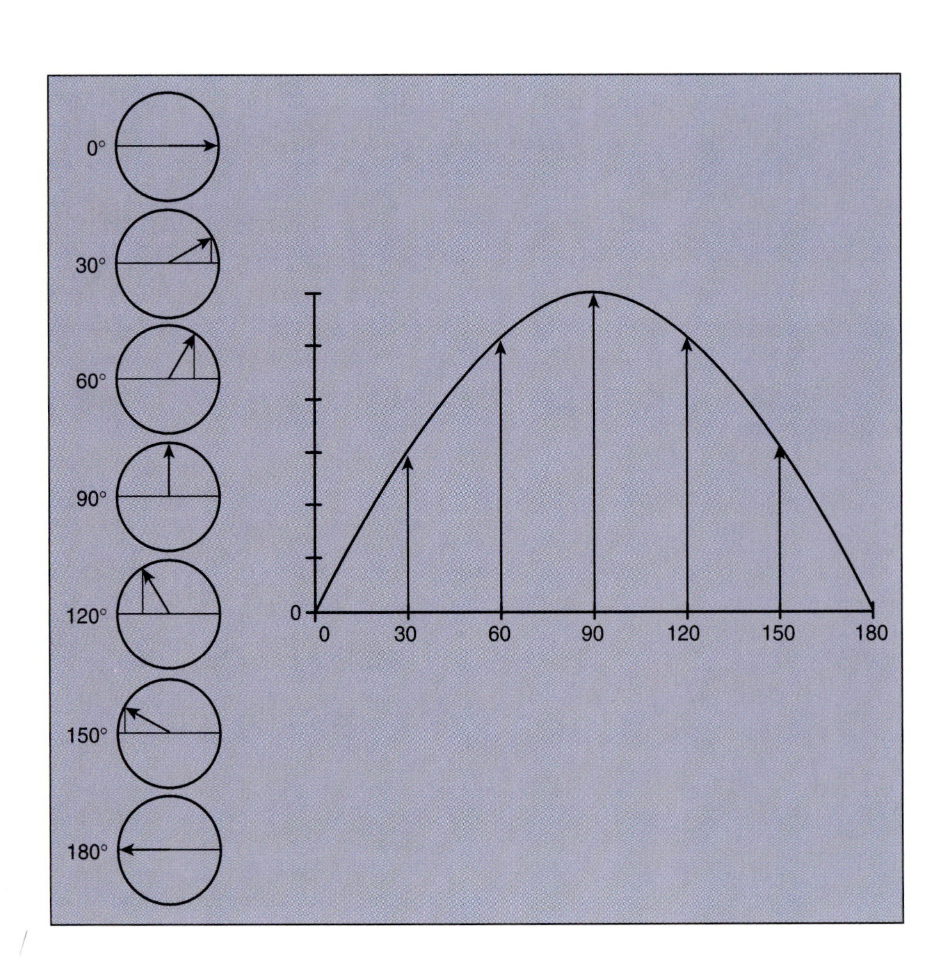

Amplitude and Frequency Modulation

OUTLINE

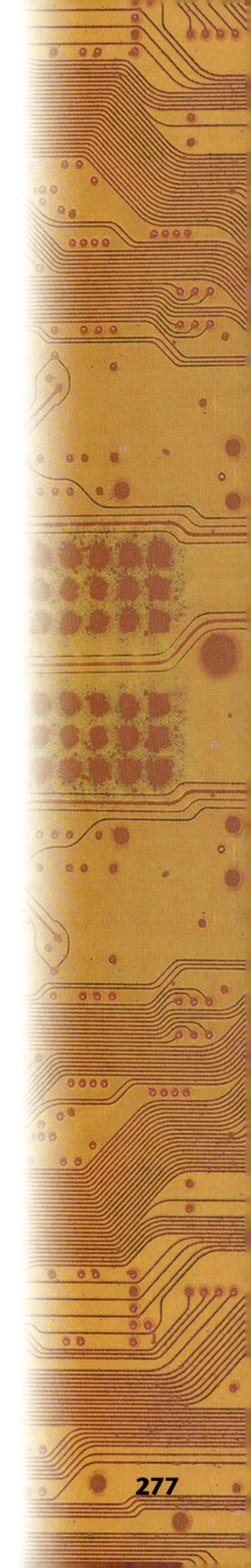

OVERVIEW

The natural desire of people to communicate between two distant points has given birth to modern-day communication systems. The advances in technology today have made it possible to reach practically any point on Earth and distant points in space. This chapter will introduce you to the basic concepts of electronic communication systems.

OBJECTIVES

After completing this chapter, the student should be able to:

1. Discuss the purpose of modulation.
2. Identify different types of modulation.
3. Define amplitude modulation and frequency modulation.
4. Explain the operation of modulators and demodulators used in AM and FM.
5. Explain the operation of AM and FM receivers in a block diagram approach.

GLOSSARY

Carrier wave The signal that is used to carry the information after it has been modulated.

Demodulator An electronic circuit that recaptures the information signal from the received signal.

Detector See demodulator.

Lower sideband (LSB) The carrier frequency minus the modulating frequency.

Modulation In electronics and communications, the process of attaching information to a signal or carrier wave.

Modulator The electronic circuit that impresses the information signal onto the carrier frequency.

Upper sideband (USB) The carrier frequency plus the modulating frequency.

COMMUNICATION FUNDAMENTALS

16.1 Information Transfer

Communication is the transfer of meaningful information (intelligence) from one point (source) to another (destination). The information to be transmitted can be in the form of audio or video signals, or it could even be digital data. Figure 16–1 illustrates the basic transfer of information from source to destination.

Signals can be transmitted from the source to the destination by two means:

- Wireless communication, where an antenna is used as a transmitting device
- Line communication, where waveguides, transmission lines, or optical fibers are used as transmitting devices

The discussion in this chapter involves wireless communication. An oscillator can be used to produce a high-frequency signal. When the output of this oscillator is fed to a transmitting antenna, it converts the high-frequency alternating current to a radio wave. A radio wave travels at approximately the speed of light, which is 3×10^8 meters/sec. When this radio wave is incident on another antenna, a high-frequency current is induced in it. This induced current is of a smaller strength but identical in shape to the current in the transmitting antenna. Thus, transfer of electrical energy from one point to the other without the use of wires or wireless communication is accomplished.

16.2 Modulation and Demodulation

Modulation is defined as the process of impressing information onto a high-frequency signal for the purpose of transmission. The information is in the form of a low-frequency signal and is called the modulating signal, whereas the high-frequency signal is called the carrier. The high-frequency signal "carries" the low-frequency signal from the source to the destination. The **modulator**, therefore, has to be present at the source, or the transmitter end, of the communication system.

Demodulation is the process of removing information from the carrier. A **demodulator** re-creates the low-frequency modulating signal at the destination, or the receiver end, of the communication system.

You may wonder at this point why the process of modulation or demodulation is required at all. Why not transmit the information directly?

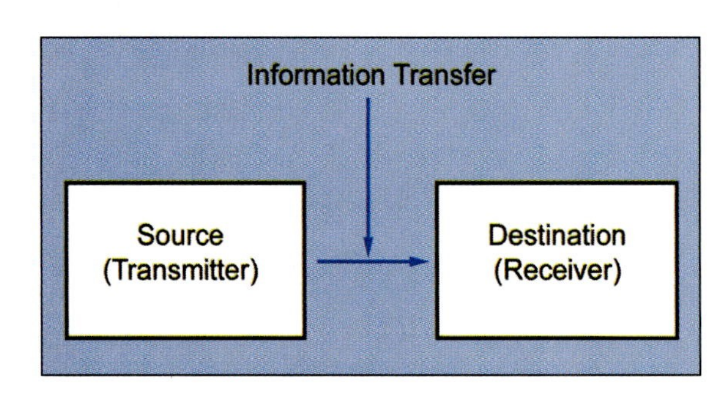

FIGURE 16–1 Transfer of information

The reason is the impracticality of transmitting low-frequency signals. Two important factors contribute to the impracticality. They are:

- Antenna length: For efficient propagation signals in the air, the length of the antenna is crucial. The antenna length depends on the wavelength[1] of the signal. High-frequency carriers have small wavelengths, and the antenna required to transmit these signals ranges over a few feet in length. On the other hand, low-frequency information has a larger wavelength, and the antenna required to transmit these signals may span many miles, rendering transmission highly impractical.
- Interference: If everyone transmitted low frequencies directly, interference between each transmitted signal would make all of the signals ineffective. With modulation, a different high-frequency carrier signal can be assigned to each transmitter, and hence, interference between transmitters is eliminated.

Table 16–1 shows the various frequencies in the spectrum. Atmospheric communications are accomplished at frequencies of 550 kHz and higher.

16.3 Basic Communication Systems

There are three basic types of communication systems, based on the characteristics of the carrier that is altered. They are:

1. Systems that use amplitude-modulated (AM) carriers
2. Systems that use angle-modulated modulation, which in turn comprises two variations:
   a. Frequency-modulated (FM) carriers
   b. Phase-modulated (PM) carriers

Table 16–1 Radio Frequency Spectrum

Frequency	Designation	Abbreviation
30–300 Hz	Extremely low frequency	ELF
300–3000 Hz	Voice Frequency	VF
3–30 kHz	Very low frequency	VLF
30–300 kHz	Low frequency	LF
300 kHz–3 MHz	Medium frequency	MF
3–30 MHz	High frequency	HF
30–300 MHz	Very high frequency	VHF
300 MHz–3 GHz	Ultra high frequency	UHF
3–30 GHz	Super high frequency	SHF
30–300 GHz	Extra high frequency	EHF

[1]You have already learned that wavelength in meters is defined by the equation $\lambda = \frac{3 \times 10^8}{f}$.

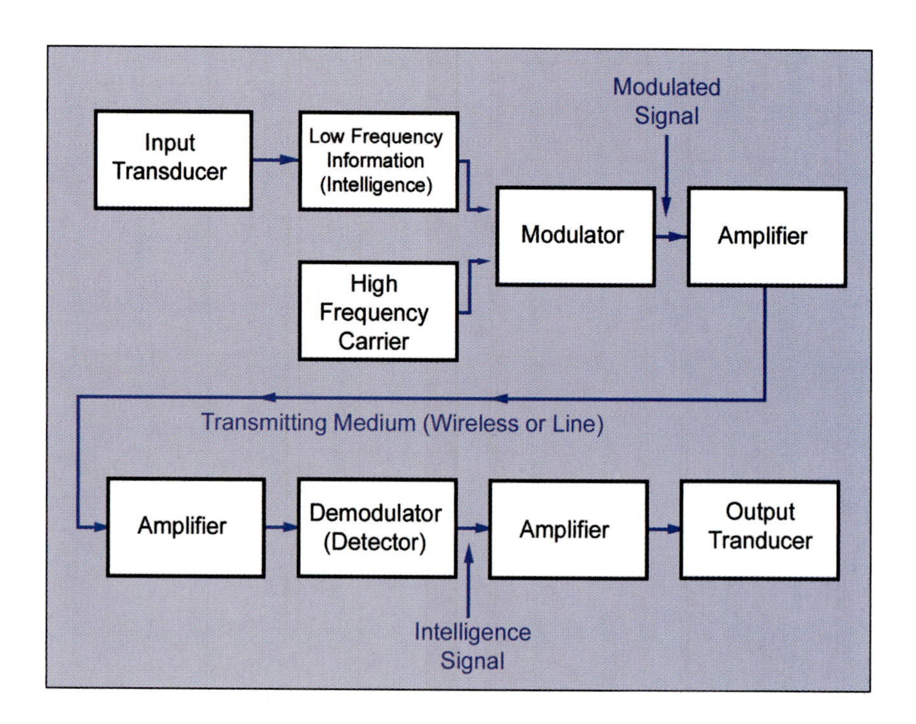

FIGURE 16–2 Basic communication system

3. Systems that use digital techniques, normally referred to as pulse modulation

This chapter focuses on amplitude modulation (AM) and frequency modulation (FM) only.

Figure 16–2 is a block diagram of a communication system. The following two sections explain how it works.

Transmitting End

An input transducer (such as a microphone or a camera) converts information into an electrical form. This electrical information is the modulating signal. The modulating signal and a high-frequency signal, called the carrier, are injected into the modulator, which produces a modulated signal. The modulated signal is amplified.

Receiving End

The receiver picks up the signal and amplifies it to compensate for the attenuation that occurred during transmission. The output of the amplifier is fed to a demodulator, or **detector**, where the information is extracted from the carrier. The demodulated signal is then fed to another amplifier, which suitably amplifies it so that it can be fed to an output transducer such as a speaker (for audio signals) or a monitor (for video signals). The output transducer converts the information from its electrical form to a physical form such as sound or picture.

AMPLITUDE MODULATION (AM)

16.4 AM Fundamentals

As the name implies, in an AM system, the amplitude of the carrier is altered by the information. In this discussion, we refer to the carrier as

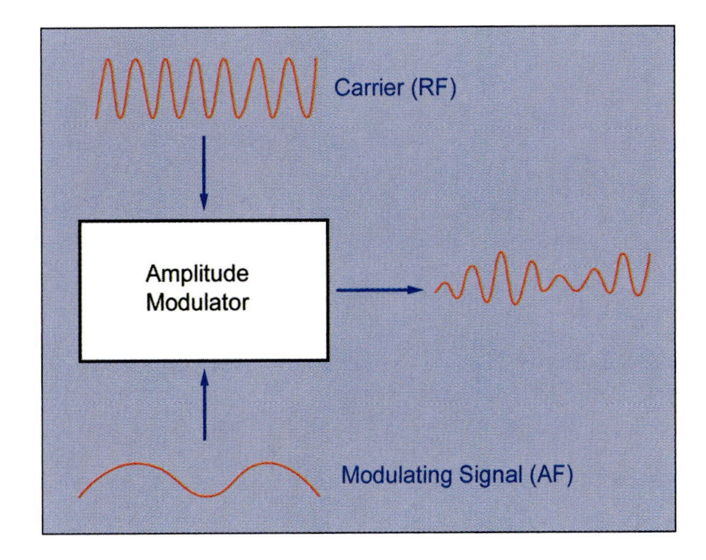

FIGURE 16–3 Process of amplitude modulation

radio frequency (RF) signals, and the information is audio frequency (AF) signals. Figure 16–3 illustrates the process of amplitude modulation, which produces a resultant amplitude-modulated (AM) wave. It can be seen that the output of the modulator produces an RF signal whose amplitude varies in accordance with the audio frequency (AF) signal.

An amplitude-modulated signal consists of three frequencies:

- The carrier frequency that is the original RF oscillator signal
- The **upper sideband (USB)**, which is carrier frequency + modulating signal frequency
- The **lower sideband (LSB)**, which is carrier frequency − modulating signal frequency

The bandwidth of an AM signal is calculated as

$$\text{Bandwidth (BW)} = \text{USB} - \text{LSB}$$

or

$$\text{Bandwidth (BW)} = 2 \times \text{Modulating signal frequency}$$

If a 2 kHz audio signal (modulating signal) is used to modulate a 500 kHz RF carrier, the output of the modulator will be made up of three frequencies:

- The carrier frequency is 500 kHz.
- The upper sideband (USB) is calculated as (500 kHz + 2 kHz) = 502 kHz.
- The lower sideband (LSB) is (500 kHz − 2 kHz) = 498 kHz.

The bandwidth is calculated as

$$\text{BW} = \text{USB} - \text{LSB} = 502 \text{ kHz} - 498 \text{ kHz} = 4 \text{ kHz}$$

or

$$\text{BW} = 2 \times \text{Modulating frequency} = 2 \times 2 \text{ kHz} = 4 \text{ KHz}$$

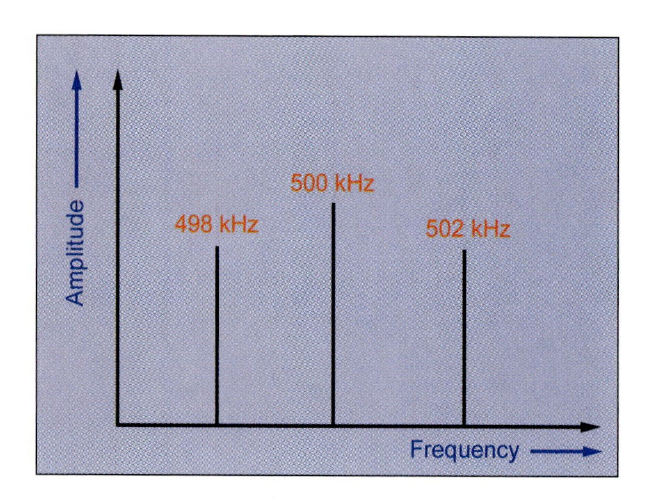

FIGURE 16–4 AM represented in the frequency domain

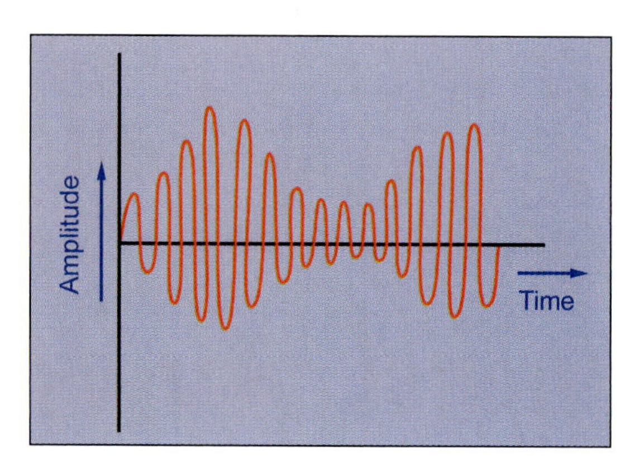

FIGURE 16–5 AM represented in the time domain

The AM signal, consisting of its carrier frequency (500 kHz) and its two sidebands (502 kHz and 498 kHz), can be represented with frequency on the x-axis, shown in Figure 16–4. An instrument called a spectrum analyzer is used to view an AM signal in the frequency domain. Figure 16–5 shows the AM signal with time on the x-axis. Oscilloscopes are used to view AM signals in the time domain.

16.5 A Typical Amplitude Modulator

A typical amplitude modulator is shown in Figure 16–6. The circuit is basically that of a class C amplifier, with a difference that the value of V_{CC} applied to the circuit is varied by the audio signal (information). The RF input (carrier) is fed to the base of transistor Q_1 via the input circuit formed by resistor R and the capacitor C.

The transformer T_1 and the capacitor C_{T1} form a tank circuit whose resonant frequency is that of the RF input signal. The transistor functions in the class C mode and is biased to be in the cutoff region of operation. The presence of the RF input causes the transistor to conduct for portions of its positive half-cycle, while the tank circuit re-creates the rest of the RF input. The frequency of the output signal at the collector of the transistor is, therefore, that of the RF signal.

The transformer T_2 controls the supply voltage V_{CC} applied to the collector. As can be seen from Figure 16–6, the audio signal is applied to the secondary of transformer T_2. The audio voltage either aids or

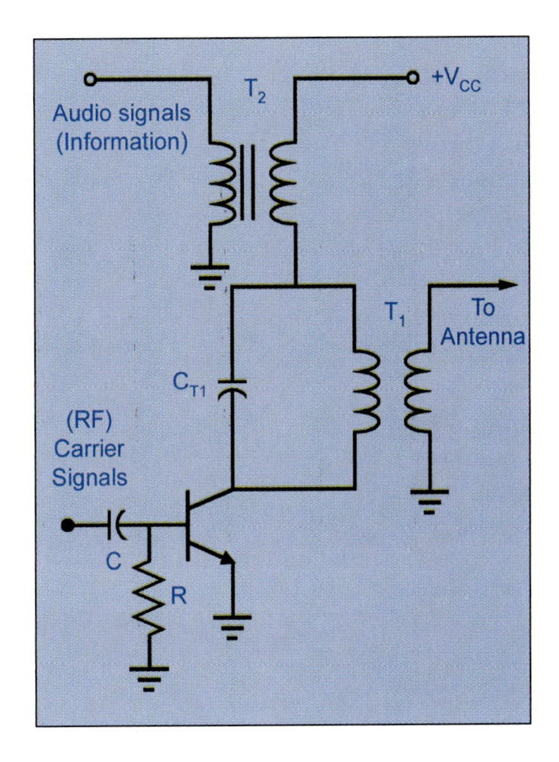

FIGURE 16–6 A typical amplitude modulator

opposes the supply voltage, depending upon its polarity. For instance, suppose that the audio signal at the secondary of T_2 is of the value 20 V peak to peak, and the supply voltage is V_{CC} is 10 V. If the audio input goes to its positive peak (with respect to ground) at 10 V, this voltage opposes the applied V_{CC} and the transistor gets 0 V. On the other hand, if the audio input goes to its negative peak of 10 V (with respect to ground), this voltage aids the applied V_{CC} and the transistor gets 20 V. Thus, the collector supply for the transistor is not a constant, but varies with the audio input. The output signal at the collector of the transistor has amplitude that depends on the supply voltage applied, which in turn depends on the audio signal. Amplitude control is hence achieved.

16.6 AM Detector or Demodulator

An AM detector, or demodulator, recovers information from the modulated signal and is an important part of a receiver. Nonlinear devices such as diodes and transistors are the heart of the detectors. A very simple diode detector is shown in Figure 16–7.

In Figure 16–7, the modulated signal is applied to the primary of transformer T. The capacitor C_1 and the primary windings form a resonant circuit that can be tuned to the carrier frequency. At the secondary end of the transformer, the diode D passes only the positive halves of the carrier. The capacitor C_2 acts as a filter, charging and discharging in accordance with the positive peaks of the modulated signal. The output of the detector thus traces the positive peaks, or envelope, of the AM signal, which re-creates the information signal.

An AM detector that uses a transistor is shown in Figure 16–8. The base-emitter junction of the transistor acts as a detector diode. The

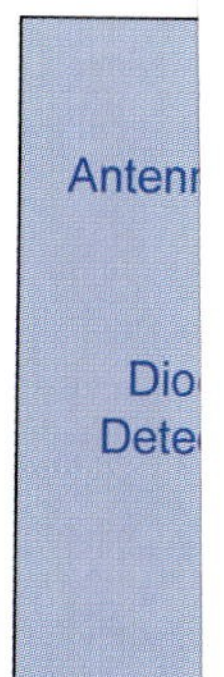

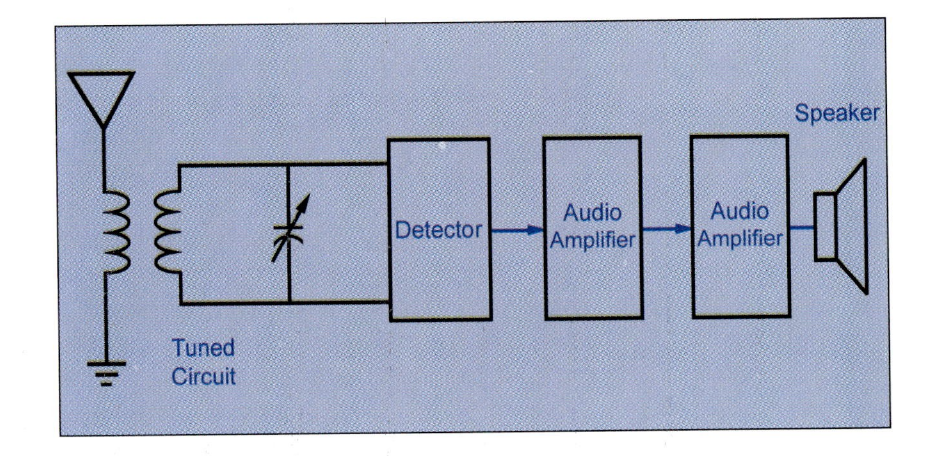

FIGURE 16–10 An improved radio receiver

the receiver will "hear" every carrier signal that is strong enough to drive the headphones. The addition of a tuned circuit (LC) can resolve this situation.

An improved radio receiver is shown in Figure 16–10. Of all the frequencies intercepted by the antenna, the desired carrier frequency is chosen, using the tuned circuit. The detector or demodulator re-creates the audio signal, which is amplified by the audio amplifier before it is fed to the speaker. It can be seen that in the improved radio receiver, sensitivity and selectivity are both improved.

Figure 16–11 shows one of the earliest types of receiver used for commercial radios—the tuned radio-frequency (TRF) receiver. The sensitivity and selectivity is greatly improved in a TRF receiver by using more amplifiers and tuned circuits. It employs three tuned circuits, which provide the desired selectivity, and four amplifiers—two radio frequencies and two audio frequencies, which provide the required gain and better selectivity.

There are, however, a few disadvantages to the TRF receiver. The three tuned circuits are gang tuned, which means that they are tuned in synchrony. The term 'tracking' is used to refer to how closely resonant frequencies can be matched at a certain setting of the tuning control. It is very difficult to match (or track) the resonant frequency of each of the tuned circuits. Also, each tuned circuit can have different bandwidths. The differences in bandwidth and resonant frequency of the receiver can pose a problem. This can be eliminated by the use of a superheterodyne receiver.

FIGURE 16–11 A TRF receiver

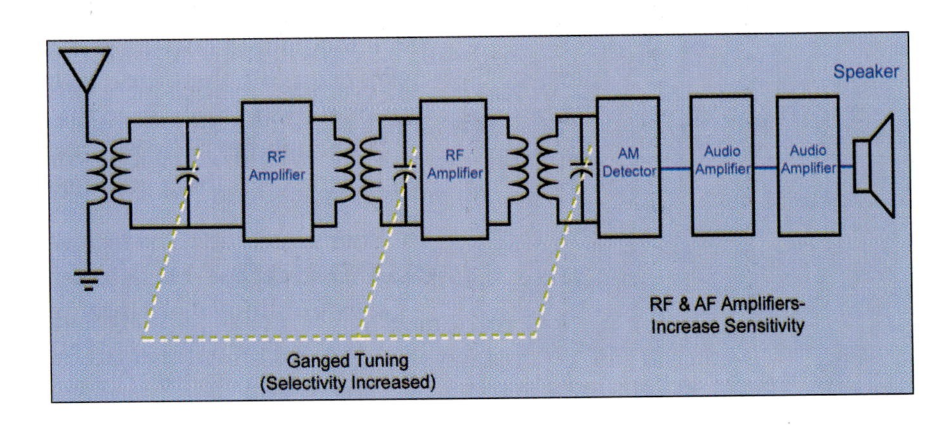

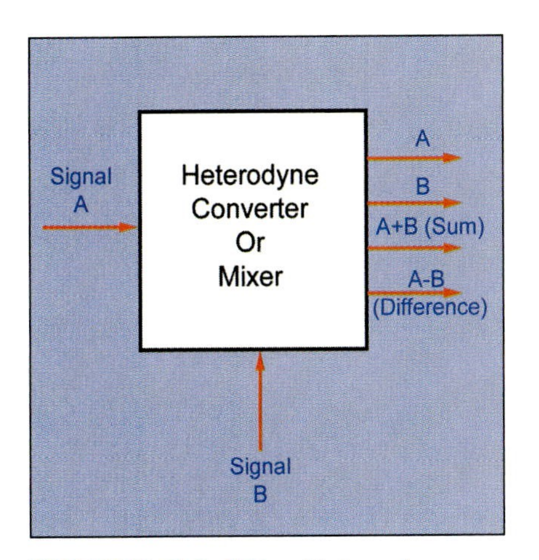

FIGURE 16–12 Heterodyne converter (also called a mixer)

16.9 Superheterodyne Receivers

In the superheterodyne receiver, some of the tuned circuits are confined to a single fixed frequency, thus eliminating the problem of tracking and changing bandwidths as seen in TRF receivers. This single frequency or fixed frequency is called intermediate frequency (IF). A superheterodyne receiver converts any received frequency to the IF by the process of mixing or heterodyning. The most commonly used IF for the AM broadcast band is 455 kHz. Different values of IF are assigned to other broadcast bands such as FM and the shortwave bands.

Figure 16–12 shows the process of heterodyning. The mixer or heterodyne converter is fed with two input signals, A and B. The output of the mixer contains the sum of the two inputs (A+B), the difference of the two inputs (A−B), and the original signals A and B. The difference signal is usually the one that is used for the IF.

Figure 16–13 shows the block diagram of a superheterodyne receiver. The mixer gets its inputs from an oscillator and the signals coming from the antenna. The oscillator is tuned to a frequency that is greater than the received frequency by an amount equal to the IF. For instance, if a carrier frequency of 1,080 kHz is to be received by an AM broadcast receiver, the oscillator should be tuned to a frequency that is equal to

$$\text{Received frequency} + \text{IF} = 1{,}080 \text{ kHz} + 455 \text{ kHz} = 1{,}535 \text{ kHz}$$

The mixer produces the sum and difference frequency at its outputs. The mixer is followed by an IF amplifier stage, which selectively amplifies and passes IF frequencies (those frequencies that are around 455 kHz). The output of the mixer produces a sum frequency of 1,080 kHz + 1,535 kHz and a difference frequency of 1,535 kHz − 1,080 kHz = 455 kHz. The sum frequency is rejected by the IF amplifier, while the difference frequency, which yields a 455 kHz signal, is passed and amplified by the IF amplifiers, and subsequently its amplitude variations are detected by the detector.

If, while the 1,080 kHz signal is being received, a 970 kHz carrier signal is also picked up by the antenna, the mixer will produce a sum frequency of 970 kHz + 1,535 kHz = 2,505 kHz and a difference frequency of 1,535 kHz − 970 kHz = 565 kHz. Both the sum and difference frequencies do not fall in the IF amplifier's pass band, so they are both rejected, thus eliminating interference from other carrier frequencies.

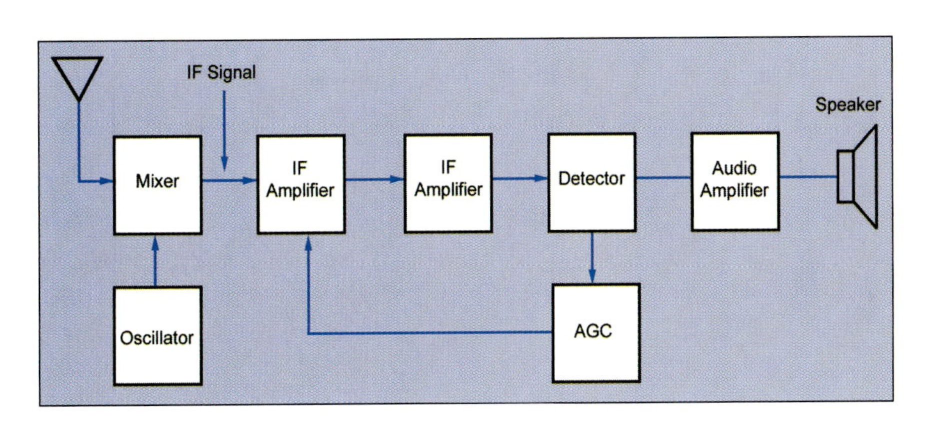

FIGURE 16–13 Block diagram of a superheterodyne receiver

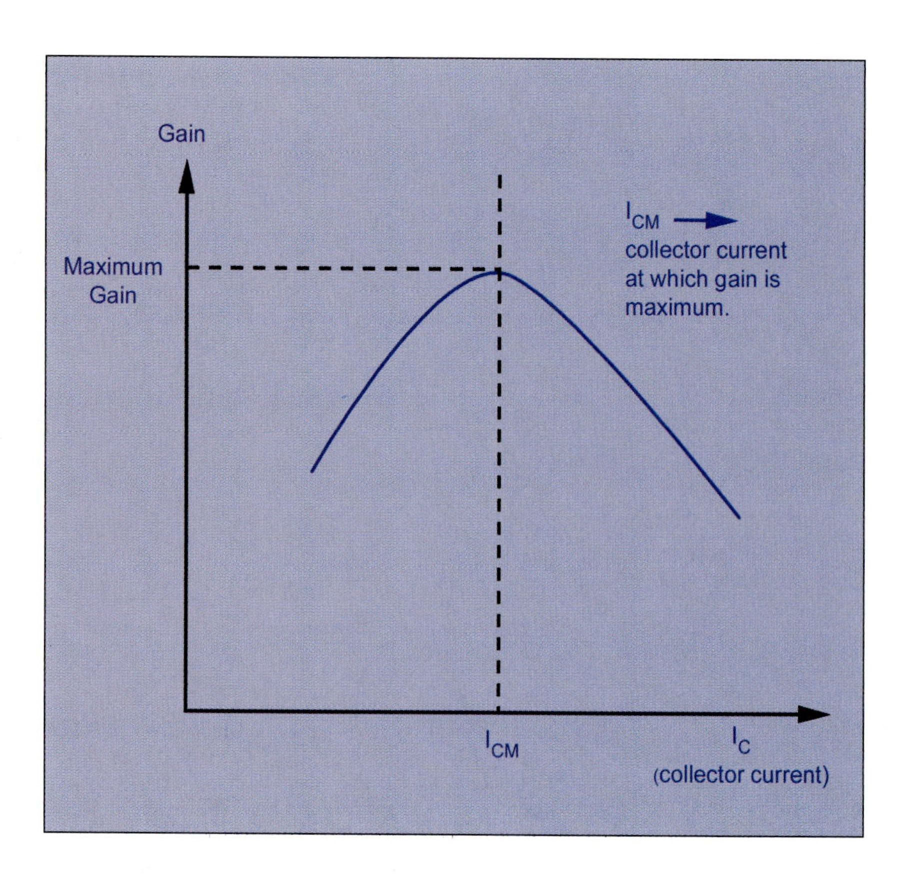

FIGURE 16–14 AGC characteristics of a transistor

The output of the detector is subjected to AGC (automatic gain control) or AVC (automatic volume control). If the gain of the IF amplifier is kept a constant, when the strength of the signal received by the antenna varies, so will the output voltage of the receiver. The purpose of AGC is to maintain a more-or-less constant voltage at the output of the receiver, and hence a constant volume. A feedback loop is formed between the detector output, AGC block, and the first IF amplifier. AGC develops a control voltage depending on the strength of the signal at the input of the detector. This control voltage is used to influence the gain of the IF amplifier. The transistor is employed as the amplifying device in an IF amplifier, and the transistor's gain is varied by the control voltage.

Figure 16–14 shows the AGC characteristics of a transistor. As indicated in the graph, maximum gain is present only at one particular value of collector current (I_C). As collector current increases or decreases from this value, gain drops. The control voltage can be used to control the bias of the transistor so that the collector current is changed. Thus, if the antenna receives a stronger signal, AGC subjects it to a smaller gain, whereas a weaker signal is subjected to a higher gain.

Superheterodyne receivers are subjected to a problem of image interference. This is because there are two values of incoming frequencies that can mix with the oscillator to produce IF. For instance, in the case already discussed, an oscillator frequency of 1,535 kHz is used to detect a 1,080 kHz signal. The difference between the two frequencies yields 455 kHz. Now let us consider that the antenna picks up a frequency 1,990 kHz. The difference signal that is available at the output of the mixer will be 1,990 kHz − 1,535 kHz = 455 kHz. This means that the receiver detects a carrier signal of 1,990 kHz, which is considered to be the image frequency of 1,080 kHz. The interference produced by the image frequency is called image interference. It can be seen that the image frequency is separated by 910 kHz (2 × 455 kHz). Image rejection is achieved by having a tuned circuit before the mixer. This tuned circuit can also have an RF amplifier stage to increase the sensitivity of the receiver as well as the selectivity.

FREQUENCY MODULATION (FM)

16.10 FM Fundamentals

Frequency modulation (FM) is the process by which the instantaneous frequency of the carrier is made to vary by an amount proportional to the modulating signal amplitude. Figure 16–15 illustrates a frequency-modulated waveform. Note that when the modulating signal is at zero amplitude, the carrier is at resting frequency, which is the carrier frequency. As the amplitude of the modulating signal increases, the carrier frequency increases, and as the amplitude of the carrier frequency decreases, the carrier frequency decreases.

Frequency modulation is preferred over amplitude modulation in commercial broadcasting networks because of its immunity to noise. AM is very sensitive to noise. This causes things such as lightning, automobile ignition, or sparking to manifest as amplitude variations on a **carrier wave**. In AM, because amplitude variations carry information, these amplitude changes are detected by the receiver and contribute to noise. In the case of FM, the noise signals can be eliminated in the receiver by using limiters that clip off amplitude increases or spikes.

FIGURE 16–15 Frequency-modulated waveform

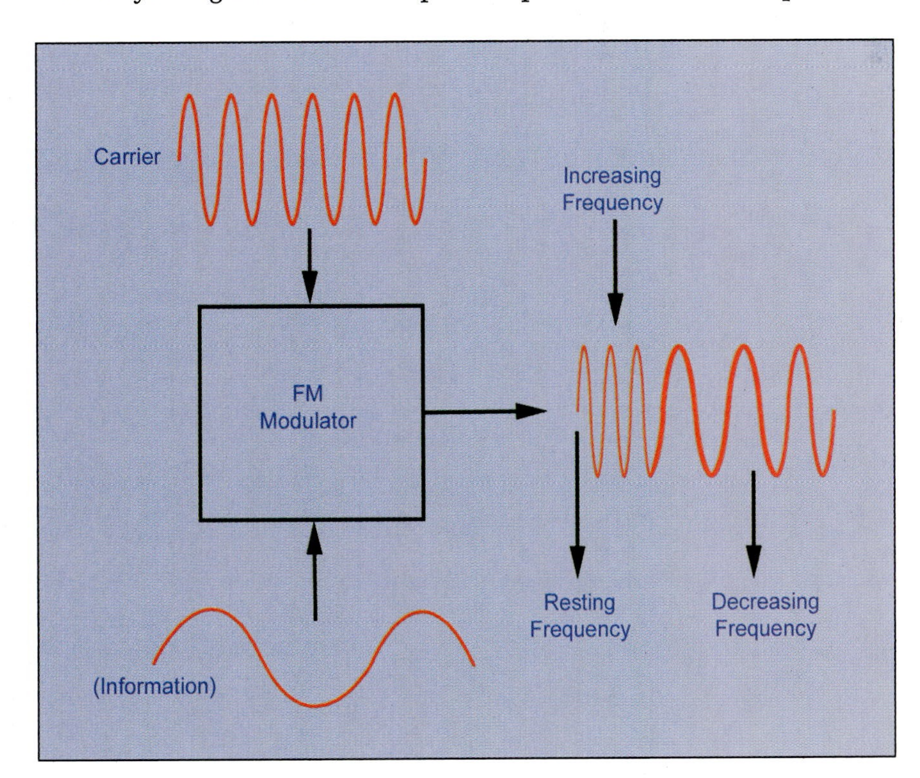

FM produces sidebands, just as AM does. But FM signals have several sidebands. Consider an FM system with a 200 MHz carrier being modulated by a 10 kHz audio tone. Figure 16–16 shows the sidebands that are produced, in a frequency domain graph.

16.11 A Frequency Modulator

In its basic form, an FM receiver consists of an LC tank circuit that is used with an oscillator. Figure 16–17 illustrates a basic frequency modulator. The audio input voltage is used to change the instantaneous value of the capacitor. The carrier frequency generated at the output of the oscillator is dependent on the change in capacitance. Thus, the oscillator generates a carrier whose frequency is proportional to the amplitude of the modulating signal (audio input).

Figure 16–18 shows a detailed frequency modulator. A Colpitts oscillator is used to generate the carrier frequencies. The values of L and C determine the frequency of oscillations. The diode D is used in reverse bias to function as a varicap or a variable capacitor. Resistors R_1 and R_2 form a voltage divider network and bias the diode. The value of the diode's capacitance is dependent on the audio input. This diode capacitance is in parallel with capacitor C, and hence the oscillator frequency is changed with the net change of capacitance (C_{D1} in parallel with C).

16.12 FM Detector or Discriminator

FM is detected by using a circuit called a discriminator. A discriminator reproduces the information (modulating signal) at the output, when fed with an FM signal. A discriminator is shown in Figure 16–19. Tank

FIGURE 16–16 FM sidebands

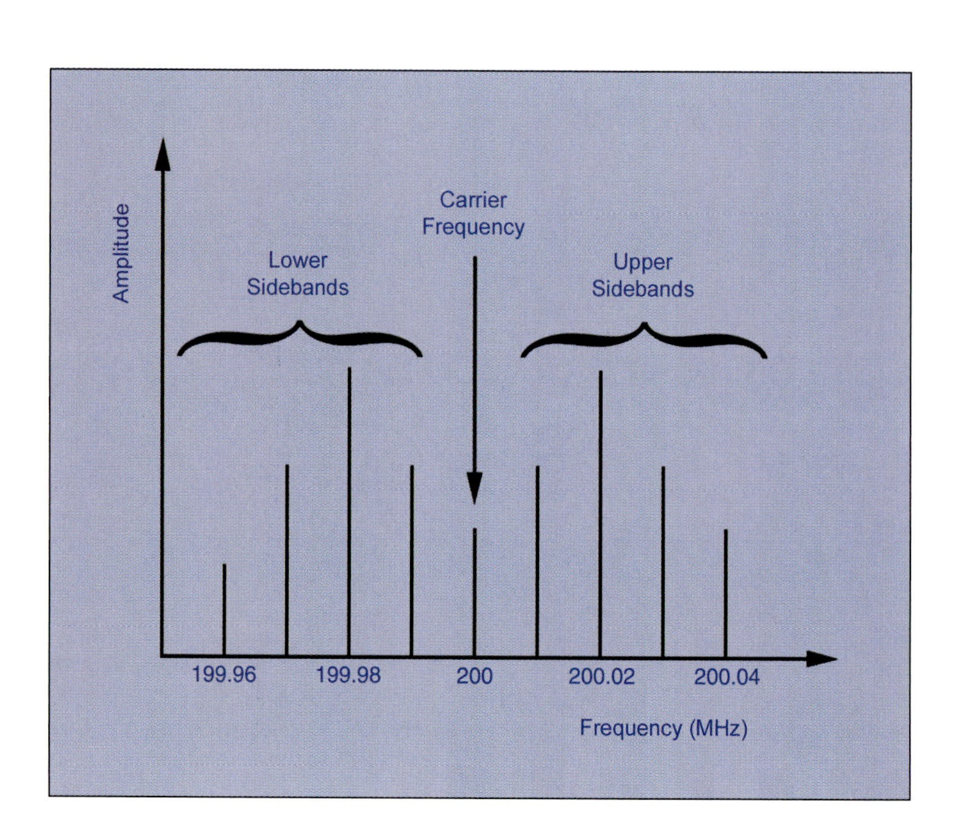

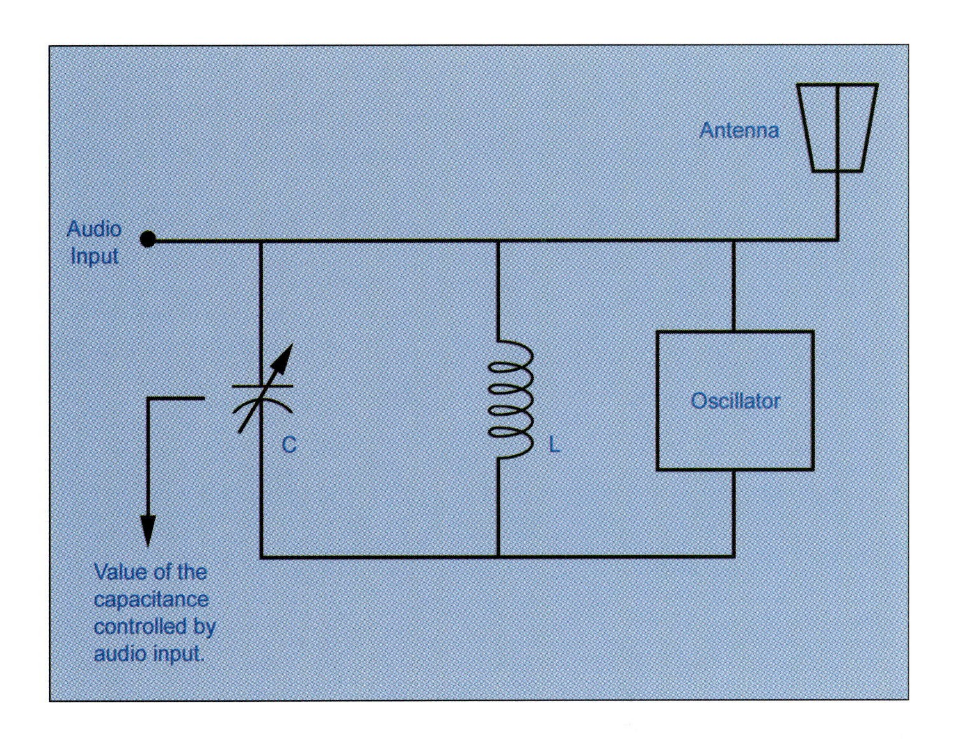

FIGURE 16–17 A basic frequency modulator

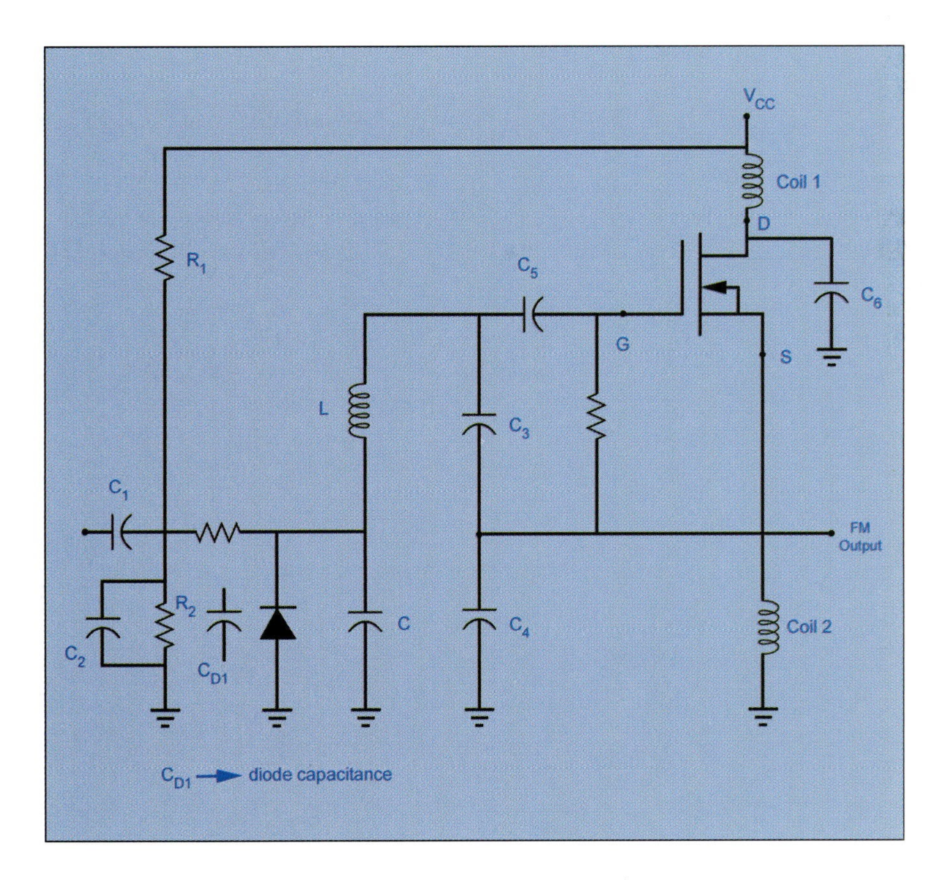

FIGURE 16–18 A complete frequency modulator

circuits L_1C_1 and L_2C_2 are present on the secondary side of the transformer. The frequency response curve for the two tank circuits is shown in Figure 16–20. As indicated in the figure, the resonant frequencies for the two tank circuits are above and below the carrier frequency f_0.

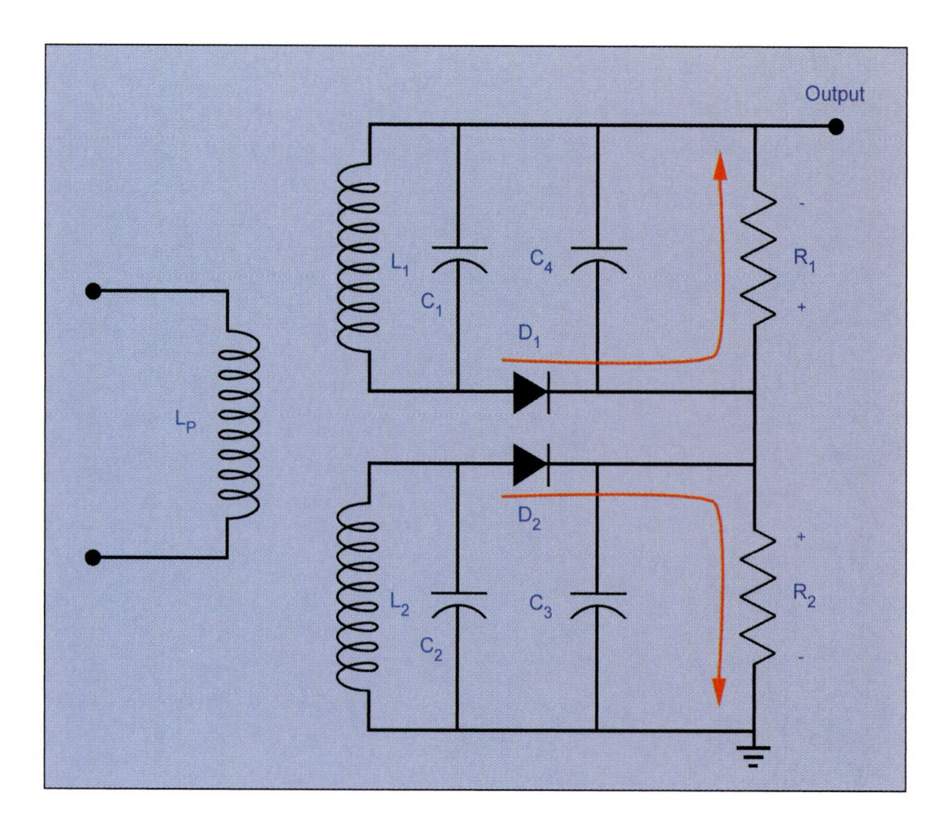

FIGURE 16–19 A discriminator circuit

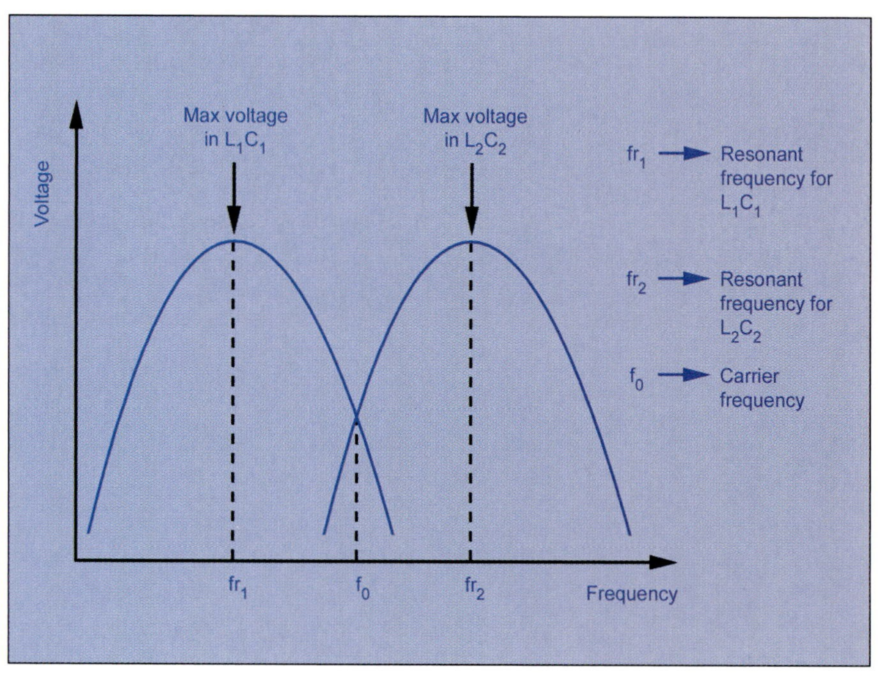

FIGURE 16–20 Frequency response curve for tank circuits

When the nonmodulated carrier, which is at the resting frequency, is fed to the primary side of the transformer, both the tank circuits have identical voltages because they are operating at the carrier frequency. The identical voltages drive the diodes D_1 and D_2 to conduct in equal amounts. Currents through the resistors R_1 and R_2 are identical in magnitude, but opposite in direction, because of the direction of the diodes. This causes the drops across the resistors R_1 and R_2 to be equal in magnitude but opposite in sense. Hence, the output voltage measured across R_1 and R_2 amounts to zero. The discriminator thus produces no output voltage for the resting frequency or the nonmodulated carrier frequency.

When the frequency of the carrier increases due to modulation, the tank circuit L_2C_2 has a higher voltage than L_1C_1. This causes diode D_2 to conduct more than diode D_1, and hence the voltage across resistor R_2 is more than resistor R_1. The result is a positive output voltage. As frequencies increase towards the resonant frequency of L_2C_2, the output voltage continues to be more positive.

On the other hand, when the frequency of the carrier decreases due to modulation, the tank circuit L_1C_1 has a higher voltage than L_2C_2. This causes the diode D_1 to conduct more than D_2, and hence the voltage across resistor R_1 is more than resistor R_2. The result is a negative output voltage. As frequencies decrease towards the resonant frequency of L_1C_1, the output voltage continues to be more negative.

This is summarized as follows:

- The output of the discriminator is zero when the carrier frequency is at rest.
- The output voltage of the discriminator increases in the positive direction as the frequencies increase.
- The output voltage of the discriminator increases in the negative direction as the frequencies decrease.

The output of the discriminator is thus a function of the carrier frequency, and the information, which is in the form of changing frequencies, is retrieved.

16.13 FM Receiver

The block diagram of a typical FM superheterodyne receiver is shown in Figure 16–21. Mixing or heterodyning is achieved just as described in the case of an AM receiver. The intermediate frequency is 10.7 MHz for the FM broadcast band. The output of the IF amplifier is fed to a limiter. The limiter is provided to eliminate noise signals that occur as spikes or amplitude variations. Limiting clips the output, which means the noise components are removed. In doing so, the FM signal does not lose any information components in the carrier, because information is coded in the form of frequency variations. Figure 16–22 illustrates the operation of a limiter.

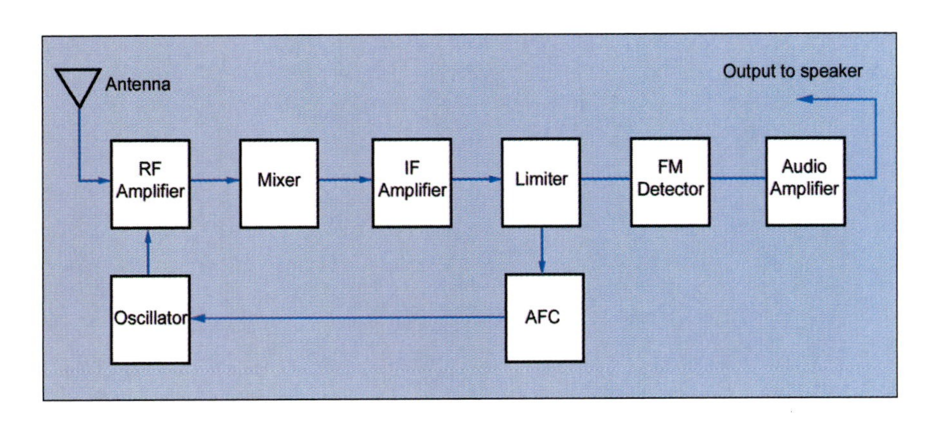

FIGURE 16–21 Block diagram of an FM superheterodyne receiver

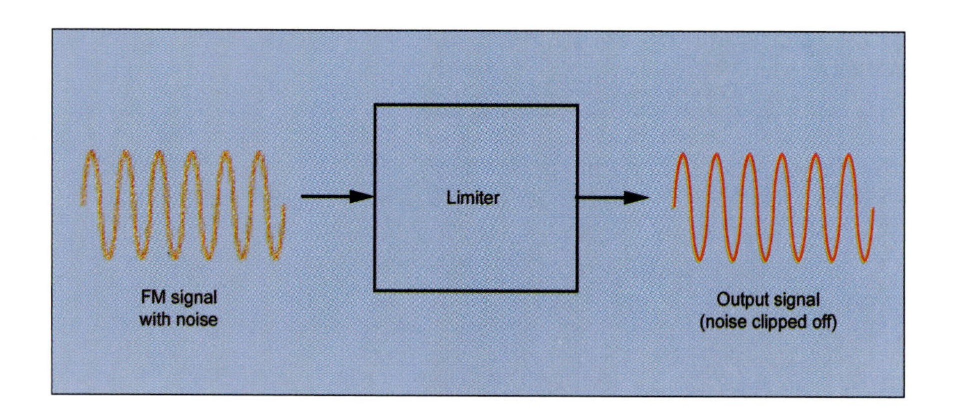

FIGURE 16-22 Operation of a limiter

Note that the process of limiting or clipping to eliminate noise could not be used in AM. In AM, the information is coded in the form of amplitude variations. The noise signals also appear as amplitude variations, and if they are clipped, relevant information may be lost.

The output of the limiter is fed to the detector or discriminator. The detector feeds its signals to an audio amplifier and a block named AFC, which stands for automatic frequency control. AFC is used to keep the IF frequency of 10.7 MHz from drifting. The detector produces steady DC voltage, which can be used as a control voltage to compensate and correct the drift in the oscillator frequency.

SUMMARY

Information of various types often needs to be transmitted from one location to another. For example:

- Voice and music from a studio to your home via a radio
- Voice, music, and/or video from a remote location to your television receiver
- Digital data from one computer to another
- Process information from an industrial process to a controller
- Control signals from a controller to a process

In each of these cases, the information can be converted to an electric signal that varies in amplitude according to the information. For example, digital data can be encoded as an on or off signal. This is a form of amplitude modulation or AM. In other cases, the information may be of a continuous or analog nature, such as audio or video. In yet other situations, the information is converted from analog to digital.

Whatever the format of the information signals, they are usually fairly low frequency—on the order of a few kilohertz up to a maximum of 100 kHz or so. For a variety of reasons, these low frequencies do not lend themselves to long-distance transmission means such as radio waves or light waves through fiber-optic cable.

To remedy this situation, the information is often impressed on a higher frequency carrier signal. This action is called modulation. This combination of the two signals is readily transmitted for long distances, depending on the type of transmission path (e.g., fiber-optic cable, radio, coaxial cable, twisted pair, etc.).

This situation is the basis for most radio and television transmissions. In this type of transmission, a transmitter impresses information onto the carrier wave, and the result is then applied to an antenna and broadcasted into the atmosphere. At the other end, a receiver demodulates the received signal and converts the information back into the original format.

Modern radio systems use either amplitude modulation (AM) or frequency modulation (FM).

In AM, the information is used to vary the amplitude of the carrier signal. In FM, the information is used to vary the frequency of the signal. Other offshoots of these two fundamental types have been developed; however, they are all based in one way or another on the two mainstays.

REVIEW QUESTIONS

1. The choice of carrier frequency makes a large difference in how far a transmitted signal will carry. FM stations in the United States use frequencies in the vicinity of 100 MHz, which is line-of-site communication. Why? What problems could occur if the signals traveled farther than line-of-site?
2. Why is a higher frequency carrier signal required for radio transmission? Discuss the pros and cons of such a system.
3. Using Figure 16–2, walk through a block-by-block explanation of a basic communications system.
4. Bandwidth is the difference between the lowest frequency and the highest frequency in a transmitted signal. For example, the bandwidth of an AM signal with a carrier frequency of 1,600 kHz and an applied audio signal of 4 kHz is 8 kHz. (1,604 kHz − 1,586 kHz). Consider what would happen to a signal if, after modulation, the carrier signal and the lower portion of the sideband were removed. What would the bandwidth be? (Such a system actually exists. It is called single sideband and is used extensively in modern communications systems.)
5. Assume the circuit of Figure 16–6 is being used to generate an audio signal for a radio broadcast. What will happen if the audio input is adjusted so that the output is clipped? (This is called overdriving.) What will happen if the audio input is decreased to a very small value?
6. Figure 16–23 is an improved version of the basic receiver shown in Figure 16–9. Note the additional components. Discuss how the additional components might help overcome the two problems of Figure 16–9 that were discussed in the chapter.
7. How might the receiver shown in Figure 16–13 be improved for additional selectivity and/or sensitivity?
8. Most of the modern music radio stations use FM. Why?
9. A certain receiver is being used to tune a 600 kHz radio station. The intermediate frequency for the receiver is 455 kHz. Another radio station is very close and broadcasts on a frequency of 1,510 kHz. Is there any possible problem? Why or why not?
10. In the circuit of Figure 16–23, the diode rectifies the input so that only half of the waveform is passed through to the capacitor and the diode. The capacitor bypasses the high RF and allows the audio voltage to develop a signal across the headphones. How does the circuit of Figure 16–9 work?

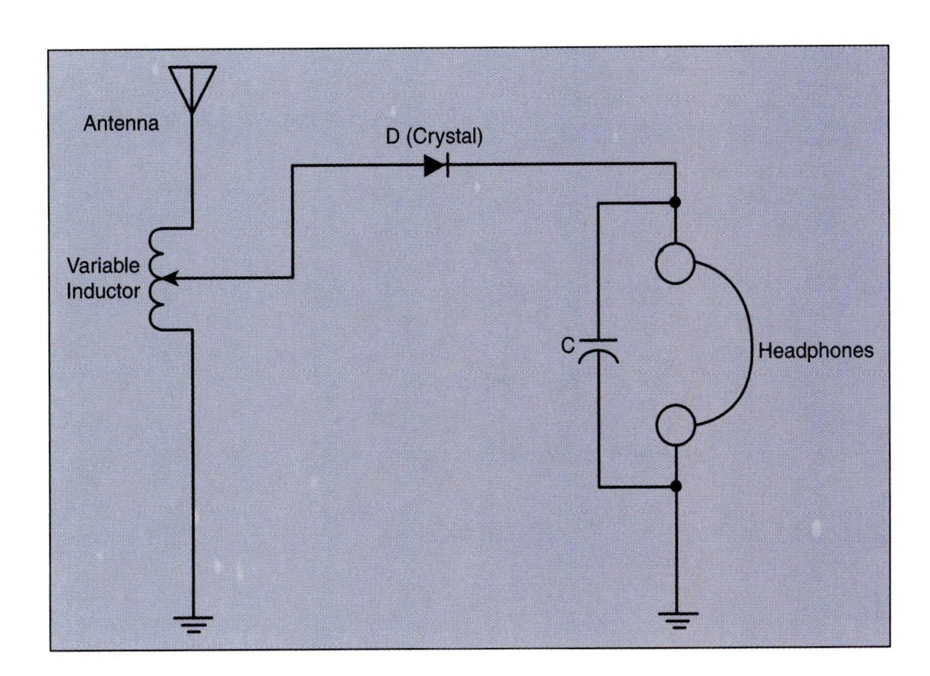

FIGURE 16–23 Improved basic receiver (crystal set)

chapter 17

Integrated Circuits

OUTLINE

OVERVIEW

From the earliest days of electrical technology until the late 1950s, electronic circuits consisted of individual components installed on a chassis or framework and wired together. Then, in 1958, a Texas Instruments engineer named Jack Kilby invented a device that has become known as the integrated circuit. Kilby's invention uses one piece of semiconductor material with many devices etched into it. Resistors, capacitors, and various types of semiconductor devices are most commonly found in an integrated circuit. In the past fifteen years, the integrated circuit has reached a level of sophistication where literally millions of components can be etched onto one chip that is smaller than a postage stamp.

Since integrated circuits were introduced in 1958, electronics has advanced at a quick pace. The use of integrated circuits makes possible applications that would otherwise be too large, too costly, or require too much cooling. This chapter will introduce you to integrated circuits and some of their more common implementations.

OBJECTIVES

After completing this chapter, the student should be able to:

1. List and define the different types of IC timers.
2. Identify the internal parts of the IC timers.
3. Describe the operations for circuits covered.
4. Describe digital signal processing.
5. Perform the calculations introduced in this lesson.

GLOSSARY

Astable A circuit that alternates automatically and continuously between two unstable states at a frequency dependent on circuit constants; for example, a blocking oscillator.

Bistable Able to operate steadily in either one of two states. Will not leave one state until triggered to do so.

Bistable multivibrator A multivibrator in which either of the two active devices may remain conducting, with the other nonconducting, until the application of an external pulse. Also known as Eccles-Jordan circuit, Eccles-Jordan multivibrator, flip-flop circuit, or trigger circuit.

Discrete circuit A circuit that uses individual component parts such as resistors, transistors, and capacitors.

Flip-flop See bistable multivibrator.

Integrated circuit A circuit with most of its components created in a single semiconductor chip.

Isolation diffusion Part of the IC fabrication process that creates insulating barriers.

Monostable Having only one stable state.

Multivibrator A relaxation oscillator using two tubes, transistors, or other electron devices, with the output of each coupled to the input of the other through resistance-capacitance elements or other elements to obtain in-phase feedback voltage.

Photolithography The creation of ICs using a photographic system.

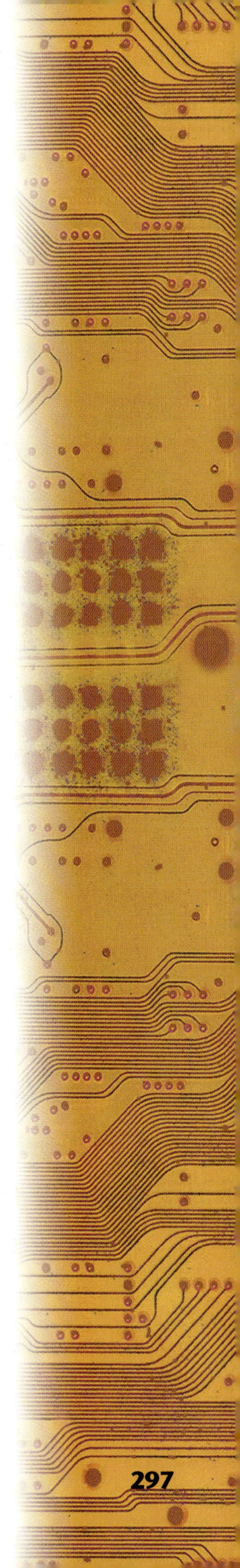

INTRODUCTION TO INTEGRATED CIRCUITS

17.1 Discrete Circuits versus Integrated Circuits

The differences between **discrete circuits** and **integrated circuits** can be summarized as follows:

- Discrete circuits use individual resistors, diodes, transistors, capacitors, and other devices to complete the circuit function. These individual parts are usually mounted on some sort of a circuit board and then interconnected using wires or solder traces.
- To accomplish the same function, discrete circuit assemblies are usually more expensive than an equivalent integrated circuit.
- Although integrated circuit (IC) assemblies do not eliminate the need for circuit boards, assembly, soldering, and testing, they do allow the same circuit to be produced for a lower cost, and a more complex circuit to be produced for the same cost.
- With integrated circuits, the number of discrete parts can be reduced.
- ICs are smaller and use less power, in addition to costing less to manufacture.
- Because the electronics involved in IC assemblies often require fewer alignment steps, integrated circuits can be set up and calibrated for a lower cost.
- Because ICs allow for fewer discrete parts in a piece of equipment, IC systems tend to be more reliable.

17.2 Schematics

The internal features for integrated circuits are seldom shown in schematics. The technician does not usually need to know the circuit details inside the IC. It is more important to know what the IC is supposed to do and how it works as a part of the overall circuit. Figure 17–1 is the normal way of showing an IC. This schematic and a few voltage specifications are all that a technician should need to verify proper operation of the IC.

17.3 Fabrication

The manufacture of integrated circuits starts in a radio-frequency furnace. The P-type silicon wafers are processed using **photolithography**. This process creates the desired P- and N-type zones in the substrate (wafer). Later in the process, the numerous circuit functions in the IC are electrically insulated from each other by **isolation diffusion** (forming an area designed to block current flow).

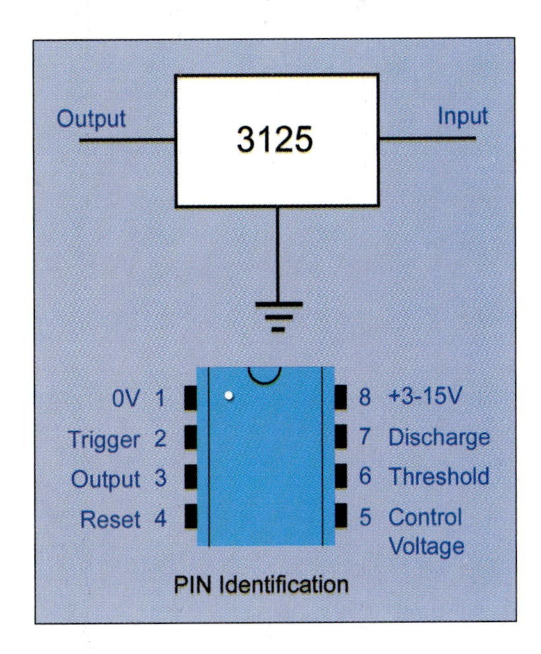

FIGURE 17–1 Normal way of showing an IC on a schematic

THE 555 TIMER

17.4 Introduction

The NE555 IC timer is very popular among circuit designers because of its low cost, versatility, and stable time delays. It allows timing intervals from microseconds to hours. Depending on the preferred output waveform, the oscillator mode requires three or more external components. Frequencies from less than 1 Hz to 500 Hz with duty cycles from 1 to 99 percent can be achieved.

17.5 Internal Circuitry and Operation

The major components of the 555 timer are shown in Figure 17–2. It houses two voltage comparators, a **flip-flop** (also called a **bistable multivibrator**), a resistor divider network, a discharge transistor, and an output amplifier with up to 200 mA current capability. The three divider resistors are 5 Ω each. This network sets the trigger comparator at one-third of V_{CC} and the threshold comparator trip point at two-thirds of V_{CC}. V_{CC} may range from 4.5 V to 16 V.

In Figure 17–2, assume that $V_{CC} = 12$. Given this, the trigger point will be 4 V (⅓ × 12 V) and the threshold point will be 8 V (⅔ × 12). When pin 2 falls below 4 V, the trigger comparator output changes states and sets the flip-flop to the high state, and output pin 3 goes high.

If pin 2 returns to a value higher than 4 V, the output stays high because the flip-flop knows that it was set. On the other hand, if pin 6 climbs above 8 V, the threshold comparator changes states and resets the flip-flop to its low state. This causes the output (pin 3) to go low and the discharge transistor to be turned on.

Notice that the output of the 555 timer is digital; that is, either high or low. When it is low, it is near ground potential, and when it is high, it is close to V_{CC}.

FIGURE 17–2 Diagram of NE555 IC timer

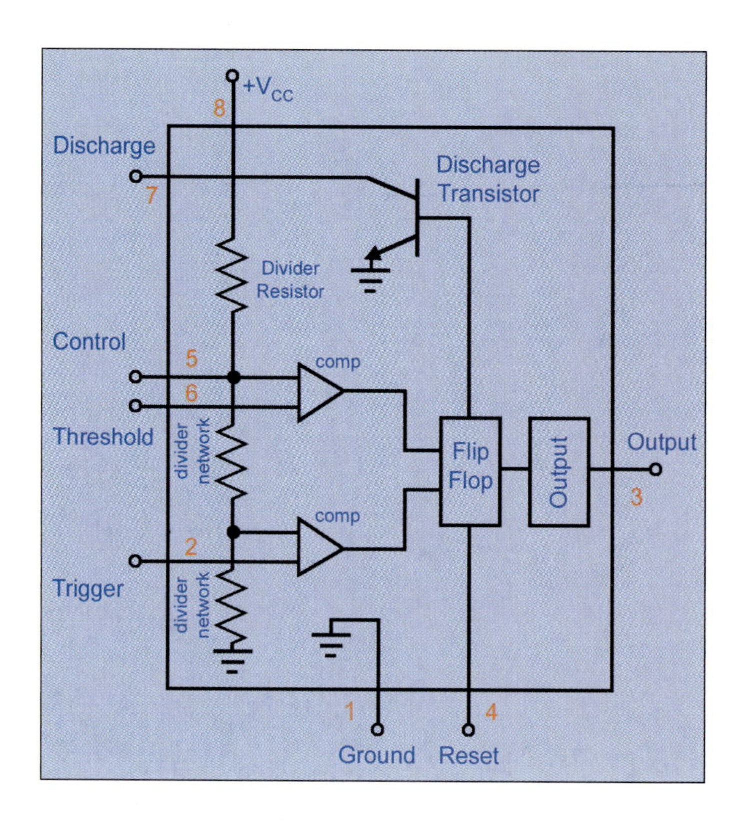

In Figure 17–2, pin 6 is usually connected to a capacitor that is part of an external RC timing network. If the capacitor voltage exceeds ⅔ V_{CC}, the threshold comparator will reset the flip-flop to the low state. This turns on the discharge transistor that can be used to discharge the external capacitor in preparation for the next timing cycle.

The reset, pin 4, allows for direct access to the flip-flop. This pin overrides the various timer functions and pins. The reset can be used to stop a timing cycle. The reset pin is a digital input and when it is taken low, it resets the flip-flop, turns on the discharge transistor, and drives output pin 3 low. The reset function is not usually needed, so pin 4 is normally tied to V_{CC}.

17.6 Monostable Mode

The integrated circuit timer is connected for the **monostable** (one-shot) mode in Figure 17–3. This mode produces an RC-controlled output pulse that goes high when the device is triggered. The timer is considered to be negative-edge triggered, which means that the timing cycle begins at t_1 when the trigger input falls below ⅓ V_{CC}.

Once the trigger input is greater than ⅓ V_{CC}, the time-out period begins, which means the trigger pulse cannot be wider than the output pulse. When the trigger pulse is wider, the trigger input will need to be AC-coupled. Using Figure 17–4, you can see the 0.1 μF coupling capacitor and the 5 kΩ resistor differentiate the input trigger pulse. The effective width of the trigger pulse is decreased by this pulse differentiation (AC coupling).

The width of the output pulse is RC controlled in the monostable (one-shot) circuit and the timing capacitor begins charging through the

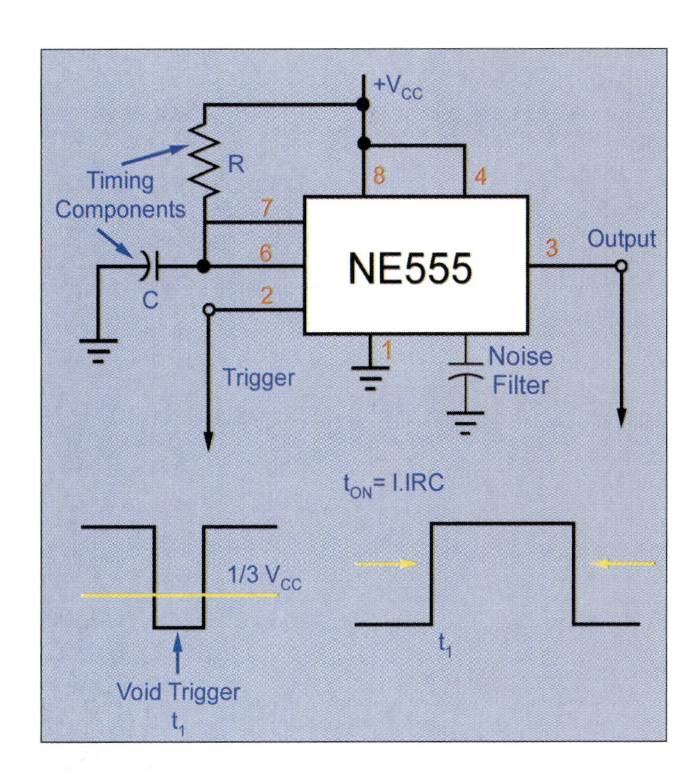

FIGURE 17–3 Monostable mode

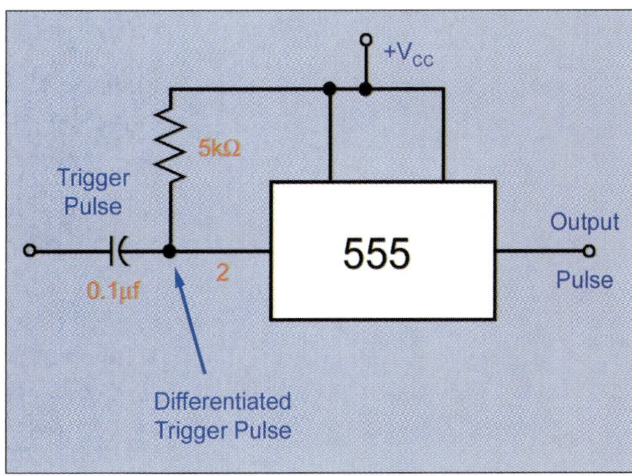

FIGURE 17–4 AC-coupled trigger pulse

timing resistor when the timer is triggered. Recall that, as the capacitor voltage reaches ⅔ V_{CC}, the flip-flop is reset. This turns on the discharge transistor, and the capacitor is emptied in preparation for the next cycle. This results in the output pulse width equaling 1.1 time constants.

The formula for finding the output pulse is

$$t_{ON} = 1.1RC$$

EXAMPLE 1

Find the output width for Figure 17–4. $R = 5\ \text{k}\Omega$ and $C = 0.1\ \mu\text{F}$.

The pulse width is equal to 1.1 time constants.

$$t_{ON} = 1.1 \times R \times C = 1.1 \times 5\ \text{k}\Omega \times 0.1\ \mu\text{F} = 0.55\ \text{m sec}$$

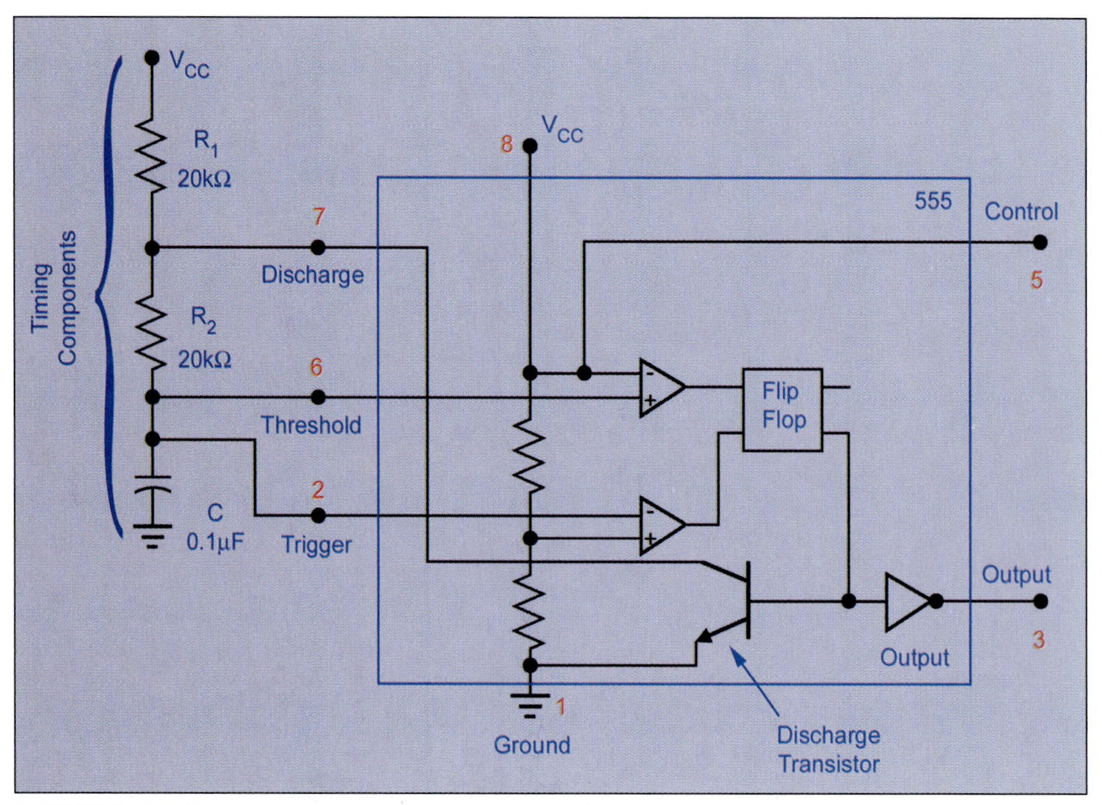

FIGURE 17–5 Free-running or astable mode

17.7 Astable Mode

An example of the timer configured in the **astable** (free-running) mode is shown in Figure 17–5, with the trigger (pin 2) being tied to the threshold (pin 6). As the circuit is turned on and the timing capacitor is discharged, the timer begins charging through the series combination (R_1 and R_2). When the capacitor voltage attains ⅔ V_{CC}, the output drops to low and the discharge transistor turns on. At this point, the capacitor discharges through R_2. The output changes to high and the discharge transistor turns off as the capacitor reaches ⅓ V_{CC}. Now the capacitor starts charging through R_1 and R_2. This cycle will continuously repeat, as the capacitor charges and discharges.

The following formula calculates the time that the output is held high in an astable circuit through two resistors:

$$t_{high} = 0.69(R_1 + R_2)C$$

EXAMPLE 2

Say that both timing resistors in Figure 17–5 are 20 kΩ and the timing capacitor is 0.1 μF. Find the time the output will remain high.

$$t_{high} = 0.69(R_1 + R_2)C = 0.69(20\ k\Omega + 20\ k\Omega)0.1\ \mu F = 2.76\ ms$$

Because the discharge path is passing through only one resistor (R_2), the time the output is held low is found by using a similar formula.

$$t_{low} = 0.69(R_2)C$$

EXAMPLE 3

Using the same data from the previous example, with the 20 kΩ resistor and the 0.1 μF timing capacitor, find the time the output will remain low.

$$t_{low} = 0.69(R_2)C = 0.69 \times 20\ \text{k}\Omega \times 0.1\ \mu\text{F} = 1.38\ \text{ms}$$

Because the resistors are equal, the time held low is half the time held high.

Because the times held high and low are different, the output waveform is nonsymmetrical. When you add t_{high} and t_{low}, the result is the total period, which is needed to calculate the output frequency. The output frequency is equal to the reciprocal of the total period, which can be found by using the following formula:

$$f_0 = \frac{1.45}{(R_1 + 2R_2)C}$$

EXAMPLE 4

Again using the data from the previous example, where the resistors are 20 kΩ and the timing capacitor is 0.1 μF, find the output frequency.

$$f_0 = \frac{1.45}{(R_1 + 2R_2)C} = \frac{1.45}{(20\ \text{k}\Omega + 40\ \text{k}\Omega)0.1\ \mu\text{F}} = 241.67\ \text{Hz}$$

The duty cycle is the portion (percentage) of time that the output is high. This is found by taking the time that the output is high and dividing it by the total period of the waveform.

The formula is

$$D = \frac{R_1 + R_2}{R_1 + 2R_2} \times 100\%$$

EXAMPLE 5

Using the information from Figure 17–5 and two 20 kΩ timing resistors, calculate the duty cycle of this rectangular waveform.

$$D = \frac{20\ \text{k}\Omega + 20\ \text{k}\Omega}{20\ \text{k}\Omega + 2 \times 20\ \text{k}\Omega} \times 100\% = 66.7\%$$

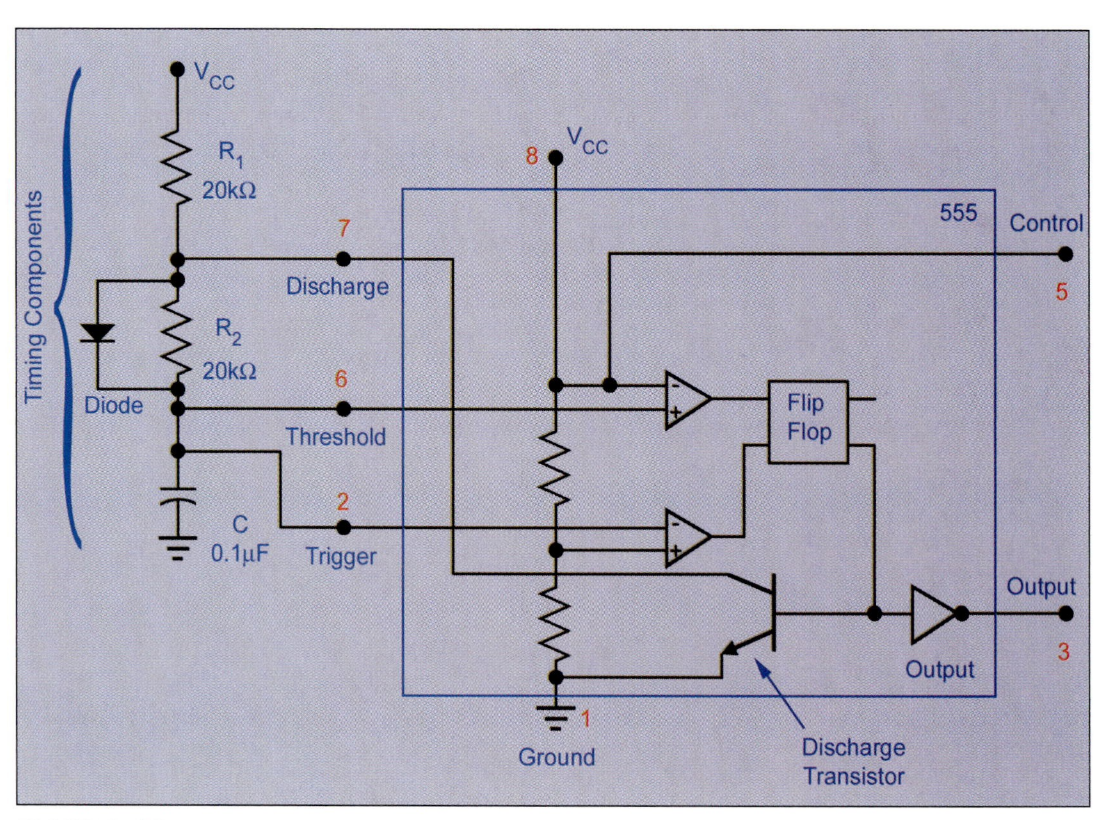

FIGURE 17–6 Bypass diode around R_2

Because the timing capacitor charges through the two resistors and discharges only through R_1, this circuit can produce only a rectangular wave. A square wave, which is a rectangular wave with a 50 percent duty cycle, cannot be produced with this circuit. As R_1 gets smaller relative to R_2, the duty cycle will get closer to 50 percent but will not reach 50 percent. Making R_1 equal to 0 Ω in the formula will result in a 50 percent duty cycle; however, this would damage the integrated circuit because there would be no current limiting for the internal discharge transistor.

An alternative circuit that allows duty cycles of 50 percent or less is shown in Figure 17–6. A diode, which bypasses R_2 in the charging circuit, has been added in parallel with R_2. The timing capacitor will still discharge through R_2 but only charges through R_1. The previous formulas are modified slightly for the altered circuit.

$$t_{high} = 0.69(R_1)C$$

$$t_{low} = 0.69(R_2)C$$

$$f_0 = \frac{1.45}{(R_1 + R_2)C}$$

$$D\% = \frac{R_1}{R_1 + R_2} \times 100$$

It is now possible to find resistor values that will produce square waves.

EXAMPLE 6

Find the resistor values that will produce a 2 kHz square wave given a timing capacitor of 0.01 μF.

Solution:
Begin with the formula for output frequency.

$$f_0 = \frac{1.45}{(R_1 + R_2)C}$$

Now, manipulate the formula to find $R_1 + R_2$, the total resistor value.

$$R_1 + R_2 = \frac{1.45}{f_0 \times C} = \frac{1.45}{2\ \text{kHz} \times 0.1\ \mu\text{F}} = 72.5\ \text{k}\Omega$$

Because a square wave has a 50 percent duty cycle, each resistor must be the same. Therefore, each resistor must be half of 72.5 kΩ.

$$R_1 = R_2 = \frac{72.5\ \text{k}\Omega}{2} = 36.25\ \text{k}\Omega$$

17.8 Time-Delay Mode

The versatility of 555 applications is shown in Figure 17–7. This configuration creates a stable timing output with the addition of a transistor and two diodes to the RC timing network. The frequency can be varied over a wide range while maintaining a constant 50 percent duty cycle. When the output is high, the transistor is biased into saturation by R_1 so that the charging current passes through the transistor and R_2

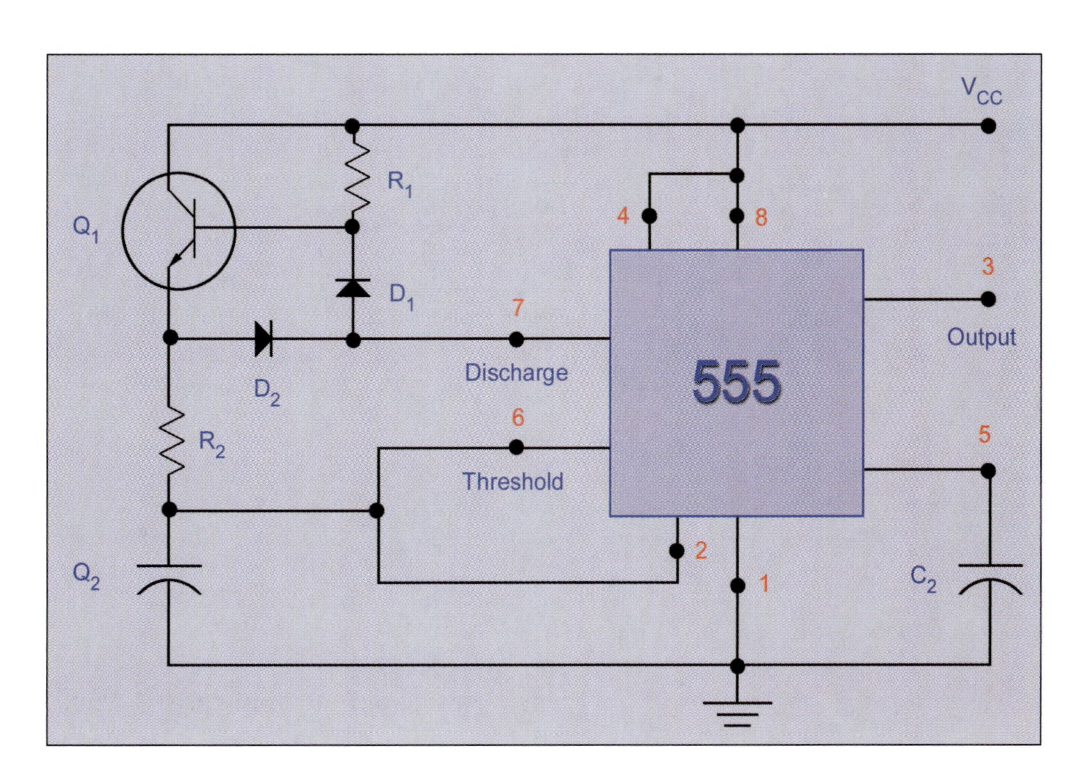

FIGURE 17–7 Time-delay mode

to C_1. When the output goes low, the 555 internal discharge transistor (pin 7) cuts off the transistor Q_1 and discharges the capacitor through R_2 and the D_2. The high and low periods are equal. The values of capacitor (C_1) and resistor (R_1) are dependent on the type or length of timing desired.

EXAMPLE 7

If $R = 260\ \text{k}\Omega$ and $C = 25\ \mu\text{F}$, the time delay is calculated by using the following formula:

$$t_{\text{delay}} = 1.1 \times R \times C = 1.1 \times 260\ \text{k}\Omega \times 25\ \mu\text{F} = 7.15\ \text{sec}$$

In this mode, if the trigger signal goes high before the IC times out, the output will not go low. Security alarms take advantage of this feature by allowing time to clear an area before the alarm is armed.

OTHER INTEGRATED CIRCUITS

In earlier chapters, you were introduced to some of the most widely used integrated circuits such as differential amplifiers and operational amplifiers. Integrated circuits can also replace transistor stages such as an IF amplifier. The use of an IC (with multiple transistors) will normally provide improved performance such as more gain, better selectively, greater noise rejection, and so on.

Linear integrated circuits that provide more than one function are usually called subsystem integrated circuits. An example of a subsystem IC is a television sound system containing a limiter, an FM detector, an IF amplifier, a regulated power supply, and an electronic volume control. This type of integrated circuit reduces the number of parts needed in the sound section of a TV receiver. With the number of parts being reduced, not only is the cost reduced, but the reliability is improved.

DIGITAL SIGNAL PROCESSING

Digital signal processing (DSP) involves changing AM analog signals (such as music) into a stream of numbers. The circuit then performs various arithmetic operations on those numbers and changes the resulting numbers back into an improved or enhanced signal. With DSP, many different things can be achieved. Some examples are:

- Eliminate unwanted signal characteristics or components (such as echoes, noise, or interference)
- Compress or expand a signal
- Enhance a signal to provide special effects (such as surround sound)
- Balance the different frequency components
- Demodulate an encoded signal

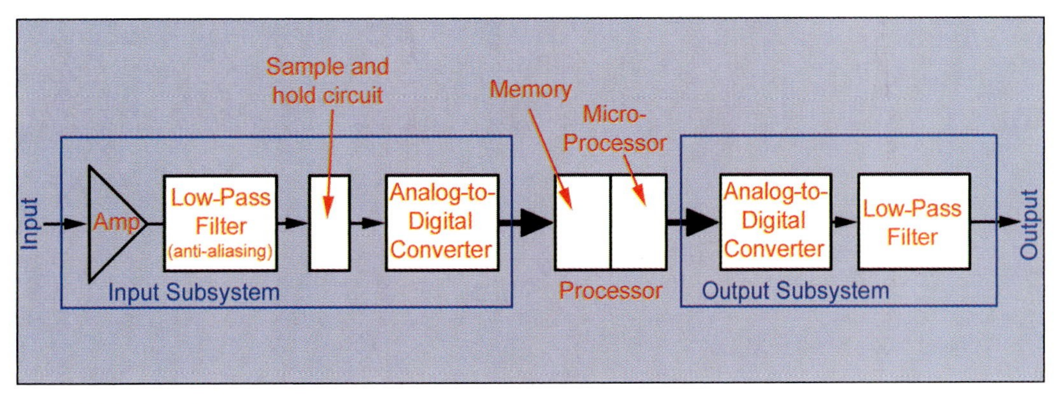

FIGURE 17–8 Diagram of a digital signal processing system

Analog circuits can also accomplish these effects. However, the DSP is preferred because the technology is small and inexpensive when compared to analog equipment.

Other benefits of the DSP include:

- Changes can be made through software adjustments rather than component changes. This reduces the cost and time spent making the modifications.
- Digital designs are more reliable, with maintenance and calibration being easier to perform and less costly.
- They are not sensitive to environmental changes (temperature and humidity) and not affected by aging, as are analog circuits.

DSP systems usually include the stages listed here and seen in Figure 17–8.

1. An amplifier
2. A low-pass filter, also call an antialiasing filter
3. A sample-and-hold circuit
4. An analog-to-digital converter
5. Memory
6. Microprocessor
7. A digital-to-analog converter
8. Low-pass filter

ANALOG-TO-DIGITAL CONVERSION

Analog-to-digital (A/D) and digital-to-analog (D/A) conversion is required in many integrated circuit applications. The digital signal processor mentioned earlier has both in its circuitry.

Analog signals are continuous; that is, their voltage changes smoothly over time. On the other hand, a discrete signal can change in voltage value at any point in time. To make digital processing possible, the continuous signal must be changed into a digital signal. Changing the continuous signal into a discrete signal uses a process called sampling. Basically, 'snapshots' or samples of the signal are taken at different points in time.

The higher the sampling rate, the better the discrete signal will represent the continuous signal. Conversely, the discrete signal will be worse as the sampling rate becomes lower. This will result in a poor quality of sound or picture.

Before the signal is received for the sample-and-hold process, it first passes through antialiasing filter, which helps remove interference and high-frequency noise. After sampling, the signal is then sent to the analog-to-digital (A/D) converter. An A/D converter converts each signal sample into a string of bits (bit is the combining of the words binary and digit).

Bits are either 0s or 1s, with 0 being called low and 1 being called high. Some examples of an output are:

- 000 (this binary number represents the decimal 0)
- 100 (represents the decimal 4)
- 111 (represents the decimal 7)

After the A/D has converted the signals to bits, they are sent to the memory and then to the microprocessor, which performs the calculations specified by the software program. From there, it is sent to the digital-to-analog converter and then on to a low-pass filter.

SUMMARY

The invention of the integrated circuit changed the world of electronics forever. Circuits and systems are routinely developed using integrated circuits that would have required an impossible number of discrete components. Integrated circuits that you have already learned about include differential amplifiers, operational amplifiers, and Darlington pairs.

The major components of the NE555 IC timer are voltage comparators, a bistable flip-flop, a resistor divider network, a discharge transistor, and an output amplifier. The network trigger comparator is set at ⅓ V_{CC}, and the threshold comparator trip point is set at ⅔ V_{CC}. The V_{CC} ranges from 4.5 V to 16 V. The output is digital.

In this chapter, you learned the different components and modifications of three different modes of the 555 timer. The three modes are the monostable mode, the astable mode, and the time-delay mode.

The stages of digital signal processing have been introduced as well as some benefits of this system compared to analog equipment. During the A/D conversion phase of the DSP, the analog signal is converted into a string of bits, which are either a 0 (low) or 1 (high).

REVIEW QUESTIONS

1. What are the advantages and disadvantages of integrated circuits as compared to discrete component circuits?
2. Describe, in simple terms, how an integrated circuit is fabricated.
3. Using Figure 17–2, explain how the inputs and outputs are related to each other.
4. Which of the three operating modes should be used (NE555) for the following?
   a. Steady square wave output
   b. A switch that can trigger a given action after a fixed time delay
   c. A circuit that can be used to determine how long a particular operation requires

5. What are the advantages of digital signal processing as compared to analog processing?
6. Using Figure 17–8, describe the basic purpose and operation of each of the blocks in the digital signal processor.
7. In a group discussion, identify and discuss at least five (5) other electronic applications that could be enhanced by the use of integratred circuits.

PRACTICE PROBLEMS

The following questions refer to the NE555 timing circuit.

1. In the monostable mode, what is the time duration for $R = 10\ \text{k}\Omega$ and 0.5 μF?
2. In the monostable mode, if the pulse width is 2 ms and the capacitor is 0.1 μF, what is the value of R?
3. In the astable mode, you have the following data: $t_{high} = 4$ msec, $t_{low} = 1.5$ msec. What is the duty cycle in percent?
4. Using the solution from problem 3, if $R_1 = 6\ \text{k}\Omega$, find the value of R_2.
5. Using the values from problems 3 and 4 and a value of $C = 0.2$ μF, what is f_0?
6. Find the values for two resistors that will produce a 2 kHz square wave with a 70 percent duty cycle. The timing capacitor is 0.01 μF.
7. You wish to use an NE555 to build an 8-second time-delay circuit. You have a 20 μF capacitor. What size resistor do you need?
8. You are analyzing an NE555 circuit (Figure 17–5) in an astable mode. The resistors are $R_1 = 10\ \text{k}\Omega$ and $R_2 = 30\ \text{k}\Omega$, and $C = 2$ μF. Answer the following:
   a. $D = ??$
   b. $t_{high} = ??$
   c. $t_{low} = ??$
   d. $f_0 = ??$
9. Answer question 8 using Figure 17–6.
   a. $D = ??$
   b. $t_{high} = ??$
   c. $t_{low} = ??$
   d. $f_0 = ??$

Microprocessors and Systems Components

OUTLINE

OVERVIEW

Early computers used vacuum tubes, were extremely large and expensive, and had limited capabilities. Today's **microcomputers** are smaller, less expensive, and much faster than older **mainframes** or **minicomputers**. These improvements have been brought about primarily by the advent of the microprocessor.

Probably the most significant use of integrated circuits has been in the development of the **microprocessor**—an integrated circuit that has the **central processing unit** (CPU) and possibly other peripheral components all on one chip. Since the initial circuits were introduced in the early 1970s, microprocessors have continued to grow in power and capability, even though the basic electrical structure and operation of the CPU has not changed significantly. At the same time, they have been decreasing in size and increasing in density.

Microprocessors composed of millions of transistors and other circuit components are now commonplace. Truly, the small laptop and handheld computers that we use today have become many times more powerful than the room-sized mainframe computers of the 1950s and 1960s.

As you work through the material in this chapter, keep in mind that computer technology is the most dynamic technology in history. It is changing and advancing so fast that almost any publication about the "state-of-the-art" is out of date before it hits print. This chapter presents many of the fundamentals upon which computer technology of today is based.

OBJECTIVES

After completing this chapter, the student should be able to:

1. List the major components of the digital computer and CPU.
2. Explain how the major components of the CPU work together to enable a program to function.
3. Explain how multiplexers function.
4. Describe the operating principles of various types of computer memory.
5. Explain the difference between volatile and nonvolatile memory.

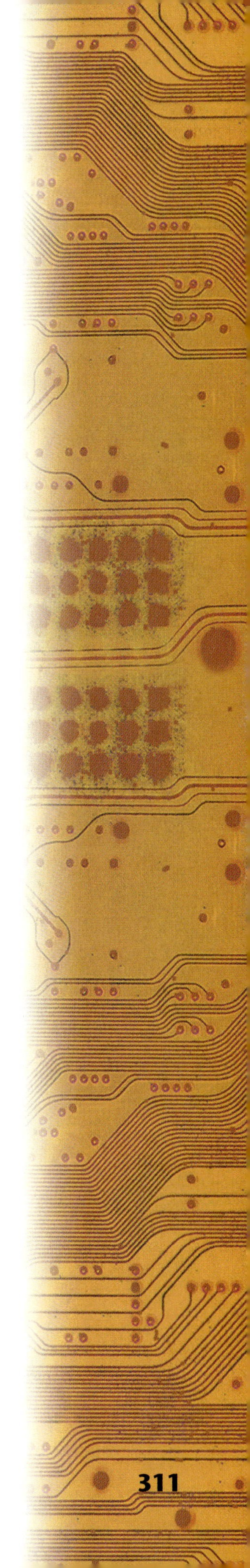

GLOSSARY

CD Compact disk. An optical media that stores information using laser-imprinted plastic media. CDs can store up to almost a gigabyte of data.

Central processing unit (CPU) The part of a computer containing the circuits required to interpret and execute the instructions.

DVD Digital video disk or digital versatile disk. A modern optical storage media that can store multiple gigabytes of data on a disk.

Mainframe computer A large computer. Older models are made of discrete components and often require cabinetry that fills several rooms.

Microcomputer A microprocessor combined with input/output interface devices, some type of external memory, and the other elements required to form a working computer system; it is smaller, lower in cost, and usually slower than a minicomputer. Also known as micro.

Microprocessor An integrated circuit that contains the entire central processing unit of a computer on a single chip.[1]

Minicomputer A relatively small general-purpose digital computer, intermediate in size between a microcomputer and a mainframe.

Multiplexer A device for combining two or more signals, as for multiplex, or for creating the composite color video signal from its components in color television. Also spelled multiplexor.

[1] Excerpted from *American Heritage Talking Dictionary.*

DIGITAL COMPUTERS

18.1 History and Modern Architecture

In addition to performing calculations (addition, subtraction, multiplication, and division), the computer can make logical decisions on how to continue as instructed by a program. The program is stored in memory and gives the computer a sequenced set of instructions.

The arithmetic/logic unit and the control unit, the two major elements of the digital computer, are usually combined together to form what is known as the central processing unit (CPU). The memory, which stores results, and the input/output (I/O) buffers are the other two major components of the computer. Although the memory and I/O buffers are sometimes contained on a single chip separate from the CPU, usually all the functions of the microprocessor are included on one chip. See Figure 18–1.

As its name suggests, the control unit controls the flow of information through the other components, whereas the ALU (arithmetic/logic unit) manipulates the data based on the instructions of the program. The program is stored in the memory, as well as other results being stored here. The I/O buffers are a temporary memory for external inputs (keyboard) and allow for outputs to external devices (printers).

The data bus (which carries and transfers data) and the address bus (which controls memory) make up the set of lines carrying information called the bus. Through the use of this basic structure, the computer manipulates bits to produce specific results.

18.2 The Central Processing Unit

Although all CPUs contain the arithmetic, logic, and control functions, they are not all the same. The one referred to as a microprocessor is a CPU contained on a single chip (shown in Figure 18–2). However,

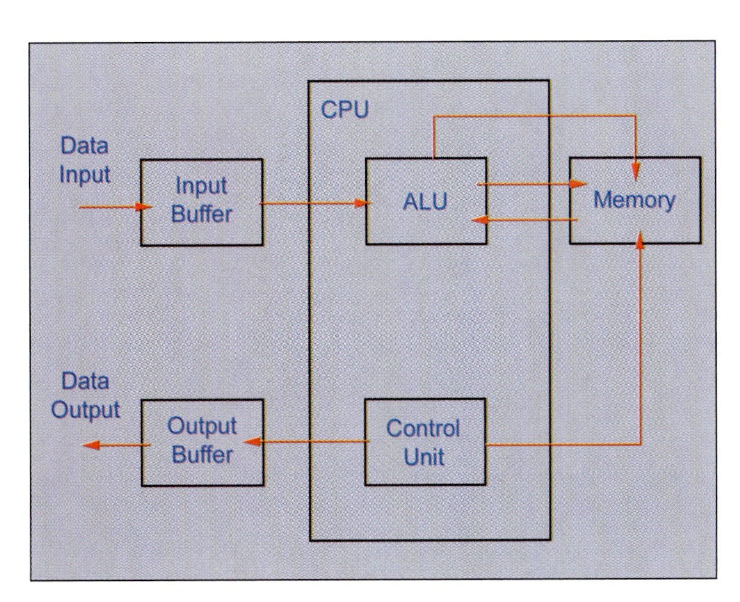

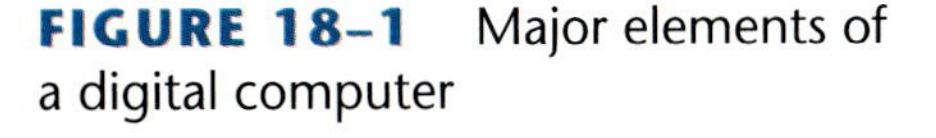
FIGURE 18–1 Major elements of a digital computer

FIGURE 18–2 The CPU

more complicated systems may involve more than one circuit board. All CPUs must contain these fundamental elements:

- The logic, control, and arithmetic functions capable of performing the instructions of the program
- A program counter, which points to the location in memory where an instruction code has been retrieved
- At least one register (accumulator), where data taken from and returned to memory will be stored
- A data counter, which points to the location in memory where data can be stored and retrieved
- An instruction register, which determines the operation to be executed as defined by the program

In general, the CPU carries out the manipulations, while the bits of information are fetched from and returned to memory. The instructions, which determine how the bit patterns are manipulated, are also in memory. These manipulations are performed in the section of the CPU called the ALU, which performs the following functions:

- Boolean operations
- Binary addition
- Complementation
- Shift of data to the right or left

Whereas the Boolean operations allow for logical decisions, the addition, complementation, and data shifting provide for any of the mathematical operations (addition, subtraction, multiplication, and division). Other functions are often included for speed, but the four functions listed allow for almost any kind of bit manipulation.

Most simple inexpensive microprocessors handle 8-bit (1 byte) to 16-bit (2 bytes) word lengths. Larger computers can handle up to 128-bit word lengths or even more. All parts of the CPU must handle the same size word lengths.

The control unit (CU) determines the order in which the ALU operates. The instruction register, which is controlled by the program, sends the instructions to the CU. The CU decodes the instructions (bit patterns) from the instruction register. With these instructions, the CU sends the appropriate sequence of signals to control the ALU logic and the flow of data through the ALU.

The data bits are transferred between the ALU and the registers by way of an internal data bus. The control unit must first understand how data bits should flow between the ALU and the registers. Once this is done and the ALU understands how to manipulate the data bits, instructions are sent to the ALU and registers from the control unit. This enables the correct signals to appear on the data bus. This allows the data bits to be sent from the registers to the ALU to be operated upon. After they have been manipulated, they are transferred back through their shared data bus. A common clock synchronizes the operations being performed.

Some computers have extra registers to allow for the holding of intermediate results and for the taking in of additional information. The purpose of the additional registers is quicker operational speed to allow for simpler programming.

Table 18–1 shows the five different status flags produced by the CPU to indicate the status or result of ALU operations.

18.3 Arithmetic/Logic Unit

Figure 18–3 shows the schematics for the pin connections of the SN74181 IC. The CPU allows two 4-bit words inputs and sixteen combinations. The sixteen combinations include both arithmetic operations and Boolean logic operations. A0–A3 and B0–B3 are the inputs for the 4-bit digital words, and S0–S3 are the control functions. Outputs are

Table 18–1 CPU Status Flags

Flag	Purpose
Carry	The carry flag is used when the entire word cannot be handled in a single operation. An example is a microprocessor that handles only 8-bit (1-byte) words but needs to add two 16-bit (2-byte) words. Adding the two lower-order bytes together and the two higher-order bytes together can accomplish this. If the two lower-order bytes produce a carry when added, the carry bit is set. If a carry is produced, one will be added to the least significant bit of the higher-order byte addition.
Interrupt	The interrupt flag helps get data into and out of the computer while ensuring that the correct sequence of events takes place.
Zero	The zero flag is set if a result is a zero.
Sign	The sign flag indicates whether a number is positive or negative. The sign flag is 0 for positive and 1 for a negative value.
Overflow	The overflow flag is used with the sign flag when multiple-byte words are processed.

FIGURE 18–3 The SN74181 IC

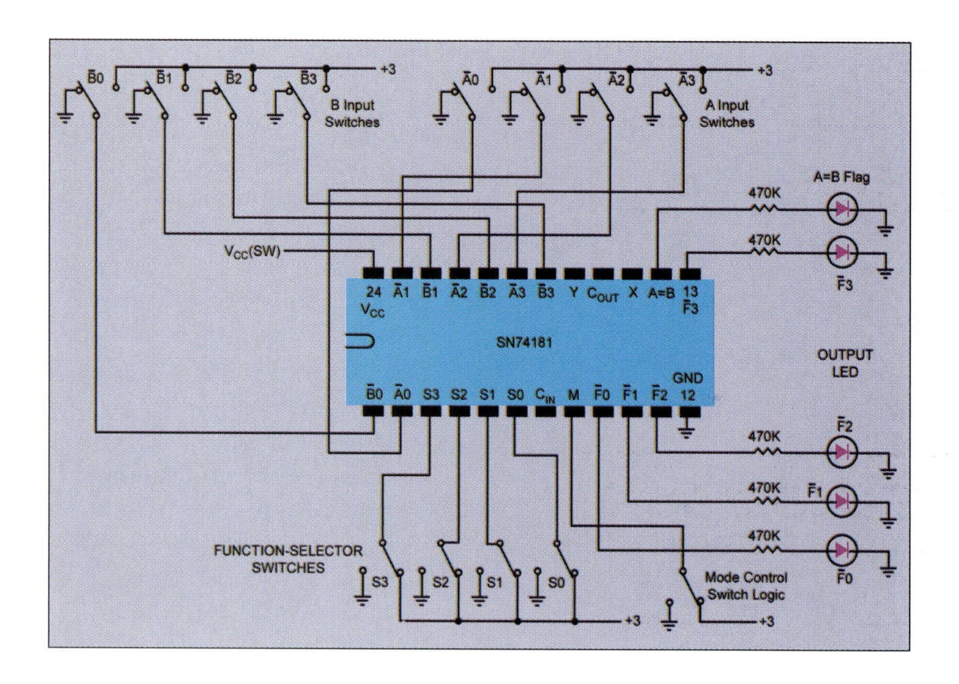

channeled through F0–F3. The ALU can be set to perform either arithmetic or logic operations by the setting of control line M. When control line M is low, arithmetic operations can be performed, and when control line M is high, logic operations can be performed. In this ALU, a flag is used when A = B.

Table 18–2 shows the arithmetic functions available for A and B with either the high levels active or the low levels active. By changing the state of the selector switches (S0–S3), you can choose the desired arithmetic functions to be performed. The desired logic outputs can also be obtained by having the mode selector switch in the logic position. There is also a table for the logic functions; it is not shown here.

Notice that if S0–S3 are set at 0001 and A is set at 0010 and B is set at 1001, the output at F0–F3 will be 1011. Although this ALU handles only 4-bit words, it is similar to ALUs found in more complex computers.

18.4 Multiplexers

Although a common bus is used to transfer data between the registers and the ALU, a **multiplexer** (shown in Figure 18–4) is used to ensure the correct interconnections. Multiplexers are used for other interconnections as well. The data bits emanating from the outputs of keyboards, adders, counters, and registers must be connected to other counters, registers, adders, and output devices at the correct time. The multiplexer conducts pulse trains from different sources, such as the ones just mentioned.

The CU directs the multiplexer selector switch. This extremely fast electronic switch can accept data from many different inputs. Its job is to connect the right devices together at the right time so that digital information can be transferred.

Demultiplexing, which is the inverse operation, is carried out at the output of the multiplexer. It receives inputs from a single line that

Table 18–2 Table of Arithmetic Operations

Function Select				Output Function	
S3	**S2**	**S1**	**S0**	**Low Levels Active**	**High Levels Active**
L	L	L	L	F = A MINUS 1	F = A
L	L	L	H	F = AB MINUS 1	F = A + B
L	L	H	L	F = AB MINUS 1	F = A + B
L	L	H	H	F = MINUS 1 (2's complement)	F = MINUS 1 (2's complement)
L	H	L	L	F = A PLUS (A + $\bar{B}$)	F = A PLUS AB
L	H	L	H	F = AB PLUS (A + $\bar{B}$)	F = (A + B) PLUS A $\bar{B}$
L	H	H	L	F = A MINUS B MINUS 1	F = A MINUS B MINUS 1
L	H	H	H	F = A + B	F = AB MINUS 1
H	L	L	L	F = A PLUS (A + B)	F = A PLUS AB
H	L	L	H	F = A PLUS B	F = A PLUS B
H	L	H	L	F = AB PLUS (A + B)	F = (A + B) PLUS AB
H	L	H	H	F = A + B	F = AB MINUS t
H	H	L	L	F = A PLUS At	F = A PLUS At
H	H	L	H	F = AB PLUS A	F = (A + B) PLUS A
H	H	H	L	F = AB PLUS A	F = (A + B) PLUS A
H	H	H	H	F = A	F = A MINUS 1

FIGURE 18–4 Multiplexer

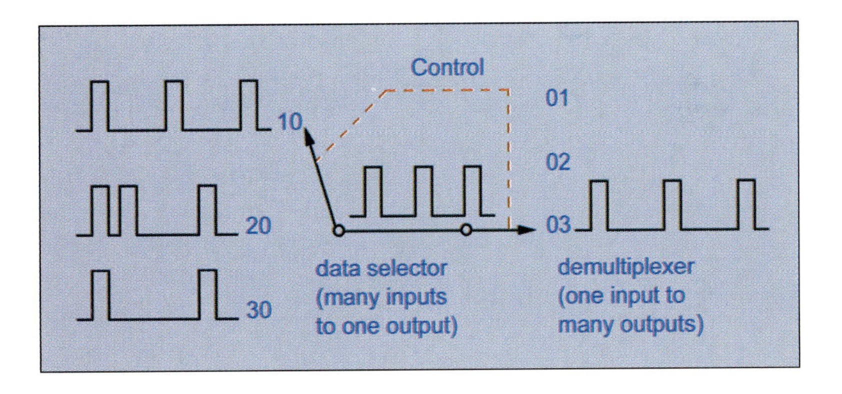

carries different types of data. It might first carry a train of pulses from source 3, then from source 1, and a different train from source 2, and so on. It directs each component of the multiplexed train to the correct destination at the appropriate time. It does this by moving its electronic switching pole to the correct output terminal at the correct time. When many remote terminals are to be used by a computer, multiplexers not only are useful but also reduce the expense of having many lines connecting to each terminal.

A simple multiplexer comprising one OR, four NAND, and four AND gates is shown in Figure 18–5. Any of the four data input lines could be connected to a single output, depending upon the two data inputs. Depending upon the combination of the two data inputs, the appropriate output will be selected, as shown in Table 18–3.

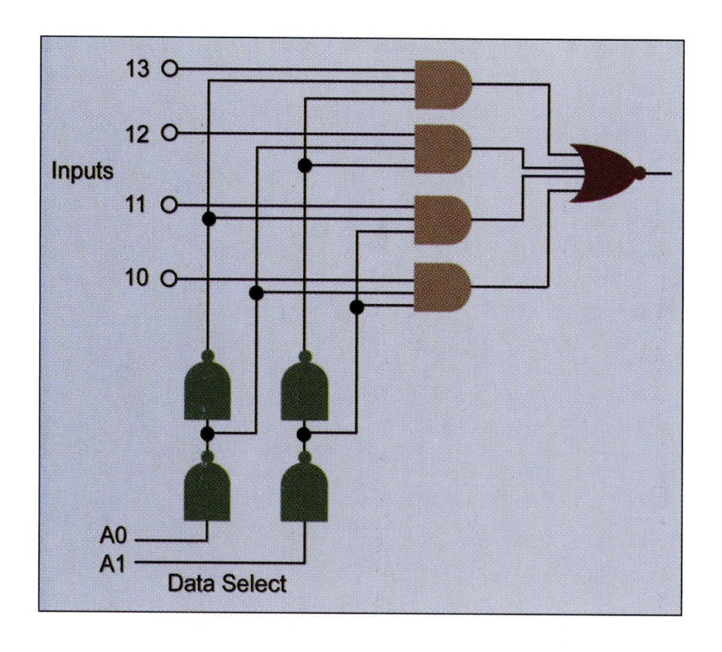

FIGURE 18–5 Four-input multiplexer

Table 18–3 I/O Table for Figure 18–5

Data inputs		Input appearing at output
A0	**A1**	
1	0	12
1	1	13
0	0	10
0	1	11

This is an example of a 4:1 multiplexer, where any one of four lines can be selected. If the number of data inputs were increased to three, it would become an 8:1 multiplexer, where eight different lines can be multiplexed onto a single line. This could be expanded to accommodate any number of lines to be connected to a single output line. The same could be done with the demultiplexer. This would allow different pulse trains to be distributed to as many outputs as necessary.

Under the direction of the control unit and using the common bus structure, multiplexing provides that the correct interconnections between all the CPU parts occur at the right times.

MEMORY TYPES

18.5 Introduction

The storage location of fixed and dynamic memories can be reached by an appropriate address. The internal bus already discussed has two parts that are associated with these memories. The address bus determines the location of the data and the data bus carries the information to and from the memory (Figure 18–6). It is important to understand

FIGURE 18–6 Bus types

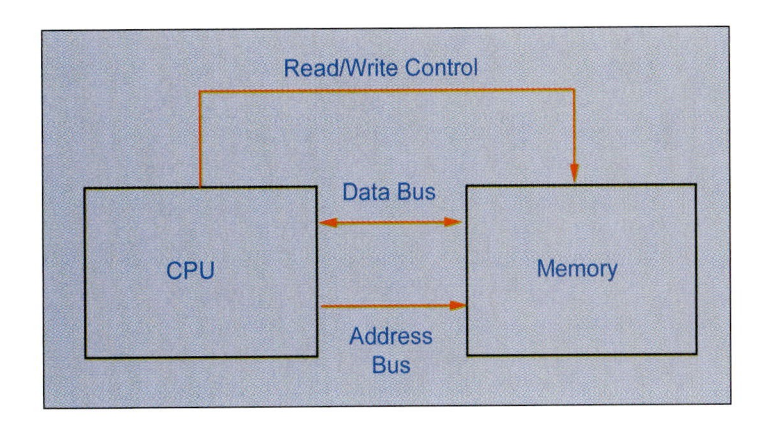

that the contents of the memory remain unchanged when it is read. The memory will change only when additional information is written into it.

Memory is categorized either as volatile or nonvolatile. Volatile memory holds data that is lost when power is turned off, and nonvolatile memory will retain the data despite the power being removed. When using fixed programs, loading memory programs, or running programs for initializing the computer, memory that will remain unchanged is necessary. These programs should remain constant no matter what manipulations have been performed and no matter what short-term changes have been made to the program. These programs must remain unchanged unless the program is rewritten for a desired change.

It is possible to control many different memory locations because the data bus and address bus are separate. For example, assume that a certain microprocessor CPU handles word lengths of 1 byte (8 bits); it could have addressing capabilities of 2 bytes (16 bits). This means there are 2^{16}, or 65,536, addressable locations. Each location is capable of holding one 8-bit word. This 65,536-word memory is what is known as 64K memory, and 16,384-word (2^{14}) memory is called 16K.

The CPU needs to know which memory is read only and which memory can be written into. This is accomplished by using a single-bit read/write control line.

18.6 Internal Memory—Core

Larger, older computers used core memory, whereas today's smaller digital computers use RAM and ROM for their memory components. Core memories allow for random access and are also nonvolatile, which gives them permanent storage capabilities. Tiny rings of magnetic oxide are mixed with ceramic material, used to strengthen the product, to form the core memory.

The rings will store a 0 when magnetized in one direction and a 1 when magnetized in another direction. The cores are built in rectangular patterns (arrays). The length of a core is the same as the word length for the computer. It will be as wide as possible. This length and width produce what is called a plane (see Figure 18–7), and these planes are stacked to increase the memory size.

FIGURE 18–7 Core planes

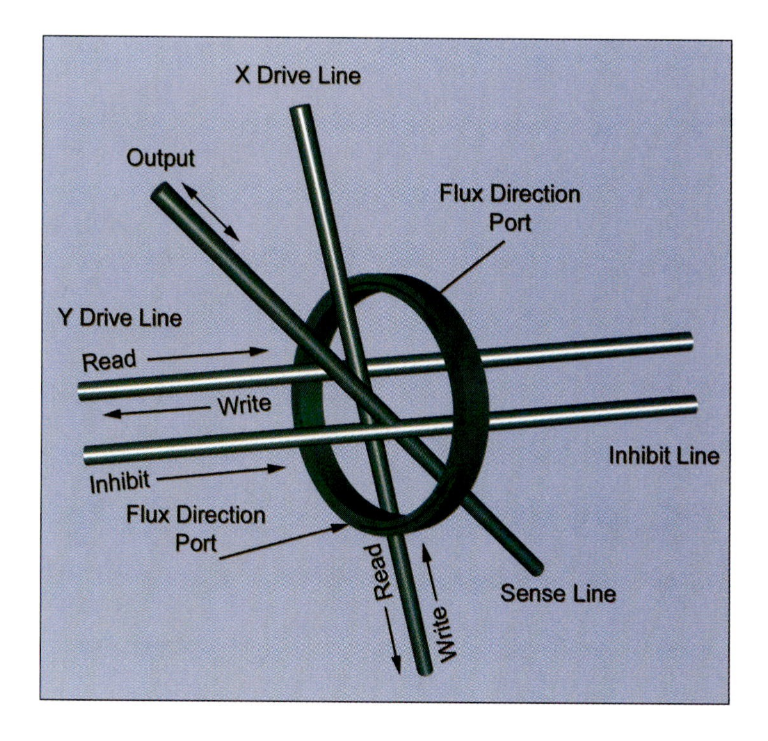

FIGURE 18–8 Magnetic core memory

As seen in Figure 18–8, wires go through each core element. These wires allow for the reading, writing, and addressing functions. Depending upon the direction of the current in the X and Y lines, the core can be read or written into. These lines also allow for addressing. The sense and inhibit lines assure the core an output and control.

Although core memory is nonvolatile, its data must be rewritten as it is read out. Because this restoring of data is done internally (inside the core), it appears that the data has been retained. Some advantages of using core memory are:

- Useful in larger computers because they can hold huge amounts of data
- Extremely reliable
- Nonvolatile

However, disadvantages of using core memory are:

- Expensive
- Complex circuit boards
- Operating speed is limited

18.7 Internal Memory—Static and Dynamic Random-Access Memory (RAM)

RAM memory chips are made of capacitors and transistors and work by storing electronic charges. The capacitor stores the charge, and the transistor can turn the charge on or off. In the RAM chips, the state of the charges can be changed, whereas ROM (read-only memory) is either permanently on or off.

DRAM (dynamic RAM) is the standard main memory and can be thought of as a rectangular array of cells. Each transistor holds a single bit of data. The capacitor holds the charge temporarily, so it must be refreshed. This means the value is read and rewritten in a clean version. The refresh speed is expressed in nanoseconds (ns). The refreshing cycles, which is why the memory is called dynamic, slow down the accessing of data.

Unlike DRAM, the SRAM (static RAM) can store data without the automatic refreshing. Static comes from the fact that nothing changes, with one exception. It will be refreshed when a write command has been performed, then the data remains until the power source is removed.

The SRAM has two transistors per bit with the second transistor controlling the output of the first. SRAM has the advantage of being much faster, although more expensive, than DRAM.

An example of a 4,096 $\times$ 1 bit SRAM is shown in Figure 18–9. As seen in the diagram, memory is a 64 $\times$ 64 array, which gives us 4,096 (64 $\times$ 64) bits. The two sets of buses address the array. A0–A5 decode the row, and A6–A11 decode the column.

To access the element in 2,1, lines A1 and A6 would be high and the others low. Another example, to access the element in 16, 32, lines A4 and A11 would be high. The column I/O forms the output, which appears at D_{out}.

The CS remains high until data is needed from the memory. When the CS is low, data can be read or written into depending upon the state of WE. Data can be written into a location, selected by A0–A11, when WE is low. When WE is high, stored data can be read and becomes available at D_{out}.

18.8 Internal Memory—Read-Only Memory (ROM)

ROM contains the programming that allows the computer to boot up, and this memory contains data that can only be read. ROM retains the data even when the power is turned off because a small battery in the computer sustains it. (Some modern ROM technologies allow information storage with no battery supply.)

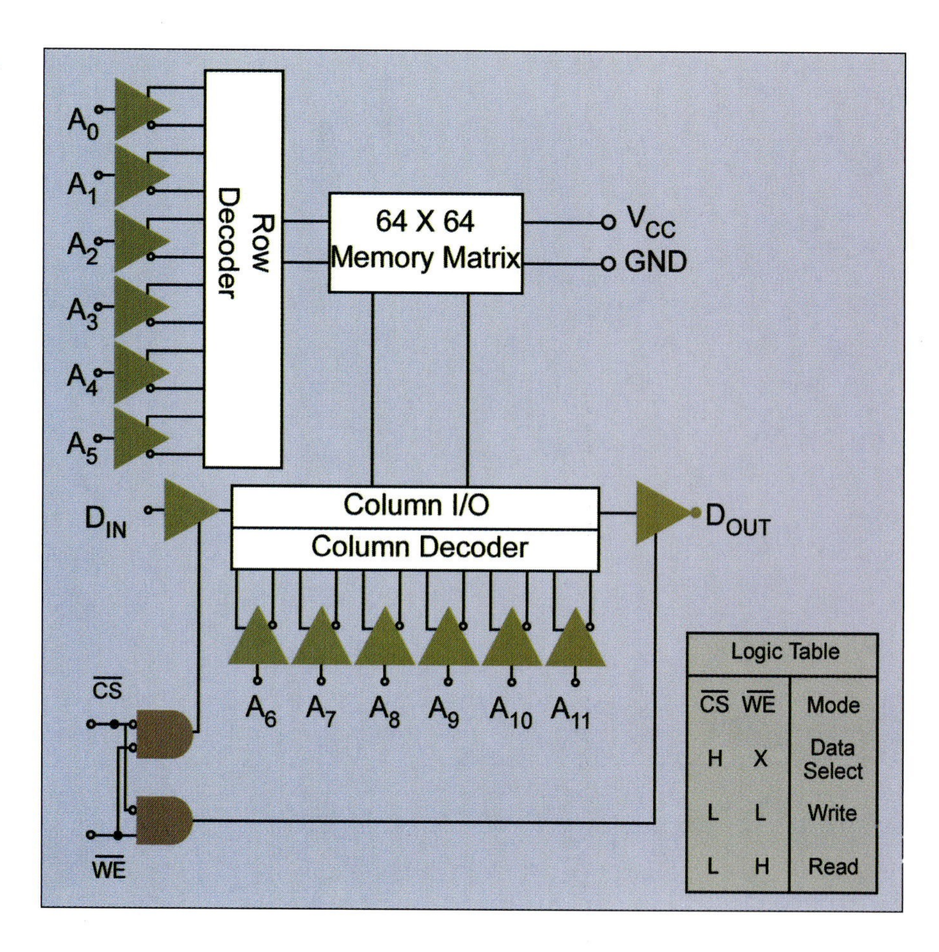

FIGURE 18–9 Block diagram of a 4,096-bit SRAM

ROMs consist of a memory matrix, as shown in Figure 18–10, with each cell containing 1 bit of permanent data. Although the types of memory cells vary, assume here that the cells are made up of diodes and switches. When the switch is open, the electric signal cannot get through (low state). If the switch is closed and the circuit is complete, the electric signal can get through (high state).

The switch is a link that is placed there during the fabrication process. Therefore, the high or low (0 or 1) state of each cell is permanent because it is set according to the needs of the program.

To read the contents of any of the rows in Figure 18–10, you must address that row. The row is addressed by applying a high state to that row. Only the content of the row that is addressed is seen at the output lines. The contents of the other rows will not affect the output. If switches A and D in row 9 are high, then the word 1001 will appear at the ABCD outputs lines.

In this arrangement, fourteen connections were needed to access the forty individual memory elements (10 words of 4 bits each). If the same arrangement were used for a typical ROM, this would lead to an impractical number of connections and wires. A typical ROM has 512×8 elements, which would require 520 ($512 + 8$) connections.

Therefore, an arrangement using decoders is used to access each element in an array. Figure 18–11 shows a 4-bit BCD-to-decimal decoder connected to the input lines of the previously shown 10×4 memory. A 2-bit decoder is used to select any of the four outputs. As is shown, this decoder is connected to the output with AND circuits.

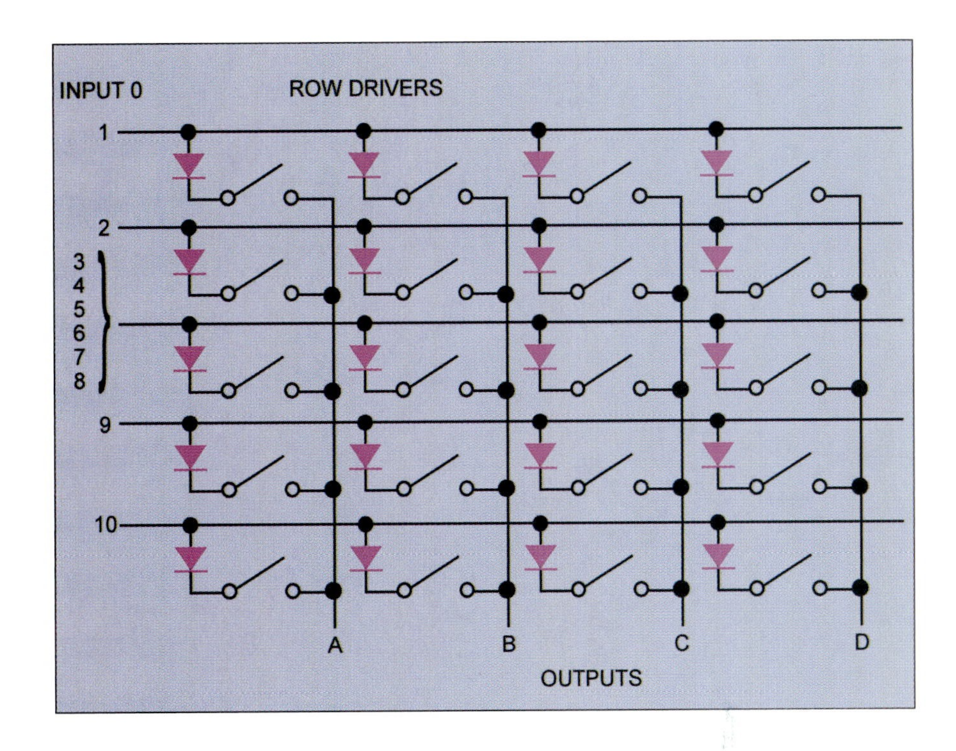

FIGURE 18–10 Structure of read-only memory

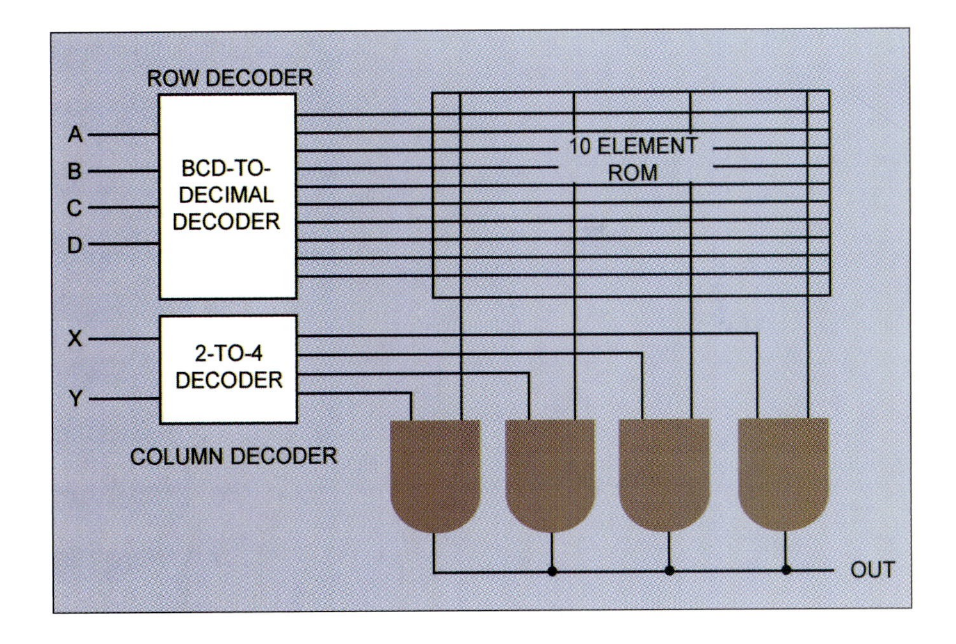

FIGURE 18–11 Decoders allow for fewer connections

Applying the 4-digit and 2-digit codes to each decoder will result in a 1 at the chosen row and column. The stored pattern, whether it is a 0 or 1, will be sent to each output line when a 1 is applied to the selected row. Then a 1 is applied to the selected column AND gate, which energizes only that AND gate. If the matrix output is 0, then the output is 0. A 1 will produce an output of 1.

With this new arrangement, the 10 × 4 memory needs only six connections instead of the original fourteen. Using this organization, it becomes practical to use ROMs with as many memory cells as required. This same structure is used with RAM.

18.9 Internal Memory—Programmable Read-Only Memory (PROM)

PROMs have the same properties as ROMs with the exception that they can be written on once. Once they have been written on, or programmed, they are just like the ROMs. The benefit of PROMs is that they can be economical in small quantities.

The memory cells of the PROM are manufactured in the closed state (1) by using a switch contact consisting of a thin alloy layer. This layer is something like a fuse. The programmer applies a current through any memory cell and blows the fuse element so the cell is now in the open state (0).

The EPROMs (erasable PROMs) do not use permanently open or closed switches. The memory can be changed electrically by exposing the chip to ultraviolet light.

18.10 External Memory—Mass Storage

Magnetic Storage

External memories are used in addition to internal memory when data is not needed for the moment but must be kept for an extended period of time. Data and programs that can be accessed quickly can be stored in external memories. Most computer systems use an external mass storage system.

Digital tape is one form of external memory. Because of the slow access to the information on the digital tape, it is used mainly for long-term storage that will be eventually read into internal memory.

The most popular external storage is the magnetic disk system, which allows for almost immediate access to data stored on it. Data can be stored onto the disks almost as quickly. They allow for large amounts of data to be stored externally. Modern magnetic disk systems can be large storage or small. Disks with capabilities up to a terabyte are becoming available. Sizes in the hundreds of gigabytes are common.

The disk or (in the case of disk systems) disks are coated with a magnetic material. The read/write heads of the computer can read stored data or they can write data onto the disk. They can range in size from the common floppy disk to the very large hard disks used with mainframe computers.

Optical Storage

Modern storage media such as CD and DVD are becoming more and more common. These two storage devices use a plastic disk. The information is stored digitally on the disk by etching the information into the material using a laser. A large variety of these two devices is available including both permanent and rewritable types.

SUMMARY

The major elements of the microprocessor are control unit, arithmetic/logic unit, input/output buffers, and memory. The control unit and the ALU are usually housed together to form what is known as the CPU. The control unit directs how the information flows through the other components. The data is manipulated in the ALU, and the results and other data are stored in memory. All of this information is carried through the bus lines, either the address bus or the data bus.

All CPUs must contain five basic elements. They must be able to perform the instructions of the program using the logic, control, and arithmetic functions, have a program counter, have at least one register, have a data counter, and have an instruction register.

The ALU must include four basic functions, but it can include more. It must be able to perform Boolean operations, binary addition, complementation, and shift of data. The control unit controls the sequence in which the ALU performs.

Multiplexers ensure the proper interconnections throughout the CPU and that they occur at the right time. They also allow for fewer wires, enabling practical applications.

Memory can be categorized as either volatile (permanent) or nonvolatile (temporary). Some of the memories discussed in this chapter were:

- Static and dynamic random access memory (SRAM and DRAM)
- Read-only memory (ROM)
- External storage (magnetic tape or disks and optical disks)

REVIEW QUESTIONS

1. Name the elements found in a microprocessor.
2. What is the purpose of
   a. The central processing unit?
   b. The arithmetic/logic unit?
   c. The control unit?
3. What are the four basic functions carried out in the CPU?
4. How does a multiplexer work? (Use a four-input multiplexer for your explanation.)
5. Explain the use of the five flags in a CPU.
6. What type of memory is a
   a. Computer floppy disk?
   b. A computer hard drive (fixed disk)?
   c. An SRAM chip?
   d. A CD?
7. What is an EPROM used for?
8. What is a PROM used for?
9. When your personal computer first boots up, what kind of memory establishes its initial startup parameters?
10. What are input/output buffers used for in a CPU?

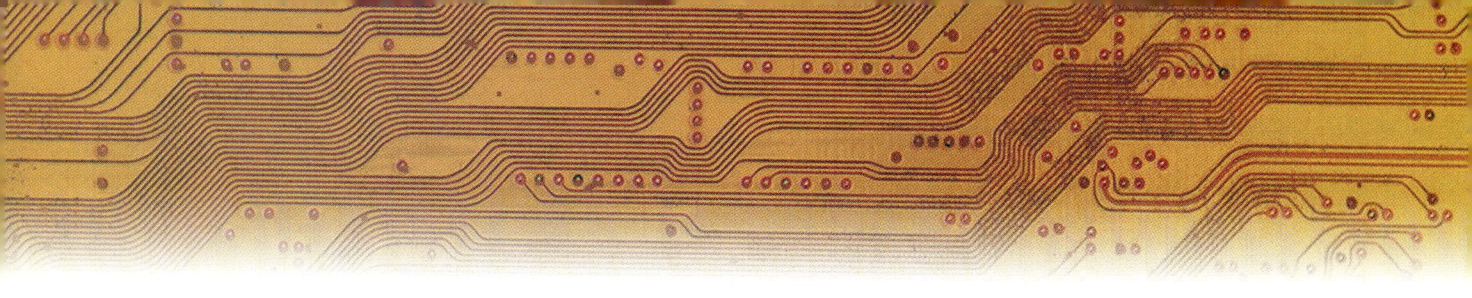

Index